FOOD SUSTAINABILITY AND THE MEDIA

FOOD SUSTAINABILITY AND THE MEDIA

Linking Awareness, Knowledge and Action

Edited by

MARTA ANTONELLI

Division on Climate Change Impacts on Agriculture, Forests and Ecosystem Services (IAFES),
Fondazione Centro Euro-Mediterraneo sui Cambiamenti Climatici (CMCC), Viterbo, Italy

PIERANGELO ISERNIA

Department of Social, Political and Cognitive Sciences, University of Siena, Siena, Italy

ACADEMIC PRESS

An imprint of Elsevier

ELSEVIER

Academic Press is an imprint of Elsevier
125 London Wall, London EC2Y 5AS, United Kingdom
525 B Street, Suite 1650, San Diego, CA 92101, United States
50 Hampshire Street, 5th Floor, Cambridge, MA 02139, United States

Notices
Knowledge and best practice in this field are constantly changing. As new research and experience broaden our understanding, changes in research methods, professional practices, or medical treatment may become necessary.

Practitioners and researchers must always rely on their own experience and knowledge in evaluating and using any information, methods, compounds, or experiments described herein. In using such information or methods they should be mindful of their own safety and the safety of others, including parties for whom they have a professional responsibility.

To the fullest extent of the law, neither the Publisher nor the authors, contributors, or editors, assume any liability for any injury and/or damage to persons or property as a matter of products liability, negligence or otherwise, or from any use or operation of any methods, products, instructions, or ideas contained in the material herein.

ISBN: 978-0-323-91227-3

For Information on all Academic Press publications
visit our website at https://www.elsevier.com/books-and-journals

Publisher: Nikki P. Levy
Acquisitions Editor: Nancy J. Maragioglio
Editorial Project Manager: Kyle Gravel
Production Project Manager: Rashmi Manoharan
Cover Designer: Vicky Pearson Esser

Typeset by MPS Limited, Chennai, India

Working together
to grow libraries in
developing countries

www.elsevier.com • www.bookaid.org

Contents

List of contributors   vii

1

1. A primer to understand the link between food sustainability and the role of the media   3

Marta Antonelli and Pierangelo Isernia

Introduction   3
References   8

2. The role of media in addressing global food sustainability: Cultural, social, and economic contexts   11

Kristen Alley Swain

The role of food journalism   12
Food imagery   14
Food information technologies   16
Food misinformation   19
Media framing of food sustainability   22
Food policy framing   26
Food on social media   29
Conclusion   42
References   44

3. Food culture and food sustainability on social media   57

Yelena Mejova

Introduction   57
Social media analysis methods   58
Social media and the individual   59
Social media and organizations   64
Case study   68

Looking forward   71
References   71

4. Food security in the Eurobarometer opinion trends   77

Arianna Marcolin and Elena Cadel

Introduction   77
Getting data on food security using the Eurobarometer surveys   78
Availability   79
Access   85
Utilization   85
Stability   86
Agency   87
Sustainability   87
Food safety   89
Discussion   96
Conclusion   98
Acknowledgment   99
References   99

5. Can we slow climate change while feeding a hungry world? Media narratives about the food − water − energy nexus   101

Kristen Alley Swain

Media framing of climate change impacts   102
Agricultural impacts of climate change   104
The food − water − energy nexus   105
Agri-food corporate social responsibility   106
Communicating about animal products   108
Food security   121
Communicating food risks   133
Conclusions   145
References   147
Further reading   159

2

6. The Chefs' Manifesto   163

Paul Newnham

How the Internet has changed food messaging   164
A short history of food messaging   165
The impact of a new platform   166
Why do we listen to people who aren't scientists or
    credible health practitioners?   166
Chefs bridging the gap   167
The global food crisis   168
Introduction to the Chefs' Manifesto   169
Building the network and guiding principles   170
Chef engagement   171
How chefs spread the message   172

7. Enabling sustainable, healthful eating in
the cafeteria setting through education and
social engagement: the SU-EATABLE LIFE
project   177

Laura Bouwman, Leah Rosen, Marta Antonelli,
Simona Castaldi and Katarzyna Dembska

Introducing the SU-EATABLE LIFE project   177
SU-EATABLE LIFE project approach   178
Project design process   179
SU-EATABLE project design and
    evaluation—stage I   183
The way forward: stage II of the SU-EATABLE LIFE
    project   188
Acknowledgments   191
References   191

8. Design thinking workshop: using
experiential learning, creativity, and
empathy to learn about the complexities of
food insecurity and sustainability   195

Sarah M. Zehr, Deana McDonagh, Jennifer Vokoun,
Francesca Allievi and Sonia Massari

Introduction   195
Presidents United to Solve Hunger Leaders Forum
    and the Universities Fighting World Hunger:

multidisciplinary and multigenerational
    experiment to experience sustainability through
    food   197
PUSH-UFWH design thinking workshop: everyone
    should be an agent of change   203
Discussion and reflections   213
Conclusions   215
Appendix   217
References   217
Further reading   219

9. Global food ecosystems: new models
to cover messages about food
systems   221

Sara Roversi

Intro   221
Value-based communication   223
Food for Climate League   223
Boot Camps   228
Hackathon   231
Conclusion   233

10. Open innovation in sustainable
corporate communication: a case study
from Italian food companies   235

Marco Bassan

Introduction   235
Sustainable development goals in the food
    industry   237
Sustainability and stakeholder engagement   238
The relationship of stakeholder engagement and
    communication process   239
Methodology   240
Case selection   240
Data collection   241
Data analysis   242
Case study   243
Skretting   245
Coop   246
Results   247
References   248

Index   253

List of contributors

Francesca Allievi Center for Sustainable Food Design, Jamk University of Applied Sciences, Jyväskylä, Finland

Marta Antonelli Fondazione Barilla, Parma, Italy; Impacts on Agriculture, Forests and Ecosystem Services (IAFES), Euro-Mediterranean Centre on Climate Change Foundation (CMCC), Viterbo, Italy

Marco Bassan Department of Business Economics, Università degli studi Roma Tre, Rome, Italy

Laura Bouwman Chair Group Health and Society, Social Sciences Group, Wageningen University, Wageningen, The Netherlands

Elena Cadel "Riccardo Massa" Department of Human Sciences for Education, Piazza dell'Ateneo lavoro Nuovo, Milano, Italy

Simona Castaldi Dipartimento di Scienze e Tecnologie Ambientali Biologiche e Farmaceutiche, Università degli studi della Campania Luigi Vanvitelli, Caserta, Italy

Katarzyna Dembska Fondazione Barilla, Parma, Italy

Pierangelo Isernia Department of Social, Political and Cognitive Sciences, University of Siena, Siena, Italy

Arianna Marcolin Department of Economics, Management and Quantitative Methods University of Milano Via Conservatorio, Milano, Italy

Sonia Massari Tourism and Service Business, School of Business, Pisa University, Pisa, Italy

Deana McDonagh Department of Education Policy, Organization & Leadership, College of Education, University of Illinois Urbana-Champaign, Urbana-Champaign, IL, United States

Yelena Mejova ISI Foundation, Torino, Italy

Paul Newnham Chief Executive Officer, SDG2 Advocacy Hub, London, United Kingdom

Leah Rosen Chair Group Health and Society, Social Sciences Group, Wageningen University, Wageningen, The Netherlands

Sara Roversi Future Food Institute, Bologna, Emilia-Romagna, Italy

Kristen Alley Swain Department of Integrated Marketing Communications, University of Mississippi, Oxford, MS, United States

Jennifer Vokoun School of Art & Design, College of Fine & Applied Arts, Walsh University, North Canton, OH, United States

Sarah M. Zehr Department of Education Policy, Organization & Leadership, College of Education, University of Illinois Urbana-Champaign, Urbana-Champaign, IL, United States

PART 1

A primer to understand the link between food sustainability and the role of the media

Marta Antonelli[1] and Pierangelo Isernia[2]

[1]Impacts on Agriculture, Forests and Ecosystem Services (IAFES), Euro-Mediterranean Centre on Climate Change Foundation (CMCC), Viterbo, Italy [2]Department of Social, Political and Cognitive Sciences, University of Siena, Siena, Italy

Introduction

Transforming food systems is one of the biggest challenges of our time. How to achieve this transformation has been interpreted from a variety of perspectives (i.e., Fanzo, 2021; Herrero et al., 2020, 2021; Loboguerrero et al., 2020). The challenges are multifaceted and have environmental, social, economic, and cultural dimensions.

From a nutritional point of view, over 4 billion people may be affected by overweight and obesity by 2035, compared with over 2.6 billion in 2020, with an increase from 38% of the world's population in 2020 to over 50% by 2035 (World Obesity Federation, 2023). NCDs continued to cause the highest disease burden worldwide. Noncommunicable diseases, for which the diet is a modifiable risk factor, caused 74% of global deaths (41 million people) and 63% (1.6 billion people) of global Disability-Adjusted Life-Years (DALYs) in 2019, especially cardiovascular diseases, cancer, chronic respiratory disease, and diabetes (WHO, 2023).

The food system, from farm to fork to disposal, increases competition over scarce land and water resources, generating anthropogenic greenhouse gas (GHG) emissions and biodiversity loss (Poore & Nemecek, 2018; Herrero et al., 2016; Springmann et al., 2018; Steffen et al., 2015). The expected increase of the global population (UN 2022) can only exacerbate these burdens, coupled with urbanization and changing dietary preferences. Food system emissions account for about one-third (Crippa et al., 2021) of the total GHG emissions from anthropogenic activities, and they are mostly related to agriculture and land use/land use change. Up to 3.6 billion people around the globe live in contexts that are highly vulnerable to the impacts of climate change (WHO, 2023). Studies have also shown that food consumption is a substantial driver of transgression of planetary limits. For instance, in the Mediterranean region, the consumption for renewable resources and ecosystem services outpaces the capacity of ecosystems to provide them (Galli et al., 2017). Preserving the domestic and global natural capital is crucial to reducing our footprints but also to improving food system resilience to provide food security in the upcoming decades (Nyström et al., 2019).

According to United Nations' projections, in a business-as-usual scenario, by 2050, food production will need to increase by 70% compared with 2009, to meet the food demands of a growing and increasingly urbanized population with rising incomes, demanding more vegetables, fruits, meat, dairy, and fish (FAO, 2019). Yield gains seem unlikely to happen without also increasing environmental burdens, even with efficiency improvements (Davis et al., 2016). Over the same period, food-related GHG emissions are expected to grow by 87%, the demand for cropland use by 67%, blue water use by 65%, and phosphorus and nitrogen application by 54 and 51%, respectively. This would exceed planetary boundaries and put at risk key ecosystem processes (Springmann et al., 2018). As several studies pointed out (i.e., Springmann et al., 2016; Willett et al., 2021), transformation to more sustainable and healthier food systems requires a global shift of diets, reductions of food loss and waste, and radical improvements in agriculture and food production practices.

In the European Union (EU), food systems are at a crossroads. Food consumption accounts for about 45% of the overall environmental and biodiversity impacts of EU consumption (Sala & Mengual, 2022). In 2020, the European Commission put forward the Farm to Fork Strategy, which is at the core of the Green Deal's target to decarbonize the EU by 2050. The Strategy calls for a broad food system transformation, with an upcoming legislative framework by 2023 (European Commission, 2020). A few studies have claimed that the implementation of the Farm-to-Fork Strategy could lead to leakage effects in terms of GHG emissions (Fuchs et al., 2020), a decrease in agricultural production (Barreiro-Hurle et al., 2021), price increases, and income losses for producers (Candel, 2022). Shifting toward more sustainable diets is key to any attempt at transforming food systems, as food patterns in the EU are already close to exceed the reference value of 2.49 kg $CO_2eq/capita/d$ used to assess the climate sustainability of diets (Castaldi et al., 2022).

A recent assessment of the carbon footprint of the EU diets also showed that, compared with the current Europeans' diet, the adoption of a desirable diet—with higher intakes of fruit, vegetables, wholegrains, low-glycemic-index cereals, nuts, legumes and fish, and lower amounts of beef, butter, high-GI cereals or potatoes and sugar—would reduce by almost 50% the carbon footprint (Giosuè et al., 2022). While reducing sugar and animal product consumption is key to increasing health, reducing consumption of animal products is the key mitigation option to reduce environmental impacts, according to Science Advice for Policy by European Academies (SAPEA, 2023). Moreover, the distribution of household carbon footprints is highly unequal within and across EU countries (Ivanova & Wood, 2020). Previous assessments of the water footprint of EU diets showed that a reduction in meat intake would contribute most to the water footprint reduction, with respect to other food groups (Antonelli & Greco, 2015). Carbon, land, and water footprint linked to the EU consumption and lifestyles were also assessed through an MRIO model by Steen-Olsen et al. (2012), who showed that all EU countries were 35% or more above the global average of land footprint, over twice the global average of carbon footprint, and about 10% of the global blue water footprint, with food trade playing a significant role. A recent study suggested that the incorporation of novel foods in European diets could reduce global warming potential, water use, and land use by over 80% (Mazac et al., 2022). The social risk associated with the EU food production and consumption was also recently evaluated. Vegetables, fruits, and rice emerged as major hotspots for social risks, in contrast to environmental assessments on food, which revealed the higher impact of animal-based products (Mancini et al., 2023).

Barriers to consumption changes are situated both at the individual and the contextual level and represent the potential entry points of targeted policies and interventions (SAPEA, 2023). At the individual level, these barriers include a perceived lack of consumer motivation and capabilities; at the contextual level, there is a lack of physical, financial, and social opportunities to acquire healthy and sustainable foods.

The book presented here analyzes the role of media in the food system transition in a constantly changing information landscape. Not only is more information readily available, but so is more misinformation and fake news. The perception of the media as biased is increasing as citizens' struggle to identify objective and trustworthy news sources of information. Against this context, the chapters presented in this volume explore the role extent to which media can fulfill this role vis-à-vis the urgency of food system transition in the face of malnutrition, agriculture, and climate change challenges. The book then explores the problems and challenges of different ways of communicating about food behavior in a sustainable and healthy way, using both traditional and less traditional methods, looking at both the supply and the demand side, to the role of companies and the media in communicating about food issues, and to ways through which to engage the individual consumers in raising their awareness on food sustainability and its implications.

The book is opened by a thorough and detailed review of the literature on food communication by Kristen Alley Swain in which she addresses, in a thorough review of the literature, two main topics of research on the media effects on food sustainability: the media framing and the role of social media in the promotion food consumption.

The first section looks at the supply side, and it starts with the contribution by Mejova, who explores the use of social media and internet data as a lens into the lifestyle behaviors of people around the world, as it captures both health-relevant behaviors and values attributed to food. Social media provides an important resource for monitoring both the behaviors and attitudes of a vast number of people, as well as the effectiveness and reach of campaigns aiming to encourage healthier lifestyles and awareness of issues concerning food, climate change, and sustainability. In this chapter, the author discusses the role of social media platforms as drivers for consumer behavior change looking on the one hand at the extent to which both organizations and grassroots activists use social media platforms to communicate their stances on sustainability issues and to create communities around them and on the other hand, how the communities around powerful influencers shape the discourse around sustainability.

Newnham reports of a very specific kind of influencers, the chef, discussing the origins, characteristics, and impact of "The Chef Manifesto Action Plan," a chef-led network bringing together more than 800 chefs from 80 countries. The internet and social media have made food and chefs' work more visible, and thereby more influential than ever. What chefs choose to profile today can directly drive future demand, as seen with ingredients ranging from quinoa to rocket. However, much of the conversation around food, nutrition, and agriculture is technical and uses a language that is often misinterpreted by people without special expertise in these fields. This is why the Chef Manifesto was born: to bridge the gap between farm and fork and to explore how Chef can help deliver sustainable food systems. This chapter details the Chefs' Manifesto Action Plan and its concrete, simple, and practical areas of actions in which chefs can work on in their kitchen, creates a clear framework for chefs to deliver on the Sustainable Development Goals. This is increasingly the focus of projects and initiatives at the European level, such as the LIFE Climate Smart Chefs project, which has developed, with funding from the European Commission, the first high-level training course for chefs that aim to combine culinary expertise and skills, with knowledge on sustainability and tools to account for the impact of what they cook.

Bassan et al. discuss the challenges of company level's communication strategy. This is a complicated process, both in terms of messages to be conveyed and the target audience to be reached. To effectively address this process, involving stakeholders can be a key action to pursue effective communication that can help to achieve sustainability. Indeed, stakeholder engagement can be useful not only in understanding needs and gathering information but also in spreading information to customers and society. Through a set case studies, the chapter highlights how some Italian food companies have used stakeholder engagement to transform sustainable communication into a competitive advantage. The analysis shows that thanks to the structured involvement of direct and indirect third parties (e.g., universities, environmental associations, internal employees), companies have also achieved a higher level of transparency and reputation toward the market. The results show that companies with a strong focus on sustainability communication are implementing an open communication ecosystem to connect in different ways to different stakeholders: it can be one to many, one to one, many to many, or many to one, depending on how much control remains in the hands of the company. From the analysis, a framework for a dynamic communication ecosystem has emerged that allows to (1) collect market needs in real time to promptly adapt its communication strategies to sustainable challenges; (2) build communication messages shared with stakeholders; (3) increase internal awareness on the issues; (4) make sustainability issues penetrate the corporate mission; (5) build the market perception of a company that is close to the social urgency.

Moving to the next section to the demand side, a first article offers a contextual analysis on what attitudes the general public holds on these issues. The article by Marcolin and Cadel draws upon the rich, and often underutilized, set of questions asked by the Eurobarometer series, over the years, about the attitudes and beliefs of European citizens regarding food security over time. Based on the assessments of 53 questions retrieved from all the Eurobarometer waves released from 2010 to 2022, the chapter found a low level of concern with Food Security than Food Safety by general public opinion with some changes over the most recent years due to the extreme natural events caused by climate change. This concern, the chapter found, has not translated into a change in the utilization of food, which is driven by diet and personal health issues, but into a higher request for actions from the European Institutions. In fact, during the decade studied, more individuals believe the role of European institutions should be strengthened to assuring sustainability of food security within the Member States. On the contrary, concern about food safety has increased. In particular, respondents are worried about additives, pesticides, and food hygiene.

The following chapters adopts less conventional and more participatory and cocreational approaches to the study of ways of communicating about healthy and sustainable-aware ways of food consumption. Bowman et al. discuss a series of activities that engage EU citizens to adopt a sustainable and healthy diet at university and company cafeterias. The EUE-funded SU-EATABLE LIFE project designed a multilevel, multistrategy approach in collaboration with universities and companies in Italy and the United Kingdom to show that we can achieve a substantial reduction in greenhouse gas (GHG) emissions and water usage through activities at the food service and consumer level. In discussing the experience, they stress how actions are possible at both the food service level and the consumers' level. As to staff and management, they were invited to revise their methods of food procurement and preparation and align their food offering with the SU-EATABLE criteria for sustainable, healthful meals. Eight, science-based criteria were identified and translated into actions, which any citizen or stakeholder within the food sector (retailers, caterers, restaurants, cafeterias, municipalities) could apply to reduce CO_2 emissions and water use, based on changes in dietary choices. At the consumption level, cafeteria customers were invited to join onsite and online learning activities (via the GreenApes mobile application) about the why, how, and what of sustainable food practices. Monthly challenges were aimed not only to emphasize cognitive capacities (i.e., educational/informative approach) but also to stimulate reflection, trigger search for information, encourage social participation, and simplify the selection and consumption of sustainable meals. The article discusses the results of a preliminary experiment launched in early 2020 at seven university and company cafeterias in the United Kingdom and Italy.

Along similar experiential learning strategies, Zehr et al. discuss the use of experiential learning in the form of a design thinking workshop during a conference to encourage attendees to think critically about food systems and the challenges of food sustainability, in particular those related to food insecurity and how they can be addressed. The workshop applied Kolb's theory of experiential learning and Mezirow's transformational learning theory to create a situation where attendees participated in a novel food experience related to sustainability and insecurity. Throughout the workshop, participants were encouraged to reflect on their assumptions about food systems and food supply chains and their role in them. Aspects of the workshop enabled participants to experience firsthand media impacts in food choices and habits and food inequalities. The workshop leaders asked participants to reflect on their emotions or feelings throughout the workshop, which fostered empathy, collaboration, and communication skills. Despite reporting knowledge about food insecurity prior to participating in the workshop, more than 80% of participants felt that their participation in the experience improved their understanding of the food system and sustainability and provided them with ideas about how to address it. Participants left with a belief that they could take action to address food complexities in their local communities and beyond.

Finally, Roversi et al. discuss the longer-term approach to education and learning adopted by the Future Food Institute that since 2014 has represented a benchmark in delivering new models to reshape and regenerate the global food ecosystem to protect the Planet, empower people, and enable prosperity, adopting a bottom-up approach, starting from education as a mean to raise awareness and shift the mindset toward global issues. It has done this using a three-pronged approach: education, community, and innovation. Education aims at fostering life-changing learning experiences, involving the global community of food system experts and innovators and at the same time local communities rooted in ancient food traditions. Community aims at embracing inclusion for diversity as an added value of learning. Caring about mutual relationships, improving listening skills, stimulating creativity, stressing critical thinking, developing abilities to codesign for prosperity, and being different but united as inspired a broader and multistakeholder community to take action on SDGs, and as a result, nudging a behavioral shift is a process that requires long-term planning. Innovation is considered as both a driver for global economic development and a catalyst for change. These three aspects represent the center of the Future Food compass, on which purpose-driven research is developed to break the silos of food systems, using a systemic approach, and building a thriving society through food, which is a perfect glue to represent culture identities, values, and the nexus with environment and territories. The chapter describes some tools that this model is using to map the transformation of the role of food in achieving SDGs, such as Food for Earth toolbox, composed of five innovative areas (food diplomacy, circular living, prosperity, climate-smart ecosystems, and food identity) and four main tools (humana communitas, metrics, models, and platforms) to study the impact of food on sustainability.

References

Antonelli, M., & Greco, F. (2015). *The water we eat: Combining virtual water and water footprints*. Springer Nature.

Barreiro-Hurle, J., Bogonos, M., Himics, M., Hristov, J., Pérez-Domínguez, I., Sahoo, A., Salputra, G., Weiss, F., Baldoni, E., & Elleby, C. (2021). Modelling transitions to sustainable food systems: Are we missing the point? *EuroChoices, 20*(3), 12−20. Available from https://doi.org/10.1111/1746-692X.12339.

Candel, J. (2022). EU food-system transition requires innovative policy analysis methods. *Nature Food, 3*(5), 296−298. Available from https://doi.org/10.1038/s43016-022-00518-7.

Castaldi, S., Dembska, K., Antonelli, M., Petersson, T., Piccolo, M. G., & Valentini, R. (2022). The positive climate impact of the Mediterranean diet and current divergence of Mediterranean countries towards less climate sustainable food consumption patterns. *Scientific Reports, 12*(1). Available from https://doi.org/10.1038/s41598-022-12916-9.

Crippa, M., Solazzo, E., Guizzardi, D., Monforti-Ferrario, F., Tubiello, F. N., & Leip, A. (2021). Food systems are responsible for a third of global anthropogenic GHG emissions. *Nature Food, 2*(3), 198−209. Available from https://doi.org/10.1038/s43016-021-00225-9.

Davis, K. F., Gephart, J. A., Emery, K. A., Leach, A. M., Galloway, J. N., & D'Odorico, P. (2016). Meeting future food demand with current agricultural resources. *Global Environmental Change, 39*, 125−132. Available from https://doi.org/10.1016/j.gloenvcha.2016.05.004.

European Commission (2020). Farm to Fork strategy. Available online at: https://food.ec.europa.eu/horizontal-topics/farm-fork-strategy_en

Fanzo, J. (2021). Achieving equitable diets for all: The long and winding road. *One Earth, 4*(4), 470−473. Available from https://doi.org/10.1016/j.oneear.2021.03.007.

FAO. (2019). *How to feed the world in 2050*. Rome: Food and Agriculture Organization. Available from http://www.fao.org/fileadmin/templates/wsfs/docs/expert_paper/How_to_Feed_the_World_in_2050.pdf.

Fuchs, R., Brown, C., & Rounsevell, M. (2020). Europe's Green Deal offshores environmental damage to other nations. *Nature, 586*(7831), 671−673. Available from https://doi.org/10.1038/d41586-020-02991-1.

Galli, A., Iha Katsunori, H., Martin., El Bilali, H., Grunewald, N., Eaton, D., Capone, R., Debs, P., & Bottalico, F. (2017). Mediterranean countries' food consumption and sourcing patterns: An ecological footprint viewpoint. *Science of The Total Environment, 578*, 383–391. Available from https://doi.org/10.1016/j.scitotenv.2016.10.191.

Giosuè, A., Recanati, F., Calabrese, I., Dembska, K., Castaldi, S., Gagliardi, F., Riccardi, G., Vitale, M., Vaccaro, O., Antonelli, M., & Riccardi, G. (2022). Good for the heart, good for the Earth: Proposal of a dietary pattern able to optimize cardiovascular disease prevention and mitigate climate change. *Nutrition, Metabolism and Cardiovascular Diseases, 32*(12), 2772–2781.

Herrero, M., et al. (2021). Articulating the effect of food systems innovation on the Sustainable Development Goals. *Lancet Planetary Health, 5*, e50–e62. Available from https://doi.org/10.1016/S2542-5196(20)30277-1.

Herrero, M., Henderson, B., Havlík, P., Thornton, P. K., Conant, R. T., Smith, P., Wirsenius, S., Hristov, A. N., Gerber, P., Gill, M., Butterbach-Bahl, K., Valin, H., Garnett, T., & Stehfest, E. (2016). Greenhouse gas mitigation potentials in the livestock sector. *Nature Climate Change, 6*(5), 452–461. Available from https://doi.org/10.1038/nclimate2925.

Herrero, M., Thornton, P. K., Mason-D'Croz, D., Palmer, J., Benton, T. G., Bodirsky, B. L., Bogard, J. R., Hall, A., Lee, B., Nyborg, K., Pradhan, P., Bonnett, G. D., Bryan, B. A., Campbell, B. M., Christensen, S., Clark, M., Cook, M. T., de Boer, I. J. M., Downs, C., ... West, P. C. (2020). Innovation can accelerate the transition towards a sustainable food system. *Nature Food, 1*(5), 266–272. Available from https://doi.org/10.1038/s43016-020-0074-1.

Ivanova, D., & Wood, R. (2020). The unequal distribution of household carbon footprints in Europe and its link to sustainability. *Global Sustainability, 3*. Available from https://doi.org/10.1017/sus.2020.12.

Loboguerrero, A. M., Thornton, P., Wadsworth, J., Campbell, B. M., Herrero, M., Mason-D'Croz, D., Dinesh, D., Huyer, S., Jarvis, A., Millan, A., Wollenberg, E., & Zebiak, S. (2020). Perspective article: Actions to reconfigure food systems. *Global Food Security, 26*. Available from https://doi.org/10.1016/j.gfs.2020.100432.

Mancini, L., Valente, A., Barbero, V. G., Sanyé Mengual, E., & Sala, S. (2023). Social footprint of European food production and consumption. *Sustainable Production and Consumption, 35*, 287–299. Available from https://doi.org/10.1016/j.spc.2022.11.005.

Mazac, R., Meinilä, J., Korkalo, L., Järviö, N., Jalava, M., & Tuomisto, H. L. (2022). Incorporation of novel foods in European diets can reduce global warming potential, water use and land use by over 80. *Nature Food, 3*(4), 286–293. Available from https://doi.org/10.1038/s43016-022-00489-9.

Nyström, M., Jouffray, J.-B., Norström, A. V., Crona, B., Søgaard Jørgensen, P., Carpenter, S. R., Bodin, Ö., Galaz, V., & Folke, C. (2019). Anatomy and resilience of the global production ecosystem. *Nature, 575*(7781), 98–108. Available from https://doi.org/10.1038/s41586-019-1712-3.

Poore, J., & Nemecek, T. (2018). Reducing food's environmental impacts through producers and consumers. *Science, 360*(6392), 987–992. Available from https://doi.org/10.1126/science.aaq0216.

Sala, S. S., & Mengual, E. (2022). *Consumption footprint: Assessing the environmental impacts of EU consumption.* Brussels: European Commission.

SAPEA. (2023). *Towards sustainable food consumption.* Berlin: Science Advice for Policy by European Academies.

Springmann, M., Clark, M., Mason-D'Croz, D., Wiebe, K., Bodirsky, B. L., Lassaletta, L., de Vries, W., Vermeulen, S. J., Herrero, M., Carlson, K. M., Jonell, M., Troell, M., DeClerck, F., Gordon, L. J., Zurayk, R., Scarborough, P., Rayner, M., Loken, B., ... Willett, W. (2018). Options for keeping the food system within environmental limits. *Nature, 562*(7728), 519–525. Available from https://doi.org/10.1038/s41586-018-0594-0.

Springmann, M., Godfray, H. C. J., Rayner, M., & Scarborough, P. (2016). Analysis and valuation of the health and climate change cobenefits of dietary change. *Proceedings of the National Academy of Sciences, 113*(15), 4146–4151. Available from https://doi.org/10.1073/pnas.1523119113.

Steen-Olsen, K., Weinzettel, J., Cranston, G., Ercin, A. E., & Hertwich, E. G. (2012). Carbon, land, and water footprint accounts for the European Union: Consumption, production, and displacements through international trade. *Environmental Science and Technology, 46*(20), 10883–10891. Available from https://doi.org/10.1021/es301949t.

Steffen, W., Richardson, K., Rockström, J., Cornell, S. E., Fetzer, I., Bennett, E. M., Biggs, R., Carpenter, S. R., De Vries, W., De Wit, C. A., Folke, C., Gerten, D., Heinke, J., Mace, G. M., Persson, L. M., Ramanathan, V., Reyers, B., & Sörlin, S. (2015). Planetary boundaries: Guiding human development on a changing planet. *Science (New York, N.Y.), 347*(6223). Available from https://doi.org/10.1126/science.1259855.

UN (2022). The state of food and nutrition security in the world.

WHO. (2023). *World health statistics 2023: Monitoring health for the SDGs, sustainable development goals.* Geneva: World Health Organization.

Willett, W., Rockström, J., Loken, B., Springmann, M., Lang, T., Vermeulen, S., Garnett, T., Tilman, D., DeClerck, F., Wood, A., Jonell, M., Clark, M., Gordon, L. J., Fanzo, J., Hawkes, C., Zurayk, R., Rivera, J. A., De Vries, W., Majele Sibanda, L., Afshin, A., Chaudhary, A., Herrero, M., Agustina, R., Branca, F., Lartey, A., Fan, S., Crona, B., Fox, E., Bignet, V., Troell, M., Lindahl, T., Singh, S., Cornell, S. E., Srinath Reddy, K., Narain, S., Nishtar, S., & Murray, C. J. L. (2019). Food in the Anthropocene: the EAT-Lancet Commission on healthy diets from sustainable food systems. *Lancet, 393*(10170), 447–492. Available from https://doi.org/10.1016/S0140-6736(18)31788-4.

World Obesity Federation. (2023). *World obesity atlas 2023.* London: World Obesity Federation.

The role of media in addressing global food sustainability: Cultural, social, and economic contexts

Kristen Alley Swain

Department of Integrated Marketing Communications, University of Mississippi, Oxford, MS, United States

A rapidly changing global climate is a grand challenge facing humanity and the vulnerable systems used to feed all people. Timely, global dietary changes could greatly reduce greenhouse gas emissions and the catastrophic loss of human life caused by climate change impacts. By the year 2050, climate mitigation strategies in the agri-food sector could reduce overall direct and indirect CO_2 and non-CO_2 GHG emissions by 40% − 70% (IPCC, 2022).

To reach this goal, GHG-intensive foods must be replaced with sustainably sourced products, which will require the support of the agri-food industry and consumers. However, climate change is often framed as a "wicked problem," one that seems impossible to solve. Wicked problems are influenced by a range of complex, interacting effects, including human values and constantly changing conditions. Addressing a wicked problem effectively often involves improving the situation rather than solving the problem, since there is no clean endpoint (Hamm, 2009).

Food ecologies and economies are vital to the survival of communities, nonhuman species, and the planet. Global food system analyses urgently call for consumers to transition to healthy, sustainable diets. Major dietary shifts have occurred since World War II, including large-scale shifts in food commodity systems and diets, especially when the private sector scales up publicly funded technological innovation, with the support of state and international policies (Moberg et al., 2021).

The rapid increase in global food production has sparked environmental and societal concerns over greenhouse gas emissions, soil quality, and biodiversity loss (de Olde & Valentinov, 2019). World food production challenges include climate change, use of resources, population growth, and dietary changes. Implicit in media coverage of these challenges are major ethical

Food Sustainability and the Media
DOI: https://doi.org/10.1016/B978-0-323-91227-3.00001-9

and political questions, such as how to uphold the right to adequate nutrition or the right to preserve conditions that support a gastronomic culture (Borghini et al., 2020).

Food supply crises triggered by pandemics, population growth, drought, and other climate change impacts continuously pose imminent threats and shocks to the entire global food system. Disruptive food technologies and novel food technologies are urgently needed to develop a food system that is more resilient, secure, safe, and sustainable. More effective public communication is needed to convey the necessity of these advances because consumers are often hesitant to accept them (Siegrist & Hartmann, 2020).

Food discussions can serve as an entry point for public engagement with climate issues and with science and technology (Schneider et al., 2019). However, few scientists, research institutions, and universities have a high profile in publicly communicating about healthy diets and scientific information about food and climate issues (Weitkamp et al., 2021). As food supply chains become more globalized, fragmented, and complex, food industry stakeholders are discussing unconventional ways of managing food chains and coping with technology innovation trends. Existing sustainability frameworks offer a holistic way to evaluate food systems but often fall short of offering clear direction about managing change (Millet et al., 2020). An agenda for food systems communication also should address food system reform, justice, and sovereignty issues because global change depends on critical responses to injustice and inequity across the food chain (Gordon & Hunt, 2019).

Proposed food system changes include vertical farming, lab meat, and diets filled with insects and seaweeds. These proposed ideas have generated a polarized understanding of food and diets, rooted in political and ethical tensions between technological and anthropological fixes. While some argue that technology can deliver clean, just, pleasurable, and affordable food that would not require future generations to adjust their dietary cultures, others argue that future generations should dramatically change what they eat and how they eat it, to achieve a sustainable diet. These arguments are based on utopian misrepresentations because food is socially constructed in the context of habits, cultural norms, traditions, geographies, and climatic conditions (Borghini et al., 2020).

This chapter examines food journalism, as well as the information technologies, imagery, and misinformation that shape public perceptions about food issues. It also explores how the media frame food sustainability and food policy, as well as how social media platforms shape food branding and promote the consumption of processed and junk foods.

The role of food journalism

Food journalism is a growing field, in which daily newspapers, television, and social media popularize food and covering food issues such as healthy diets and sustainable tourism (Fusté-Forné & Masip, 2018). Narrative storytelling with a lively and vivid conversational style can help reduce negative perceptions about agricultural and food technologies, even in stories containing scientific information (Yang & Hobbs, 2020).

Growing media corporatization and commercialization have pushed food justice facts and issues to the fringes. News gathering and gatekeeping processes are engaged in a continuous tussle for prominence, and stories must be delivered under short deadlines. The news worthiness of articles is often measured in light of news values including currency, immediacy, and the impact of food-related problems on specific populations (Thakurta & Chaturvedi, 2012). Many journalists connect climate change stories to natural disasters that harm food crops—including hurricane intensity, drought, flooding, and wildfires—without mentioning how GHGs spur climate change. However, "slow journalism," an idea borrowed from the slow food movement, offers a new way for journalists to develop in-depth, ongoing climate change reporting that is less dramatic and personality-driven and connects with consumers and communities on a local level and over a longer period (Gess, 2012).

US newspaper coverage of climate change rarely mentions food system contributions, despite increasingly solid evidence of the importance of agri-food contributions to climate change. The few stories that do mention food or agricultural contributions often focus on agri-food issues or food-related animal contributions. Earlier coverage addressed food system contributions to climate change, mainly in terms of individual behavior but later expanded to address business and government responsibility. These trends indicate that nutrition experts should proactively disseminate information about food and climate change to the US media (Neff et al., 2009).

Goody's (1982) framework proposes five phases of sustainable food-related behaviors: production, distribution, preparation, consumption, and disposal. Romanian journalists reporting about food-related topics lack a solid understanding of the sustainable food field, sufficient research, and adequate sourcing. As a result, the coverage often contributes to the spread of misinformation. The stories often focus only on the consumption of fresh fruits and vegetables, framed as the only healthy foods, and as recommended diets for specific underlying health conditions such as obesity, diabetes, hypertension, and associated diseases (Marinescu et al., 2021).

Media coverage about major climate-related events can influence consumer behavior and awareness about the impacts of their everyday choices. Opinion leaders and policymakers could leverage media coverage to engage population segments that are not already knowledgeable and proactive in their everyday food choices. For example, coverage of the 2015 release of Pope Francis' encyclical "Laudato si—On Care for Our Common Home" about environmental issues and climate change, as well as the 2015 United Nations Paris Agreement, had the greatest impact on people who responded positively to them but were unaware of related discussions before they saw the stories (Ricci & Banterle, 2020).

Story frames and images can promote certain individual actions to help mitigate climate change while ignoring others such as reducing meat consumption. For example, Australian television images of black balloons and turning off light switches were part of the country's first climate change campaign. Journalistic coverage of the campaign emphasized household electricity, while technical literature highlighted the "carbon dioxide equivalence" spelled out in the Kyoto protocol. These divergent messages led to a mismatch between causal factors as understood by climate scientists and the causal factors perceived by the public (Russell, 2015).

Television cooking shows increasingly highlight the origins of foods and the ethics of food production and consumption. For example, two Australian food TV shows illustrate complex dynamics between the media and food industries, their influence on consumer behaviors and business practices, their explanations of the meanings and significance of foods, and their ability to open new markets and marketing strategies for food products and experiences. One of the TV shows is a niche lifestyle program that advocates for small food producers, while the other focuses on televisual marketing strategies of a major supermarket (Phillipov, 2016).

Providing food information to consumers can influence their expenditures on environmentally friendly food products. Public debate about sustainable food consumption in newspaper coverage predicts expenditures among regular supermarket shoppers who buy newspapers. Stories mainly influence consumer expenditures on organic, wholegrain, and low-salt products, indicating a preference for healthier foods. However, media coverage about buying less food of animal origin does not affect expenditures, except for fish purchases. Food information presented uncritically in coverage promotes higher expenditures on certain foods, while stories that present information contextualized as highly structured debate reduce expenditures of certain unsustainable foods (Bellotti & Panzone, 2016).

Food imagery

Digital photography shapes food cultures, including images of food used in photographic exchanges on social media. Users post food images to promote their identity and to interact with the online community (Walsh & Baker, 2020). Food experience is omnipresent, with the increased use of social media platforms where users share food images or video accompanied with #hashtags. Readers must tap into their visual cognition and perception of what the social media food images represent. Food images can simulate food taste, and user hashtags can affect food taste perception. For example, images of a Malaysian dessert provoked different user perceptions of cake appearance, flavor, texture, and hedonic value. An analysis of Instagram hashtags that depict taste perceptions of the cake before and after tasting it showed that many initial perceptions were inaccurate. Improved technology interactivity may be needed for visual images of food on social media to consistently trigger accurate food taste perceptions (Harun et al., 2020).

Photovoice, a technique used in community-based participatory research to document and reflect reality, can illuminate barriers to healthy eating. Instagram can serve as a tool to help communities share food images and discuss key health issues (Yi-Frazier et al., 2015). In one study, participants presented their own photos, experiences, and insights about food justice to policymakers. Although the participants were all overweight, lived in supportive urban housing, and had serious mental illnesses, their photo presentations reflected realistic optimism toward their roles as community change agents and influencers despite systemic barriers (Weinstein et al., 2021).

The global "food porn" phenomenon popularizes the pleasure of cooking and eating, through the production and circulation of food images in social media networks (Junqueira, 2019). Food porn presents food provocatively, in a style similar to glamor photography or pornography. This exaggerated style of contemporary food imaging has become normalized in recent years (Taylor & Keating, 2018). The professional roles of food visual designers include food photographers, food stylists, and prop stylists. Common themes in food and beverage photography and styling include emotional appeal, key trends, career opportunities, required skills and abilities and interventions, and ethical considerations (Cankul et al., 2021).

Design, styling, and photographic tropes used in still images of food communicate many cultural meanings associated with food. Some images use creative disruption to create a sense of intimacy, draw the gaze, and activate desire-based triggers to engage the viewer (Taylor & Keating, 2018). Photo translation of 2 million geotagged food images on Twitter in 17 food categories reflected unique appearance and other food characteristics in food images across six different regions (Okamoto & Yanai, 2019).

Unlike other social media platforms, Instagram deemphasizes textual description in favor of image use. Many food companies use Instagram as a communication channel within a green marketing digital mix, to promote environmental-oriented claims, raise awareness about their products, and position themselves as good corporate citizens. For example, Instagram posts from Croatia's first organic food store bio&bio framed environmental claims in terms of food products, food processes, food images, and environmental issues (Šikić, 2021). Similarly, another study found that clean eating food images in top Instagram posts that included the hashtags #eatclean and #cleaneating encourage this lifestyle. Idealized images of "clean" foods extol purity and often contrast them with foods perceived as defiled. These idealized clean eating posts aim to garner esteem and attention, while generating a sense of community through food media and promoting responsible food consumption (Walsh & Baker, 2020).

"Foodstagramming," the act of taking pictures of food and sharing them on Instagram or other social media platforms, has become a global consumer trend (Wong et al., 2019). It often involves photo sharing or the consumption of food images among solo diners. These "food selfies" can capture everyday social practices in motion and reveal hidden or ignored aspects of food practices (Middha, 2018). For example, in Instagram posts about a Brooklyn ramen shop, perceived territoriality and social distancing increased foodstagramming among solo diners, especially among introverted customers. Sometimes foodstagramming reflects food neophobia, the fear of new foods, or culinary tourism (Handayani, 2021). Self-expression, enrichment of dining experiences, social connection, favorable travel outcomes, virtual community engagement, and special occasion memories predict foodstagramming among tourists (Wong et al., 2019).

Foodstagramming might trigger a new gastronomic paradigm, intensified by social media influencers' global socialization of gastronomy and their growing replacement of food critique professionals. In light of Kuhn's scientific knowledge paradigms, del Moral (2020) argues that the first gastronomic paradigm occurred during the French Revolution. It was based on food smell and taste, and then the changing flavor of sauces shifted gastronomic preferences to a second paradigm. The third and fourth paradigms emphasized gastronomic touch, sight, and sound. Examples include the worldwide diffusion of Spanish tapas, globalization of ethnic and fusion cuisines, the increase of sustainability in cooking with development of local, vegan, and paleo cooking, comfort food, and smart food.

Visual design cues can influence food-related behaviors, through a range of psychological processes. These include attention, affective, cognitive, and motivational reactions to food images, food perceptions, and food-related attitudes (Vermeir & Roose, 2020). In light of the Theory of Planned Behavior, a person's intention to post food photos on social media depends on their subjective norms, perceived behavioral control, self-expression, supportive interaction, and utilitarian attitudes. Market mavens with knowledge about many kinds of food products, places to eat or shop, and other facets of food markets also contribute to intentions to post food photos. Enriched dining experiences can promote a need for self-expression and the intention to post food photos along with positive restaurant reviews on social media (Javed et al., 2021).

Sharing one's own eating and cooking experiences on social media can influence others' judgments about the person posting the content and about the intended audience. Food pictures can communicate a profile owner's characteristics. For example, posting images of gender-stereotyped dishes can shape impressions that others form about the sex of the profile owner. Posting photos of dishes that are stereotypically masculine dampens the attribution of feminine traits, regardless of the profile owner's gender. When food images promote feminine or masculine stereotypes, these gendered audience impressions influence their desire to interact with the profile owner (Cavazza et al., 2020). Turkish and American college students perceive certain foods as masculine and feminine, but the degree of masculinity or femininity they ascribe and their food consumption intentions differ by gender and country (Ekebas-Turedi et al., 2021).

Excessive food photo posting is seen as an undesirable behavior. Third-person user perceptions include beliefs that they personally post food content on social media less frequently than others do, especially undesirable content. They also believe that others are more influenced by social media food content than they are (Pham et al., 2019).

Food information technologies

Providing relevant food information to consumers is challenging and complex. Consumers perceive a growing need for information about food, partly because of publicized controversies about food adulterations and fraudulent investments in various food sectors (Felicetti et al., 2020).

Virtual Reality interventions that stimulate personal efficacy about eating also promote beliefs that can help VR users initiate and maintain sustainable food choices. For example, Meijers et al. (2022) created a VR supermarket, in which participants saw interactive pop-ups displaying environmental or health impact messages when they picked up products. The impact messages stimulated both efficacy beliefs and proenvironmental food choices, and heightened self-efficacy sparked positive spillover effects, up to 2 weeks after the VR experience. The effectiveness of the impact messages did not depend on whether they had a health or environmental appeal or whether they were presented text only or text with a visual.

As technology unlocks new opportunities to improve food supply chains, agri-food businesses have a vested interest in exploring Augmented Reality technology. AR involves overlaying virtual objects on a real-world environment. AR applications in the food industry include food process efficiencies, food decision-making, food marketing, food training, and food safety (Rejeb et al., 2021).

The Internet of People (IoP), an open source and decentralized technology infrastructure, consists of interconnected peer-to-peer networks that host individual profiles, reputation, and identity information. The proliferation of mixed-reality headsets as a next-generation IoP interface provides many opportunities for diet tracking and interventions.

Mixed-reality headset-mounted cameras for computer vision-based detection of diet-related activities, and the display of visual real-time interventions can be used to support healthy food choices. Current neural networks already enable accurate food item detection in real-world environments. For example, smart vending machines can improve consumer choices of beverages to reduce sugar and calorie intake and promote healthier food choices to reduce saturated fat consumption (Fuchs et al., 2020).

Smart e-health systems with powerful knowledge bases can provide suggestions of appropriate foods to individuals. Using an AR-based wearable food-monitoring system to make nutritional database information available to consumers in supermarkets and other locations, a Google Glass front-facing camera can provide reverse food image searches and text mining (Jiang et al., 2018). Similarly, Chang et al. (2021) developed an innovative, deep learning−based intelligent calorie management system named iBuffet. Deep learning and Internet of Things (IoT) technologies provide a calorie management system that includes intelligent buffet table modules, a mobile device application, and a cloud-based platform. IoT is the network of physical objects embedded with sensors, software, and other technologies for connecting and exchanging data with other devices and systems over the Internet. The iBuffet user records daily meal records, which are sent to a cloud-based platform to calculate calorie values.

While some agri-food businesses still have not developed mobile apps that utilize advances in Internet of Food technologies, others have created mobile apps that help consumers get better food information and improve their everyday food practices. Mobile food apps address particular dietary needs more effectively when they are ubiquitous, reduce knowledge gaps, and promote context awareness (Felicetti et al., 2020). Most communication technology tools designed to promote good health habits, physical activity, and weight loss only target adult audiences and fail to capture the attention of children. Augmented reality, serious games, and other computational techniques can provide a child-friendly experience for nutrition education and obesity prevention (Briseno et al., 2019).

Automated food and drink recognition methods connect to cloud-based lookup databases, such as food item barcodes, previously identified food images, or previously classified near-infrared spectra of food and drink items databases. These automated methods can match and identify a scanned food or drink item and report the results back to the user. The processes involve the acquisition, processing, segmentation, and classification of food images, as well as feature extraction and volume estimation (Knez & Šajn, 2020). However, these methods have limited value unless they can provide more details about ingredients, nutrients, portion sizes for a meal, and interrelationships among different foods. Boulos et al. (2015) argue that a more useful application of these food recognition systems could help users make healthier food and drink choices for their particular health condition, age, body weight, Body Mass Index, lifestyle, and preferences.

Food object recognition systems have mobile applications that can provide objective measurements of eating activity and help users manage dieting and health conditions or analyze eating patterns. Food recognition processes allow the apps to record, suggest, and provide clinical responses (Knez & Šajn, 2020). For example, Rewane and Chouragade (2019) developed a mobile app called "Food Nutritional Detection, Visualization and Recommendation for Health Monitoring" that identifies food items from images taken and shows their calorie content. In addition to mobile phone apps, wearable gadgets and AR have been used for food identification and estimating calorie counts. Other audiences for food recognition systems are farmers and other food producers. Born in the early 1980s as a multilingual agricultural thesaurus, the linked open database AGROVOC is used for indexing, retrieving, and organizing data in agricultural information systems and web pages (Caracciolo et al., 2013). It is a structured collection of terms, definitions, relationships, and concepts that represent anything related to food and agriculture.

Biomedical vocabularies, predictive modeling, and food standards all play an important role in providing relevant health information. However, there are few links among existing food ontologies because they each were developed for specific applications. One rule-based food recognition system is the Unified Medical Language System, which links biomedical vocabularies to enable computer system interoperability. Another computer resource assigns semantic tags from four food ontologies to food concepts embedded in recipes (Popovski et al., 2019).

Public communication about ecological, active, and intelligent food packaging can promote innovative ways to implement stricter requirements for food security and food safety. iFoodCloud, a knowledge base that systematically collects and analyzes public opinion on food safety issues in China from over 3000 public sources, informs food safety emergency response (Zhang et al., 2023). Wireless sensor networks, cloud computing, and machine learning have improved agri-food sustainability communications and logistics. Blockchain technology enables the production of high-quality food with minimal social and environmental consequences and the remote management of the location and conditions of food shipments and products. It also can help customers make better food-buying decisions (Belaud et al., 2019; Kramer et al., 2021; Marinagi et al., 2014; Pappa et al., 2018; Pérez Perales et al., 2019; Saurabh & Dey, 2021; Verdouw et al., 2018).

Food traceability systems also can help address food safety issues by sending information back to food producers. Consumers then need to have access to trustworthy food information from food producers and social networks, so they can knowledgably assess their own risks, especially when the media exposes food scandals. In light of social embeddedness theory, consumer social activities affect their risk perceptions about traceable food. Consumer risk perceptions are shaped by interpersonal, organizational, and social relationships. Specifically, interpersonal relationships tend to foster lower food risk perceptions, while organizational and social relationships promote higher levels of consumer risk perceptions. In turn, this higher perceived risk is amplified by a "ripple effect" (Yan et al., 2019).

Mobile apps linked to nutritional information databases are one of the most useful tools for providing healthy recipes to consumers (Huamaní-Cahuana & Cabanillas-Carbonell, 2021). For example, a mobile app created by the French nonprofit Open Food Facts enabled consumers to identify ultraprocessed foods among more than 145,000 packaged products (Open Food Facts, 2018). Contributors to the database use their phones to scan barcodes of food products at home or in stores and take pictures of ingredient lists and nutrition facts tables (Gigandet, 2012).

Uncontrolled or unmonitored food consumption and chronic stress promote obesity. Not knowing when to stop eating or how much food is too much can lead to long-term health issues. iLog, a system that monitors and creates awareness of food intake, indicates personal emotional state, and classifies eating behaviors including normal and stress eating. A deep learning model, mobile platform, and computer system for iLog use smart glasses to automatically detect, classify, and quantify foods on a user's plate with 98% accuracy and average precision of 86% (Rachakonda et al., 2020).

The agri-food industry uses thousands of chemical compounds to add color, stabilize, texturize, preserve, sweeten, thicken, add flavor, soften, and emulsify processed foods. According to the World Health Organization, these additives can contribute to top health risks including heart disease, hypertension, cholesterol, asthma, diabetes, allergies, and Alzheimer's. FoodWiki, a safe food consumption mobile system, provides traditional syntactic-based searches and semantics for rich searching to identify additives and other health and food safety issues linked to various ingredients (Çelik, 2016; Ertuğrul, 2016).

Media attention about smart farming innovations can multiply their impact on global sustainability. Information and Communication Technology (ICT) automation tools utilize technologies such as smart ear tags for livestock, smart sensors, the Internet of Things (IoT) applications, artificial intelligence, deep learning, big data, and robotics (Akkem et al., 2023). Mahfuz et al. (2022) identified smart tech stories that explain how farmers use ICT tools to automatically collect, process, and analyze data and for auto-identification and remote monitoring of livestock feeding behavior, environmental conditions, animal welfare, reproduction, health problems, nutrition and food quality, labor management, and other farm operations.

AI adoption across the food chain has increased agriculture income, enhanced competitiveness, and reduced costs. Deep learning algorithms of artificial neural networks have improved water resource management, yield prediction, price/demand forecasting, energy efficiency, fertilizer/pesticide usage, crop planning, personalized advice, and consumer behavior predictions (Ganeshkumar et al., 2023). Micro, small, and medium-sized agri-food enterprises are adopting Fourth Industrial Revolution digitization (aka Industry 4.0 or 4IR) solutions that promote stronger interconnectivity and smart automation. Industry 4.0 practices also provide more robust cybersecurity, enhanced food product information, more effective communication, and extended shelf life of products (Fernandez et al., 2021).

Food misinformation

Fake news is false information broadcast or published as news, often for fraudulent or politically motivated purposes. It includes misinformation, disinformation, opinion masquerading as fact, and other nontruthful communications (Rowe & Alexander, 2017a). Social media amplifies the influence of misinformation about food, in the form of "alternative facts" and "fake news." Misinformation and spin are age-old tactics in food policy-making (Perl et al., 2018). In some countries such as Thailand, the government focuses more on criminalizing food disinformation than debunking it (Pitigraisorn, 2021). Another approach in mitigating online disinformation is the increasing use of machine learning and semantic technologies to detect fake news about food (Demestichas et al., 2020).

Individuals often select stories from among numerous articles based on perceived self-relevance. However, selective exposure to news is not necessarily accompanied by critical judgment of news credibility (Huang, 2020). In democracies where ideas can openly compete in the marketplace for attention, misinformation generated intentionally or unintentionally can spread rapidly. Fake news can delay or prevent effective health care, in some cases even threatening the lives of individuals. While the rapid spread of misinformation dates back to the earliest days of scientific medicine, the internet has allowed instantaneous communication and powerful amplification of false information. Social media is a highly effective tool for propagating health-related and food/nutrition misinformation. Sensationalized food news easily crowds out accurate scientific information, which may be considered dull and difficult for people to understand (Wang et al., 2019).

To gain traction on social media platforms, some food product brands, ideas, and campaigns have created and disseminated compelling disinformation. For example, online disinformation tactics have contributed to brand monetization within an evolving GM foods narrative. A small group of alternative health and proconspiracy sites has received more total engagements on social media than mainstream media outlets that cover GM food topics (Ryan et al., 2020). False science news often takes the form of exaggerated or misleading reporting about research, reporting of fabricated or fraudulent research, misleading press releases, and communication about web-based scientific fantasies. When they can serve as expert interview sources, nutritionists and other scientists try to counteract this reporting by making readers more aware of food misinformation and more skeptical of false claims such as nutritional "cures," medical "breakthroughs," "miracle foods," and alarming health scares (Rowe & Alexander, 2019).

Social media channels enable the rapid spread of food-related misinformation, which reduces food literacy, healthy eating, and well-being (Steils & Obaidalahe, 2020). Diet influencers on social media are unreliable sources of food advice because they typically share misleading information about nutrition (Vasconcelos et al., 2021). Biases that can affect the quality of online food literacy include vividness, mindset, sociocultural bias, and cognitive dissonance. Lack of control over the quality of shared information and the risks of knowledge distortion affect whether food literacy is constructed positively or negatively on social media. Consumers contribute to food literacy through evaluation, adaptation suggestions, and critiques (Steils & Obaidalahe, 2020).

In an environment of internet-mediated democratization of expertise, every food blogger, social media user, and website owner is a potential subject-matter "expert." This trend is problematic for health communicators because of the declining public trust in science and the declining influence of official scientific information gatekeepers (Rowe, & Alexander, 2018). In this posttruth era, characterized by a less-skeptical view of facts, public trust in scientific and other expertise is evaporating, and scientists working in food and health no longer have influence in public discussions (Rowe & Alexander, 2017b). Health communicators also face growing challenges in delivering balanced, accurate, and credible food information that can compete with misinformation (Rowe & Alexander, 2017a).

Social media platforms present many pitfalls for communicating food risks and benefits, including inaccuracy, distrust, and lack of source credibility (Rutsaert et al., 2013). Expert messages must break through social media clutter to effectively address food-related misinformation. However, experts cannot effectively correct misinformation unless audience members first have a basic level of news literacy (Vraga et al., 2020). Wide diffusion of online fake news about food is making it more difficult for the public to recognize reliable food information and is deteriorating the relationships between public institutions and food industries. Social influence, lifestyle change motivation, and other individual psychological characteristics can play a role in promoting belief in online fake news about food (Castellini et al., 2021).

Food rumors on social media often include attractive headings and false but convincing scientific claims. Even though users are getting better at identifying misinformation, most people trust food information from social media influencers more than from the government. Food rumors often promote hope or fear (Pitigraisorn, 2021). However, "likes" and other social media popularity cues do not affect food risk misperceptions or the perceived credibility of experts (Bode et al., 2021).

The complexities of nutritional advice open the door to confusion, misinformation, and misinterpretations. For example, humans consume thousands of daily food combinations every day, which makes it difficult to identify a single dietary component's potential impact on health outcomes. The association between one nutritional factor and a clinical outcome does not constitute a cause – effect relationship. Similarly, contrary to popular belief, omega-3 fatty acids do not provide cardiovascular benefits (Monnier et al., 2020).

Rather than rely on expert advice, consumers with limited knowledge about the environmental impact of their food-purchasing decisions often resort to perceptions and heuristics to guide them instead. For example, with access to no information besides that contained on a food product's front label, consumer beliefs about the product's raw materials, transportation, and manufacturing processes influence whether they classify the food item as having a low or high carbon footprint. Brand reputation also plays a role in reducing consumer food categorization biases (Panzone et al., 2016).

To gain sufficient knowledge about carbon footprint information, most consumers need repeated exposure to the information before they will change their behavior. In one experiment, food product carbon labels with goals on virtual supermarket baskets prompted sustainable consumption behaviors and learning about carbon footprint, but only when consumers visited the online supermarket multiple times. Carbon footprint information on food products alone did not lead directly to more sustainable consumption behavior, regardless of whether it was presented numerically or as a five-color image of carbon footprints (Kanay et al., 2021).

When consumers guess the carbon footprint, calories, animal welfare impacts, and food safety of commonly eaten foods, most overestimate values and cannot accurately estimate carbon footprint and calorie content. However, these estimates are usually accurate for meat products (Miller et al., 2021). Even consumers with environmental knowledge have perceptual biases about sustainable food consumption. A "negative footprint illusion" can occur when a food choice that is both unsustainable and sustainable is thought to be less environmentally impactful than a sustainable food choice alone. For example, many pre-service science teachers in Turkey had perceptual biases about carbon footprint. After these student teachers evaluated the environmental impact of three-meal menu food types—a standard menu, a standard plus sustainable menu, and a standard plus unsustainable menu— they developed the false perception that the sustainable-addition menu had a lower environmental impact than the standard menu alone (Ateş, 2020).

Another form of "fake news" is media coverage perceived as biased. Describing solutions to social problems is an accepted and institutionalized aspect of US news media routines. Solutions-oriented news coverage can be included in food crisis stories to help citizens engage in policy deliberation. Still, some journalists argue that including solutions-oriented information in stories might create the perception that their coverage is unethical or biased. Journalists covering hunger crises have described an internal tension between maintaining a neutral, unbiased position and writing in a way that supports engagement and action. Ultimately, any journalists covering famines, food riots, and other international humanitarian agricultural crises must decide whether to include solutions-oriented information in these crisis stories (Kogen, 2019).

Another type of food bias in media coverage is speciesism, the bias against other animal species. Speciesism is a major food ethics concern because nonhuman animals suffer massive harm within the industrial farming complex. These animals are confined throughout their lives, and a high proportion are killed while still infants or juveniles. In coverage of the use of nonhumans as a food source in the New York Times (USA) and El País (Spain), Khazaal and Almiron (2016) found that both newspapers played a major role in concealing the nonhumans' cruel reality. El País conveyed crude speciesism, while the New York Times provided a camouflaged, more deceptive style in covering the topic.

Media framing of food sustainability

Media frames are storylines that provide meaning by communicating how and why an issue should be seen as a problem, how it should be handled, and who is responsible for it (Asplund et al., 2013). Frames become embedded within a text, make themselves manifest, and influence thinking. Media framing can create an awareness of the world that people experience, as well as aspects of existence that remain unexperienced, unexplained, and hidden (Tankard, 2001). Analyzing problem frames in media coverage can reveal underlying biases and values that determine environmental and food-related decisions. Entman (1993) identified various uses of framing and created a framing model widely applied across academic disciplines. It identifies several common emergent frames: problem definitions, explanations, evaluations, solutions, characteristics, causes, and consequences. For example, a solutions frame in stories that addresses links between climate change and technological innovation, individual lifestyle choices, or policy action could promote food sustainability solutions (Lee et al., 2014).

Sustainability messages are too abstract for many people unless the content can help them develop practical solutions and understand complex issues such as food sustainability, nutritional challenges, agricultural system sustainability, and food waste and food loss (Massari et al., 2021). Building upon prior experiences with sustainability can promote a better understanding of various trade-offs within food systems. Many consumers understand sustainability in terms of perceived trade-offs associated with using different agricultural resources over short-, medium-, and long-term periods (Yamashita et al., 2017).

Goody's five-phase framework highlights how stories frame food production, distribution, preparation, consumption, and disposal practices. In general, the news media in developing countries often highlight the affordability of fresh fruits and vegetables, while media in richer, developed countries emphasize long-term environmental benefits of sustainable food (Diaconeasa et al., 2022). Media coverage and social media framing shape consumer attitudes about sustainability, within audience segments that have different personality traits and worldviews. Consumers who care for the community, value diversity and gender equality, make decisions based on cooperation, and engage in other activities that reflect altruistic values are more likely to engage in prosustainable behaviors. They also have a stronger environmental self-identity, often characterized by biocentrism, ecofeminism, ecospirituality, or deep ecology. Consumers with egoistic values have weaker personal norms, which predicts a lower level of effective prosustainable behaviors that can reduce their carbon footprint (Ajibade & Boateng, 2021).

Some governmental entities use media framing about food as an agenda-building tool. For example, in partnership with local print media, the Singapore government used food to construct a national identity. Food-related stories about this initiative reflect self-improvement, ethnic culture, and cosmopolitan attitudes. All these elements are touchstones of Singapore's government-approved national identity. The government promoted more press coverage of cosmopolitan and foreign foods than local foods, to help situate the country as a global hub (Duffy & Ashley, 2012).

Media framing also can help translate new science to public audiences. Supply and demand factors can affect coverage of new food technologies, and the media can influence public perceptions and consumer behavior related to foods made with new technologies. News reporting about new agricultural and food technologies, especially biotechnology, has been characterized by agenda setting, framing effects, and the social amplification of risk (McCluskey et al., 2016). However, journalists often do not interview scientists to explore underlying causes of problems. In covering a crisis, the media often focus on its impacts without highlighting the role of climate change or other contextual causes of the situation. For example, news coverage about Brazilian droughts and floods conveyed the direct impact of these events but did not adequately highlight private sector risk management and how it can help to protect food supply chain disruptions and aid in climate change adaptation (Brito et al., 2020).

Many media frames in food sustainability stories have been positive representations. Stories about green lifestyles project an environmentally friendly existence. However, Craig (2019) found that media representations of sustainable everyday life across traditional media channels do sometimes connect home, lifestyles, and the local community to broader political, social, and economic contexts of sustainability. Online news stories about 3D food printing have framed food product fabrication technologies in positive ways, as futuristic, creative, healthy, efficient, and sustainable. These frames reflect broad food culture preoccupations, including time-saving convenience; novelty, entertainment, and leisure pursuits; effective food production and distribution; health and nutrition; environmental impacts; and global food security (Lupton, 2017).

In addition to sustainability framing in news coverage, a variety of prosustainable food arguments have been presented within recent American sustainable agriculture documentaries including "Food, Inc.," "The Garden," "Fresh," and "Farmageddon." Pilgeram and Meeuf (2015) argue that these documentaries fail to merge environmentalism and environmental justice. They frame sustainable food production around the normative issue of "good food" for capitalist consumers, while ignoring questions about community and cultural conceptions of farmers. The films also tend to promote the white family farm as the norm, which limits a reimagining of alternative food systems, the roles of diverse actors within those systems, and possibilities for concerned eaters beyond "voting with your fork." One exception was "The Garden," which suggested that cinema could help to radically reenvision participation in alternative forms of agriculture. Environment and sustainability documentaries—including films about food production, food processing, and sustainable (and unsustainable) patterns of food consumption—often communicate value judgments implicitly and explicitly. Most of these documentaries are reflexive and performative, conceptualized as having moral agency (Gottwald & Weder, 2021).

Some news coverage blames individuals rather than policies or systemic issues for the negative impacts of the agri-food industry. For example, coverage of nutritional ways to prevent bowel cancer often uses the "lifestyle" frame, laying responsibility for increasing dietary fiber at the door of the individual rather than looking at broader social, economic, or political drivers of dietary change. Similarly, news coverage of nutrition often highlights diet-related health problems such as obesity, type 2 diabetes, and coronary heart disease. In international news, these health problems are often framed as personal, lifestyle issues rather than problems requiring policy or structural change.

Nutrition research and media coverage often emphasizes ingredients or food components more than overall diet. The media often report diet-related research simplistically, without providing context. For example, in an analysis of UK news media coverage and British Medical Journal articles about dietary fiber and bowel cancer risk, the media typically disseminated research findings from press releases to national media to local media and shifted the "geographies of responsibility." The coverage of diet and bowel cancer research often highlighted technologies, news conventions, and media framing routines as decided by key gatekeepers (Wells, 2017).

While the media often portray unhealthy diets as the responsibility of individuals, this coverage also tends to ignore the impact of broader societal factors including food system processes, diversity of actors involved in food production, consumption, and distribution, and the impact of food systems on economic, social, and ecological sustainability. Food and diet topics continue to gain prominence in media coverage, in part because of public health problems, sedentary lifestyles, concerns about dietary habits, and the impact of food production on climate change. For example, Latvian print and digital media coverage of dietary practices used biomedical, psychological, and social practice frames.This coverage emphasized individualization, self-disciplining, gendering, and medicalization (Orste et al., 2021).

The American public holds mixed views about whether the government should combat childhood obesity. Media framing influences citizen views about the causes of childhood obesity and appropriate strategies for addressing the problem. US news coverage typically blames the food and beverage industry for childhood obesity, while framing solutions in terms of restrictions on the food and beverage industry and individual behavioral changes. US television news has more often focused on individual behavior change as a solution, while newspapers often identify system-level solutions that affect neighborhoods, schools, and the food and beverage industry (Barry et al., 2011).

Many specialized farming magazines provide informative climate-related articles for farmers and agricultural extension officers that frame climate change in terms of conflict, scientific certainty, economic burden, and action (Asplund et al., 2013). Western news typically frames edible insects as the solution to challenges caused by unsustainable global food systems. Some of this coverage frames insects as a "superfood," a marketing term for nutrient-packed foods promoted to consumers with promises of health, well-being, and beauty. However, the Western increase in food demand is causing negative social, environmental, economic, and cultural consequences for many companies that produce superfoods, often located far away from where the superfoods are consumed. Tensions arising from the media double-framing insects as both a superfood and an apolitical solution to climate change challenges could push insects toward an unsustainable future, unless sustainably driven actors can produce these foods on a local level (Schiemer et al., 2018).

Even when the media focuses on a specific news development, social media users often shift the story discussion to broader food-related frames. For example, online news article headlines, tweets, and Facebook comments on news stories about the release of Canada's Food Guide (CFG) online in 2019 highlighted nine prominent frames: food guide, impact, health, sustainable plant food, who will use, Canadian culture, food and consumption practices, meat, and dairy. Online news headlines alluded to potential impacts. Meanwhile, Facebook comments framed the CFG in terms of broader issues including health, foods, food consumption behaviors, sustainable plant-based foods, and meat. Tweets discussed intersects between health and food and consumption practices (Woodruff et al., 2021).

Media framing of sustainability initiatives and outcomes, through the perspective of historical change, sometimes leans toward just-so accounts and overly simplified stories. These narratives often highlight agrarian and environment successes, amid the grim realities of globalization and its impacts (Zimmerer, 2007).

Some stories romanticize farming to encourage public support for agriculture. For example, on Australia Day in 2011, a supermarket giant announced that it was slashing the price of its own-brand milk to a considerably discounted $1 per liter, a move immediately matched by its rival. In an unexpected media backlash, the dairy industry sought to expose the threat of major supermarkets to Australian dairy livelihoods. During these events now dubbed the "milk wars," media coverage of milk pricing—and the dairy industry campaigns that underpinned it—sought to galvanize media and public support through specific images and discourses associated with the "Australian farmer." The narratives and images the dairy industry used frequently and that had most cut-through with media were strongly invested in Australian pastoral sentimentality. These stories showed traditional, "small-scale," multigenerational family farmers being tragically forced off their land by the supermarket duopoly. The coverage revealed the power of romanticized depictions of farming as a media strategy for galvanizing public support for farmer livelihoods. However, the supermarket campaigns coopted these depictions to silence critique. Alternative food advocates all over the world use pastoral sentimentality as a means to encourage urban consumer support, but these depictions often do not effectively oppose corporate power in the food system (Phillipov & Loyer, 2019).

The growing need for a global "Plan B" has sparked agricultural research on long-duration space missions. In space, food production must withstand extreme conditions, including poor soil, lack of gravity, and radiation. These new technologies, funded by NASA and SpaceX, include genetic engineering, digital agriculture, 3D bioprinting, and synthetically grown meat. Shaw & Soma (2022) found that news coverage of food technologies developed for space largely has been uncritical, portraying them as inevitable and a positive good for humanity.

Even with broad coverage about the feasibility of long-duration space missions and future human settlements on Mars, few stories have addressed the societal implications of space-based food production innovations on Earth and how these discoveries might address systemic and structural inequalities in the food system. News coverage of 3D food printing has framed food product fabrication technologies in positive ways, as futuristic, creative, healthy, efficient, and sustainable. These frames reflect broad food culture preoccupations, including time-saving convenience; novelty, entertainment, and leisure pursuits; effective food production and distribution; health and nutrition; environmental impacts; and global food security (Lupton, 2017).

Food policy framing

Media framing can have powerful effects on agri-food policymaking. For example, media framing, knowledge, power relations, and public participation all play important roles in decision-making about changes to the urban edge of agricultural lands. The media often present two opposing issue frames about these choices: providing food vs. balancing complex needs to address housing and poverty. Common decision frames include balancing the complexity of conflicting needs, expert claims, and institutional knowledge. Opponents of decision-makers often rely on a knowledge base that emphasizes food security concerns, scientific knowledge, and written policy (Rose et al., 2016).

The news media cover a wide variety of rural planning and other agri-food policy issues, including food security, agrarian community welfare, farmland and landscape protection, protection of "community" in the face of agricultural globalization, dissenting views of about farming systems, global food commodity production, and conflicting land use priorities and values. Rural planning dilemmas, as reflected in media coverage, often occur on the periphery of urban areas where there are competing notions of farming and landscapes, as well as different rural lifestyles. Rural food policymaking and other governance often reflect preferences about food consumption, farming systems, and high-yield agriculture practices (Butt, 2019).

The way audiences make sense of scientific knowledge about food is important for food policymaking. Conflicting media frames, rooted in differing dietary recommendations, confuse consumers and cause distrust. Mostly issued by agri-food government agencies, official dietary guidelines are often shaped by power plays among interest groups, food controversies, emerging expertise, and companies that produce sugar, dairy, and animal products. They have considerable influence over food preferences and consumer purchasing, which can present ethical challenges. Most dietary guidelines around the world address nutrients, processes, and meals. Nutrient-oriented guidelines emphasize individual health and foods that contain one or more "healthy" nutrients. This framing implies that individuals can take responsibility for their own long-term health and interpret their own nutrition data. Process-oriented guidelines, designed by nutrition scientists, categorize foods and ingredients according to minimal, refined, and reconstituted production processes. Meal-oriented guidelines emphasize preparation of fresh food for shared meals, as formulated by politicians, social scientists, and fresh food producers (Korthals, 2017).

Palm oil, the world's most popular vegetable oil, is widely criticized for creating environmental, socioeconomic, and health sustainability challenges. While environmentalists push for elimination of palm oil as a food ingredient, palm oil producers are active in shaping consumer opinion over the benefits and the opportunities of palm oil use. The palm oil debate has intensified among other agri-food chain stakeholders. Food manufacturers and retailers typically adopt a purely informative approach on its sustainability certification and avoid engagement about sensitive issues. One study found that among 463 European palm oil agri-food chain companies, 43% had an active Twitter account, including world palm oil producers, bakery and chocolate manufacturers, and food retailers. Among tweets from palm oil agri-food chain actors, all promoted sustainability. Health was the most mentioned sustainability dimension (Ruggeri & Samoggia, 2018).

Individuals must understand food information before they can make informed decisions about food risks and food intake recommendations (Gaspar et al., 2018). The greater the knowledge gap about food supply chain issues within public audiences, the greater the importance of media in framing public opinion about those issues (Demke & Höhler, 2019). A health message or story with a regulatory focus can lead someone to avoid junk food consumption if it frames personal behavior in terms of gains instead of losses. Perceived health risk can amplify health consciousness, which promotes the avoidance of junk food. In addition, using the correct type of health information or health claims for relevant consumer segments, in a message with a disease prevention or health promotion focus, can effectively encourage consumers to avoid junk food (Shimul et al., 2021).

Food policymaking coverage and social media discussions often accommodate a diverse range of stakeholder interests, including those with localized concerns about how food should be produced and traded. However, social media influencers and people quoted in news stories can impede efforts to achieve more open, sustainable, and integrated food chains. Inadequate public information, misinformation, and poorly informed public opinion are constant challenges for food policy analysts. The volume of erroneous evidence and difficulties encountered in differentiating opinions from verified, factual food policy inputs have increased significantly in recent years. Many models and practices for food policymaking were developed during an earlier era, when a sharper duality between fact and fiction existed. Since then, policymakers have had to grapple with "truthiness," using outdated policymaking models that do not account for social media impact and other forces affecting policy processes (Perl et al., 2018).

Food supply chain actors perceive and explain agricultural markets based on their perceptions of various threats and opportunities that market volatility creates for farms and agri-food firms, as well as the reasons for volatile prices (von Zazie et al., 2010). Amid fears of vulnerabilities related to climate change, price volatility, and global food shortages, an increasingly conflicted policy field has emerged. Competing narratives in the media have provided rival orientations about food policy, land rights, the environment, and justice (McCarthy & Obidzinski, 2017).

No "silver bullet" message or policy solution exists to shift food choices toward sustainable healthy diets. However, in light of global knowledge about the transition to healthier diets and sustainable food systems in different countries, it is clear that simultaneous action among the public sector, private sector, and governments is needed to achieve success (Clark et al., 2020). Although many sustainability stories about the agri-food industry do not mention actionable measures or producer responsibility, some do discuss a broader context for specific sustainability challenges. For example, UK coverage of marine plastic pollution has discussed policy efforts and behavioral contributions to the problem, including the impact of consumer use of plastic straws and food packaging eventually eaten by small marine animals (Keller & Wyles, 2021).

Food politics has become a dominant feature of popular media, including television, social media, cookbooks, and advertising. According to media production staff, reality TV contestants, celebrity chefs, food producers, and food retailers, food politics is shaped by consumers and consumption ethics, as well as political and economic imperatives within media and food industries (Phillipov, 2017). American support for climate policy has been mainly understood in a political affiliation context, with Democrats supporting such policy and Republicans not. With the rise in antielite populist sentiment, populist sentiment about climate change policy operates inside and outside the electoral sphere. Public reactions to climate change messages from Pope Francis and the American Association for the Advancement of Science (AAAS), two global elites, were linked to the perceived elitism of a climate change messenger, as well as anger and fear responses, in shaping support for climate policy (Myrick & Evans Comfort, 2020).

Media coverage, public policies, online discussions, and marketing channels have all influenced the development of the organic foods sector. Organic food markets have evolved through relationships among policymakers, farmers associations, advocacy groups, and actors along the food chain. The dynamics of this process have been unpredictable, as they have depended on factors such as maintenance of family farms, environmental protection, gastronomic heritage, fairness in the food chain, and export promotion (Darnhofer et al., 2019).

The decision to spend more on organic fruits, vegetables, eggs, and meat/dairy products is often motivated by perceived freshness and quality, as well as concern for their own health and that of their loved ones. They seek information about organic foods primarily from experts—including dietitians and physicians—as well as from family members, social media content, and websites managed by institutions (Wojciechowska-Solis et al., 2022). The news media drives public opinion on organic food by determining discussion topics. Perceived salience of organic food coverage in news stories significantly influences their salience in reader comments, but not vice versa (Danner et al., 2022).

Some media content has promoted policy support for farmers, such as coverage that promotes consumer attachment to rural identity and a shared national sense of "who we are" (Fountaine & Bulmer, 2020). However, agricultural politics are increasingly contested. The practice of growing grain crops for bioenergy instead of food has emerged as a prominent controversy in media coverage of agri-food policy. Public debates about this issue increasingly spotlight ethical concerns about responsible land use. Although gaining independence from fossil fuels is an important climate mitigation strategy, biofuels production competes with other forms of land use and may compete, directly or indirectly with food production, causing food insecurity, malnourishment, and food poverty. Bioenergy production also may lead to deforestation and other land use changes that can raise GHG emissions (Gamborg et al., 2012).

Media agenda building is a process in which stakeholders try to embed their interests into media coverage, to influence policy decision-making, or to influence what types of policy outcomes will be produced. Some agri-food industry media agenda-building efforts have attempted to neuter policies that promote the health of people and the planet. For example, the baby food industry engages in diverse political practices to foster favorable policy, regulatory and knowledge environments, including lobbying international and national policymakers, generating and deploying favorable science, leveraging global trade rules, and adopting corporate policies that counter government regulatory action. Such practices have been a major impediment to global implementation of the International Code of Marketing of Breastmilk Substitutes and other policy actions designed to protect, promote, and support breastfeeding (Baker et al., 2021).

Although social media hype about agri-food controversies can strongly affect policy, it is almost impossible to predict. For example, an analysis of five cases of peak social media activity in the Dutch livestock sector identified four dimensions of social media hype: peak patterns of activity, issues and frames, interaction of actors, and media interplay. Stakeholders instigate and frame social media hype about agri-food controversies. These hype posts usually revolve around activism, scandals, and conflicts and use organizing concepts, a judgmental tone, and a hashtag. Peak activity, typically focusing on a few themes, is recurrent (Stevens et al., 2018). Individuals and organizations already invested in an issue are more likely to use specialist hashtags on social media (Maye et al., 2021). The Instagram profiles of Portugal's National Health Service and Brazil's Ministry of Health showed that their national health promotion agendas, including food consumption advice, often competed with political topics, and their interaction with followers were limited (Pinto et al., 2021).

Food on social media

Social media uses and gratifications

Ten uses and gratifications motivate people to use social media: social interaction, information seeking, pass time, entertainment, relaxation, communicatory utility, convenience utility, expression of opinion, information sharing, and surveillance/knowledge about others (Whiting & Williams, 2013). Adolescents consume food media on different social media platforms mainly for education, social utility, and entertainment, and their motives varied across different platforms. They watch online TV cooking shows mainly for companionship and entertainment, while they use online food media more often for information, inspiration, and social interaction. Some seek online food media to improve health and fitness or to find appetizing and esthetically pleasing food content (food porn). Friends and family, as well as existing food preferences, can also influence adolescent uses of online food media (Ngqangashe et al., 2021).

One of these social media uses, social interaction, can be seen when people post about eating situations, refer to foods and beverages they consume, or express emotions about food based on whether they are talking about breakfast, lunch, dinner, or snacking (Dondokova et al., 2019).

Information sharing on social media promotes sustainable food consumption. Although sustainable food consumption behavior acculturation often occurs differently across various ethnic groups, social media interventions can promote this acculturation by providing information that consumers need before they are asked to engage in sustainable behaviors (Choudhary et al., 2019). Popular food bloggers engage in information sharing about cocktails, cooking, special diets, and culinary travel within their own food subcategories on Twitter (Hepworth et al., 2019). Another food information sharing source on social media is the celebrity chef, a rising phenomenon of contemporary popular culture reflected in the popularity of TV cooking programs, bestselling cookbooks, and chef biographies and autobiographies. The celebrity chef has a growing influence on consumer food habits and choices through their various roles, including media performer, writer, entrepreneur, role model, and rebel (Giousmpasoglou et al., 2020).

When people share information about themselves on social media during a natural disaster, their aggregated posts highlight in real time how they are responding to food-related threats during the crisis. Social media analyses detect dietary patterns during crises and inform public health social media campaigns that advise people about stocking up on healthy, nonperishable foods ahead of natural disasters. In tracking high-frequency Twitter mentions of foods/beverages during four hurricanes, grains are the most frequently mentioned food group prehurricane, and dairy was most frequently mentioned during hurricanes. The top five most mentioned foods and beverages overall are milk, pizza, turkey, oranges, and waffles, and the foods most frequently mentioned in the tweets are typically high-calorie and protein-dense (Turner-McGrievy et al., 2020).

Information seeking on social media is an important factor in designing effective health campaigns. For example, efficient use of social media is key to raising awareness about balanced nutrition and food waste (Özkaya et al., 2021). Seven factors predict whether people seek out information about healthy eating: individual characteristics, perceived hazard characteristics, affective response to the risk, felt social pressures to possess relevant information, information sufficiency, one's personal capacity to learn, and beliefs about the usefulness of the information in various channels. People who put more effort into health information seeking and processing are more likely to develop long-term, risk-related understanding, attitudes, and behaviors. Healthy eating campaigns should help people anticipate and cope with the potential volatility of their behavioral changes (Griffin et al., 1999).

Learning about peer behavior on social media can promote healthy eating habits. For example, social media users have been persuaded to consume a larger proportion of low-calorie foods than high-calorie foods, after they were exposed to socially endorsed images of low-calorie foods on social media (Hawkins et al., 2021). Instagram and Facebook are among the most popular platforms that influence adolescent healthful eating behaviors—defined as fruit and vegetable intake—as well as unhealthful eating behaviors prompted by fast-food advertising. Online forums serve as accessible channels for eating disorder relapse prevention among youth. Social media influences on teen eating habits include visual appeal, content dissemination, socialized digital connections, and adolescent marketer influencers (Chung, Vieira, et al., 2021).

Another aspect of social media learning about others is the ubiquity of food-related norms about neighborhoods and other geographic locations of interest to consumers. For example, the "healthiness" sentiment of food-related tweets is sometimes associated with neighborhood characteristics such as income, race, age, population density—and obesity-related health outcomes in those communities. Foods high in saturated fat such as pizza, bacon, and fries are mentioned more often in less-affluent areas. Food-related tweets often mention activities such as eating, drinking, and cooking, locations where food is consumed, positive attitudes such as affection, cravings, and enjoyment, and negative attitudes such as food dislike, personal struggles, and complaints (Vydiswaran et al., 2020). Users often promote their own tweets by using real geographic references based on food experiences there and add hashtags referring to the place where the food mentioned in the post was produced or consumed or another category of food, such as wine. For example, tweets use the word "food" with different connotations in different regions of Italy. In posts about Sicily, "food" is often paired with general-use hashtags to accompany shared text or multimedia posts. For tweets about Tuscany, "wine" is often paired with "food" and geographical references (Platania & Spadoni, 2018).

Social media information sharing can improve local food systems. For example, Italians rarely use social media platforms to exchange surplus food or goods because they lack knowledge about this practice, and this type of online market is uncommon. To address this need, social media platforms could promote and facilitate social food-sharing practices (D'Ambrosi, 2018). Information sharing on social media can have both good and bad effects on healthy eating. Food consumers carry food messages, which reflect both healthy and unhealthy food choices, food consumption, and nutrition knowledge. Food content on social media is often both virtuous, leading to an increase in consumer knowledge and information, and bad in terms of leading to unhealthy food consumption practices. The types of food content on social media include user-generated content, information, risk communication, and digital marketing (Ventura et al., 2021).

The characteristics of Instagram including its immediacy, usability, and visual appeal, make it a useful intervention tool for sharing information about healthy eating among teens. For example, high school students exposed every day for a month to the Instagram Teen Food healthy eating page had an increased intention to avoid junk food consumption and a reduction in their consumption of unhealthy snacks (Capasso et al., 2020). One form of information sharing that promotes engagement with climate change and other issues is experiential-based learning, in which students produce social media content and podcasts that reflect their insights. This approach gives youth the skills needed to handle the complexity of the food system, as well as a sense of global citizenship, a feeling of belonging to a broader community, and strategies for increasing awareness of food security and sustainability (Allievi et al., 2018).

Social media influencers have shared content about green food options, food choices, zero waste, and other sustainable lifestyle topics, in order to promote more sustainable consumption patterns. Yildirim (2021) found that female microcelebrities promote sustainable food consumption patterns far more often than green male social media influencers do. Expression of opinions on social media is a critical factor in constructing meanings about food.

Expression of opinions on social media is a critical factor in constructing meanings about food. Food personalities, celebrity chefs, and other food influencers create, curate, and share the meanings of good food on social media using "good food" grammars. However, these grammars often reflect white, heteronormative middle- and upper-class privilege (Goodman & Jaworska, 2020). The notions of good food often revolve around clean eating or clean lifestyles that combine healthy, "free from" diets with fitness regimes.

Clean eating, the dietary practice of consuming foods deemed to be pure and healthy, is often framed as a form of moral food consumption that embraces particular foods while eschewing others (Walsh & Baker, 2020). Twitter users associate healthy food with a healthy lifestyle, diet, and fitness. Foods they typically associate with the #healthyfood hashtag are vegan, homemade, and organic (Pilař, Stanislavská, & Kvasnička, 2021; Pilař, Stanislavská, & Kvasnička, et al., 2021).

Similarly, Instagram users associate healthy food with healthy lifestyle, fitness, weight loss, and diet and with vegan, homemade, clean foods, and plant-based foods (Pilař, Stanislavská, & Kvasnička, 2021; Pilař, Stanislavská, & Kvasnička, et al., 2021). Users express their attitudes toward food innovations within a local or regional context, as shown by user sentiments in tweets. Food marketers can use Twitter to identify regions of interest, to identify local consumer perceptions and attitudes toward their products (Pindado & Barrena, 2021). Food cultures evolve when people share opinions, information, and knowledge about food on Twitter (Platania & Spadoni, 2018). Virtually all contentious agri-food issues discussed in German social media posts are framed in a two-sided way. However, the productivity-driven agri-food industry is generally framed as a negative development, with posts favoring natural food production (Boehm et al., 2010). Virtually all contentious agri-food issues discussed in German social media posts are framed in a two-sided way. However, the productivity-driven agri-food industry is generally framed as a negative development, with posts favoring natural food production (Boehm et al., 2010).

Social media conversations about veganism or that challenge meat and dairy-based diets can help people reconfigure their daily eating habits and explore the entire sphere of eating-related practices, including food production, distribution, purchasing and cooking food, and sharing stories and experiences of veganism. These discussions create wider communities that support and facilitate sustainable eating (Laakso et al., 2021). Overall, consumers who regularly read news stories via social media are more likely to pursue a vegetarian diet, if they also have vegetarian friends or family members. This is because news consumption via social media increases understanding of food issues, and building this knowledge base encourages readers to pursue a vegetarian diet (Kley et al., 2023).

Information seeking, as a social media use, can be quantified using social media analytics and Twitter API tools. These tools can be used to identify which topics that social media users interact with, including nutrition, population nutrition health, interdisciplinary health collaborations, alcohol consumption, dieting, and eating-out-of-the-home behaviors (Stirling et al., 2021). Twitter highlights social representations of food trends in different regions of the world (Pindado & Barrena, 2021).

Most people seek food information mainly through online sources. The proliferation of social media applications including online communities, social networking sites, and blogs, provides numerous means for receiving and providing food information. Structural changes in social media—including an increasing capacity for storing, retrieving, and reusing information—offer new opportunities for communicating food risks and benefits. The increased availability of social media apps and other digital tools has facilitated the growth of interactive communication among consumers, creating opportunities bloggers, recommender systems, and other third parties to become food information providers (Corvello & Felicetti, 2014).

The benefits of food literacy include nutrition literacy and healthy cooking behavior. Information seeking within social media groups can boost food literacy. For example, members of the Facebook group "Homemade Food for Babies" often use group content including recipes, guidelines, and advice to enhance their perceived food literacy. However, those who participated more actively and longer in the group did not demonstrate a higher level of food literacy. Even so, their motivation to participate, use expert advice, and try group recipes promoted healthy cooking behavior (Cupar & Juric, 2019).

A key gratification of social media use is relaxation, and many posts about foods and beverages reflect the quest for relaxation. For example, coffee consumption tweets often convey favorable emotions, wellness, energy, a positive state of mind, and an enjoyable and trendy lifestyle. Coffee is one of the most popular beverages in the world, and an increasing scientific consensus asserts that coffee consumption has beneficial effects on the human body. Many coffee-related tweets also express a positive perception of coffee benefits, especially as related to mental and physical well-being (Samoggia et al., 2020).

As a social media gratification, online social interaction plays a role in peer influence. Social media, peers, and siblings play a prominent role in shaping children's food-related consumer socialization. Social media exposes children to the food and drink products that children frequently use and enables them to interact with peers about food and share their food-related experiences. Peers influence children's healthy eating behavior more than siblings, but siblings often influence children to eat together with the whole family (Ragelienė & Grønhøj, 2021).

Food branding

Social media usage has a stronger impact on consumer satisfaction and sustainable food-purchasing behaviors than price satisfaction, packaging quality, and perceived brand quality (Yakubu et al., 2022). Opportunities for food and beverage companies to actively interact with current and prospective customers outweigh the potential pitfalls of social media interaction (Meixner et al., 2013). However, many food companies still have not utilized the potential of social media to promote their brands, even when these companies are upstream in the value chain, are invisible to the public, or are located in rural peripheral areas or urban cores (Carpio et al., 2020).

Similarly, most health promoters have been unable to reach and engage people on social media to the extent that Big Food brands and lifestyle personalities have. To be effective, they need to use strategies associated with higher social media engagement, to provide food-related health content. In an analysis of the most engaging posts from Facebook and Instagram's 10 most successful nutrition and food-related accounts, lifestyle personalities recorded the highest absolute engagement, while health promoters recorded the highest engagement relative to follower count. Strategies associated with higher Facebook engagement included the use of hashtags and post engagements prompted through announcements. On Instagram, higher engagement was associated with higher caption counts, providing health information links, prompting engagement through strategies that require action, and using humorous strategies. Strategies associated with lower

Instagram engagement included reposted content, general encouragement to eat strategies, encouragement to exercise strategies, not inducing any emotion/hedonic sensations, and providing a negative tone (Barklamb et al., 2020).

More than 80% of young adults use social media at least once a day, yet health and nutrition campaigns struggle to gain traction with young adults. In one analysis, the most engaging Facebook and Instagram posts by 10 food industry and lifestyle brands and six health promotion organizations included photos, videos, and links to purchasable items. These engaging item posts often featured body image messages and content that induced positive emotions. Facebook interactions were lower when posts used pop culture, relatable content, storytelling, or visually appealing graphics in their posts. The more that young adults interact with Instagram food posts, the less weight they lost (Klassen et al., 2018).

An example of Big Food marketing ubiquity is the baby food industry, whose global milk formula sales grew from US $2 billion in 1978 to $56 billion in 2019. This boom occurred because of widening geographical reach and marketing practices targeting the Global South. This trend has raised serious concerns about its impacts on child and maternal health, including breastfeeding. To grow and sustain milk formula consumption, the baby food industry markets these products through health systems, mass media, and digital advertising to promote product innovations backed by corporate science. In the realm of social media, the baby food industry's "first-foods systems" campaign reached diverse markets, to drive milk formula consumption on a global scale. Then the industry broadened its products beyond infant formula, to include follow-up, toddler, and specialized formulas for a wider range of age groups and conditions (Baker et al., 2021).

Small-scale food producers are increasingly using social media marketing, as well. Female and younger farmers, as well as incorporated agribusiness farms, are more likely to invest in digital than traditional media advertising. However, the beneficial effects of digital media advertising are not equal across all types of farms. Family farms and farms located in rural areas tend to benefit more from digital advertising (Chung, Chang, et al., 2021).

Consumers are more likely to purchase local food when it is labeled in ways that display specific sustainability attributes (Kusá et al., 2021). Four message frames in advertising and labeling have increased the consumption of local food: gain, nonloss, nongain, and loss (Carfora et al., 2022). Proenvironmental consumers are more persuaded by messages formulated in terms of gains and nongains, while health-conscious consumers are more persuaded by messages framed in terms of losses or nonlosses.

Some food and beverage brands have distinct personalities on social media that generate millions of likes and retweets. The personification strategies they use include humor, trendy language, absence of food product mentions, trending topic references, current event references, internet meme references, and niche audience targeting. The brands that spent the most on advertising have similar or fewer retweets per follower than the brands that spent relatively little on advertising (Greene et al., 2022). Parasocial relationships between audiences and celebrities on social media can promote positive public attitudes about food sustainability and activist behaviors. For example, celebrity use of first-person pronouns on Twitter led to more positive attitudes about climate change, and celebrity fear appeals were more effective than hope or emotional appeals in driving audience participation in climate activism (Park, 2020).

Many food companies use Instagram marketing to increase users and the word-of-mouth effects generated by hashtags. Among Instagram users, food-content quality is based on perceptions of accuracy, relevance, conciseness, and information usefulness. Source trustworthiness and hashtag scalability boost credibility and perceived information usefulness. Those who find food information useful and intend to share Instagram food content are more likely to follow food information advice (Lee et al., 2021). Instagram food ads endorsed by celebrities generate more pleasure and arousal in audiences than food ads endorsed by food experts do. Food ads that use complex visuals generate more pleasure and arousal than simpler visuals. Food brand marketers are often more successful if they use celebrities and increase visual complexity in ads by adding more objects, colors, objects, or textures, and by incorporating asymmetric elements (Kusumasondjaja & Tjiptono, 2019).

A livestreaming environment can enhance food and beverage advertising effectiveness. Livestreaming platforms such as Twitch can amplify influencer marketing of foods and beverages, as compared with asynchronous social media such as YouTube. Through a survey of Twitch users, most recalled seeing at least one food or beverage ad on Twitch. After observing an advertised product, 14% reported craving it, and 8% reported purchasing one. In chat rooms, 56% saw conversations related to food, and 25% participated in such conversations.

Of the Twitch users who also used YouTube, 65% reported negative emotions when encountering advertising on YouTube, as compared with 40% on Twitch. A higher proportion felt Twitch's advertising primarily supports the content creators, while a higher proportion feel that YouTube's advertising primarily supports the YouTube platform. Overall, food marketing exposures on Twitch are noticeable but less bothersome to users and influence their consumption and purchasing behaviors (Pollack et al., 2021).

Perceived informativeness, dynamism, and enjoyment motivate consumers to visit food retail chain pages on Facebook. Facebook users also visit food retail chain pages to obtain information about discounted items or to consult recommended recipes, enter contests, or learn about new food products available in stores (Ladhari et al., 2019). Some food companies use digital tools to discover how they are perceived on social media. For example, Tagxedo can be used to reveal the distinctive image of a company, while How sociable can track its social media key performance indicators, such as digital traffic and visibility (Carpio et al., 2020).

Consumer behavior is the most unstable and unpredictable part of the entire food supply chain. Measures to influence consumers to switch to more sustainable consumption are complex. The most important predictors of sustainable organic food consumption are support and guidance from news media, government, and educational and research institutions, as well as the educational level, income status, and age of consumers (Huynh et al., 2021). Acceptance of organic foods varies by culture and region. For example, most Indonesian Twitter users accept organic foods as healthy, while the rest feel it is too expensive (Rintyarna, 2021).

The desire to avoid health risks also predicts organic food consumption (Giampietri et al., 2020). Some healthy foods are promoted on social media by consumers rather than marketers. For example, most tweets about kale highlight its popularity as a healthy food and its antiinflammatory and immunity-boosting properties. However, these tweets rarely refer to the taste of kale, a mention that could shape healthier consumer behaviors (Kāle & Agbozo, 2020).

After years of discussions with experts on the proper definition of the term "organic food," the US Department of Agriculture published its proposed set of standards for the certification of these products, as well as draft legislation. The proposal was the first time the US government had attempted to define the organic food category (Lezaun, 2017). Credence attributes are qualities of a food product that consumers cannot ascertain, even after they purchase it. For example, since consumers cannot tell if a product is organic or "rainforest-friendly" even after buying and tasting it, they must rely on the reputation of the product's certification to ensure that the product carries the positive value they paid for (Pascucci, 2010).

A major attitude – behavior gap among environmentally responsible consumers is the difference between their food-purchasing attitudes and their actual buying behavior. The main barriers to purchasing sustainable, organic food are price, lack of immediate availability, sensory criteria, and lack or overload of information. Low-involvement barriers include a lack of fit with well-established consumption routines, as well as a lack of transparency and trust in organic food labels and certifications. Low-involvement reasons to purchase familiar food products include daily purchase routines and relatively low prices. These routines often result in fast, automatic, and subconscious decisions. In contrast, sustainable food consumption behavior involves decisions based on slow, logical, conscious, and arduous rationales. Considering the strong psychological barriers among consumers and the continuing low market share of organic food, proactive policy measures would need to require more transparent labeling and certifications and promote consumer and corporate responsibility and provide incentives for sustainable production (Terlau & Hirsch, 2015).

Consumers expect that leaders will ensure the safety and healthiness of foods and beverages and protect food prices. These expectations outweigh their own willingness to act. However, food labeling and economic incentives can motivate individual involvement in addressing food-related health concerns (Tepper et al., 2020). Using food labels to convey food safety information supports premium prices for food safety. Consumers are willing to pay a price premium of up to 169%, for food products that are treated to resist a specific food-borne risk factor and that are certified to be safe, tested, or inspected by public or third parties. Food safety labels are insufficient when food products are affected by hazardous and risky events, but this labeling could lower the mismatch between objective scientific-based risks and subjective perceived risks (Santeramo et al., 2021).

Similarly, "traffic-light" labels can improve consumer knowledge about food product characteristics, including calories, carbon footprint, whether the product is healthy or unhealthy, and whether it is good or bad for the environment. Traffic-light labels also improve consumer ability to rank products by calories and carbon footprint, rather than use metrics. The labels also can help them more effectively classify food products as unhealthy or good for the environment. Traffic-light labels ultimately can help consumers reduce their carbon footprint, by making it easier for them to understand carbon footprint numbers to inform more sustainable food choices (Panzone et al., 2020).

Sometimes marketers use words used to characterize organic foods and other healthy dietary choices to sell unhealthy, nonfood products. For example, e-cigarettes that were introduced as a less harmful alternative to combustible cigarettes now associate e-cigarettes with healthy foods in marketing campaigns. This messaging can mislead people to assume that vaping is a healthy practice. Both marketers and nonmarketers post tweets that include vaping keywords along with healthy food-related labels including natural, vitamin, vegan, or organic. Most of these tweets represent vaping as harmless, health-enhancing, and compatible with a healthy lifestyle, rather than refer to it as a smoking-cessation device (Basáñez et al., 2018).

Junk food marketing

The global consumption of ready-to-eat food products with low nutritional value is increasing. This trend highlights the danger of hedonistic advertising messages in shaping unhealthy habits. Unhealthy eating contributes to many medical conditions, including malnutrition. Beyond individual nutrition and physical activity, government policymaking, commercial activities of large food companies, and food production sustainability practices also contribute to eating habits (Marinescu & Rodat, 2020).

To promote both environmental sustainability and healthier lifestyles, advertising appeals must emphasize the quality of food products, because this perception influences individual dietary goals. Food ads that focus on the environment and appeal to self-improvement goals are more effective when they also appeal to the ways that individuals perceive themselves in relation to others. The same approach is effective for framing food ads that appeal to health concerns (Banovic & Barone, 2021).

The lower the nutritional value of a food product, the greater the proportion of positive emotions that emerge from exposure to its advertisement, especially when the ad uses hedonistic language (Blasco & Jiménez-Morales, 2021). Unconscious instinct can amplify ad persuasiveness. According to the insurance hypothesis, humans have developed an instinct and physiological urge to seek high-calorie foods and can detect the threat of food scarcity. Regional consumption of energy-dense foods contributes to climate change and health problems. In turn, climate change – induced food scarcity increases preferences for these foods, especially among women. Food scarcity highlighted in the media can induce overconsumption of high-calorie foods among people who are not at risk of famine. Thus, climate change communications should not discuss food scarcity issues (Folwarczny et al., 2021).

Big Food corporations often capitalize on nutritionism, the reduction of food's nutritional value to its individual nutrients. This messaging has enhanced their power and position in global processed and packaged food markets. Through lobbying and participation in nutritionally focused public – private partnerships, Big Food corporations try to directly influence policy agendas, governance, media coverage, and public opinion. Through market dominance in the nutritionally enhanced foods sector and participation in nutrition-focused, policy activities in agri-food supply chains, Big Food corporations are using public outreach and the media to present their views about the nutritional aspects of their products (Clapp & Scrinis, 2017).

Sporting events allow food companies to reach massive numbers of consumers. Many transnational food companies conduct major marketing campaigns via global sporting events such as the Olympic Games, FIFA World Cup, European Football Championships,

British football, and the English Premier League. Despite their advertising success, public health advocates have not scrutinized most prominent sports marketing and sponsorship portfolios that promote unhealthy foods (Ireland et al., 2021).

Food industry magazines often focus on ultraprocessed foods, as well as health-based marketing of snacks and "protein-rich" products. Proactive food marketing approaches include large investments in health-related promotion of ultraprocessed foods, but consumers who try these products often lack access to sufficient information about them, cannot assess their quality, and cannot compare them with alternative products. Sometimes, defensive food marketing strategies reflect a "Red Queen" effect, in which food companies promote health-related actions to keep up with competitors. In turn, these competitive strategies can undermine public health efforts to support objective product comparison and healthier choices (Cuevas et al., 2021).

Nine prominent Australian ultraprocessed food industry actors, including major trade associations, used Twitter to influence food and health policy debates. Hunt (2021) identified seven overarching lobbying strategies in these tweets: coopting public health narratives; opposing regulation; supporting voluntary, co- or self-regulation; engaging policy processes and decision-makers; linking regulatory environments to the need for ongoing profitability; affecting public perceptions and value judgments; and using ignorance claims to distort policy narratives.

Globally, the increased consumption of ultraprocessed food has contributed to an uptick in adverse health outcomes. The types and quantities of ultraprocessed foods sold worldwide have expanded, representing a transition toward a more processed global diet, especially in highly populated, middle-income countries. Ultraprocessed food sales also are growing rapidly in other regions because of the industrialization of food systems, technological change and globalization, market growth, political activities of transnational food corporations, and inadequate policies to protect nutrition (Baker et al., 2020).

Many developing countries are facing a rapid increase in the prevalence of overweight and obesity status, which has nearly reached the levels seen in high-income countries. The rise in obesity is caused by a major nutrition shift triggered by the higher affordability and consumption of heavily processed and otherwise unhealthy foods that contain high amounts of added sugar, fat, and salt. Obesity prevalence is also exacerbated by increased access to large food retailers (Big Food) and fast-food restaurants. Easy access to Big Food stores and fast-food restaurants increases the prevalence of overweight and obesity status significantly, even among those with lower socioeconomic status (Otterbach et al., 2021).

Eating out and take-out meals from restaurants can greatly increase daily caloric intake, but it has steadily increased in many countries over the last few decades. Some nutrition interventions have persuaded individuals to reduce their self-reported frequency of eating away from home and school and on the weekends (Wu et al., 2022). When people eat out, they often do not see or understand nutrition information, which can lead to a lack of consumer control and fewer healthy and sustainable food choices.

However, restaurants can use digital tools to provide enhanced dish information, online apps, or web-based platforms to provide better information. Quick and clear restaurant information should be easily accessible to consumers and highlight its relevance to good dietary choices (Bray et al., 2020). Plamondon et al. (2022) found that individuals exposed to both nutritional and environmental menu information more frequently chose meals with the highest scores for greenhouse gas emissions, nutritional quality, and eco-efficiency. However, exposure to this information did not affect the amount of food consumed. The menu information also persuaded diners to reduce food waste while they were eating out. Further, Huang (2020) found that persuasive messages that reference the environment can influence buffet diners' intentions to reduce food waste.

A major consequence of unhealthy food consumption is the prevalence of type 2 diabetes. Public perceptions about diabetes shape policy interventions, and media framing of diabetes has influenced policymaking largely because it has shifted public opinion about different ways to combat the disease. Through a content analysis of media framing of diabetes in the *New York Times* over a decade, Stefanik-Sidener (2013) found that nearly a third of the articles failed to differentiate between type 1, type 2, and gestational diabetes or to note differences in the root cause and treatment of each. An examination of frames used for each type of diabetes showed that the overall dominant frame across types was either a medical or behavioral frame, with insufficient use of a societal frame. More societal-level framing of diabetes could help the public see the wider consequences of the disease and increase the likelihood of public support for policy solutions to combat it.

Food producers add more than 3000 chemical compounds to processed foods, to add color or flavor, stabilize, texturize, preserve, sweeten, thicken, soften, or emulsify. Some governments have reduced ingredients in manufactured foods that have potential side effects or pose health risks such as heart disease, cancer, diabetes, allergens, and obesity. Advising consumers about what and how much they should eat, as well as what not to eat, can help them avoid the effects of unhealthy ingredients (Çelik, 2016). Local zoning and other land use policies are another controversial strategy to improve community food environments. For instance, US news coverage of efforts to restrict fast-food restaurants typically frame proposed zoning policies in one of two ways: either to improve community health in urban, racially diverse communities, or to protect community esthetics or local businesses in small towns or majority-white communities. Health-focused zoning policies are often subjected to more criticism than other policy proposals and generally are less likely to succeed (Nixon et al., 2015).

Food and drug government agencies generally do not use social media to warn consumers about the health outcomes of consuming certain unhealthy foods marketed through social media platforms. However, these warnings are needed because consumers generally prefer unhealthy and nutrient-poor food items promoted on social media than healthy and nutrient-rich foods (Aldossari & Al-Mahish, 2021).

These warnings might effectively guide consumer food choices only in specific situations. For instance, Instagram users do not care whether the sources of online food and beverage product reviews are consumers or influencers. Knowing the information source does not affect whether Instagram users buy food and beverage products impulsively. The perceived value of products is more likely to predict impulse buying behavior (Prabowo & Alversia, 2020).

Some health professionals serve as expert opinion leaders on social media, to provide dietary recommendations about fresh, processed, and ultraprocessed meat products. A survey of a health influencer's followers showed that credibility, trust in the content he shared, and perceptions about the informative value of his advice predicted consumer motivation to acquire more knowledge about the endorsed meat products, develop a favorable predisposition toward them, prefer them over their options, and modify their behavior by purchasing them instead of their usual foods (Segovia-Villarreal & Rosa-Díaz, 2022).

The omnipresent marketing of processed foods drives dietary choices and brand loyalty. In recent years, market data indicate a shift in food marketing expenditures to social media. Digital marketing platforms have greater potential to influence young people, given their unique peer-to-peer transmission and youth susceptibility to social pressures. Among 15 of the most popular, energy-dense, nutrient-poor food and beverage brands on Instagram, each used at least six different marketing strategies but often adhered to an overall theme such as athleticism or relatable consumers. There was a high level of branding, although not necessarily product information, and very few health claims. A high frequency of the targeted and curated posts manipulated consumer emotions rather than present information about products (Vassallo et al., 2018).

Young adults are constantly exposed to high-calorie, nutrient-poor food and beverages through advertising that influences poor food choices and negatively impacts health. In the absence of an environment that allows healthy food to be affordable and convenient, constant exposure to unhealthy food ads impedes healthy eating behaviors and promotes guilt. Young adults often discuss different aspects of the food marketing mix—promotion, product, price, and place—and can more readily recall high-calorie, nutrient-poor food ads than healthy food ads (Molenaar et al., 2021).

The influence of junk food marketing on adolescents is pervasive. When social media food marketing influences adolescents, this content often provokes social comparison, emotional engagement, and attaching symbolic meanings to foods (van der Bend et al., 2022). Food marketing, extensive exposure to social media, messages from peers, and digital marketing disguised as entertainment all take advantage of adolescents' still developing and hypersensitive reward responsivity to appetitive cues—and make them more vulnerable to junk food marketing appeals.

Disproportionate junk food advertising appeals targeting Black and Hispanic youth can exacerbate health disparities within their communities. All of these health challenges highlight the need for policies to expand food marketing restrictions, beyond existing policies that restrict TV advertising for children (Harris et al., 2021).

Adolescents are exposed to unhealthy food ads at an average rate of 4.8 per hour they spend on Facebook. A Chrome browser extension, AdHealth, estimates adolescent exposure to unhealthy food and beverage online advertisements and collects data about the ads, ad type, and length of time the ads are viewed. The WHO-Europe nutrient profile model classifies advertised food products as healthy or unhealthy. A review of AdHealth food advertising data for 4973 Facebook ads viewed by adolescents found that 98% of the ads promoted unhealthy foods, a third featured promotional characters, and a third featured premium offers (Kidd et al., 2021). Among 6 million adolescents following 27 of the most highly advertised junk food and sugary drink brands, a higher percentage followed the brands' accounts on Twitter than on Instagram, especially for the brands that spent more money on TV ads. However, on Instagram, a higher percentage of the adolescents followed sugary than low-calorie drink brands as compared with Twitter (Rummo et al., 2020).

Global calls to action have been issued in recent years, to protect children and adolescents from exposure to the marketing of unhealthy foods and beverages. However, social media advertising policies, as well as advertising policies related to food, obesity, and other areas impacting public health, generally lack comprehensive restrictions on the advertising of unhealthy foods. In comparison, 92% of social media platforms have policies restricting the advertising of alcohol, tobacco, gambling, or weight loss (Sacks & Looi, 2020).

Ultraprocessed food product marketing is used extensively on Facebook pages. Some of the most liked and talked about Facebook pages for ultraprocessed foods are Coke, McDonald's, Burger King, Kibon, Outback, and Garoto. Facebook marketing techniques for these brands often include videos, promotions, celebrities, photos, user conversations, presence of brand elements, and links (Horta et al., 2018).

Adolescents interact with food brands in ways that mimic interactions with their friends on social media, including brands that promote unhealthy foods. Most Black and non-Latinx White adolescents prefer Instagram food ads that have more likes and comments, especially "heavy" users who spend more than 3 hours daily on social media versus "light" users who spend less than 3 hours daily. This pattern demonstrates the power of adolescent social norms in shaping behavior. Heavy social media users are six times more likely to comment on ads compared with light users. There are calls to expand the 2019 Children's Online Privacy and Protection Act to online advertising restrictions, to include adolescents aged 12 − 16 years (Lutfeali et al., 2020).

Some food companies have leveraged the COVID-19 pandemic to market unhealthy food and beverage products to youth, people facing increased stress and hardship, and other vulnerable populations on social media. These products include snacks, nonalcoholic beverages, fast food, and confectionary products. This marketing has increased brand loyalty and encouraged consumption. Known as "COVID washing," this practice occurred during low levels of regulation and advertising monitoring. In an analysis of social media posts from 20 junk food companies, 27% of all early-pandemic posts that marketed unhealthy food and drink brands were COVID-19-themed, 70% referenced COVID-19, and these pandemic mentions peaked during nationwide lockdown restrictions.

Nearly half of the fast-food brands used COVID-19-themed posts. Fast-food brands had the highest number of posts overall during the pandemic and the highest consumer engagement. The most common theme, in over a third of all social media posts referring to COVID-19, was drawing upon feelings of community support during this challenging time. Suggesting brand-related isolation activities was also common (23%), as well as the message that "consumption helps with coping" (22%). The least ethical marketing posts promoted excessive junk food consumption or targeted children (Gerritsen et al., 2021). Another recent "whitewashing" tactic among some food companies is referring to moral values such as patriotism and social justice in their marketing efforts, which can amplify their health claims about foods or ingredients (Boeuf, 2019).

Similarly, some agri-food companies market both healthy and unhealthy foods that allow consumers to "buy in" to climate mitigation. This approach has been criticized as "green-washing," a false veneer of democratization via consumer choice that may curtail more progressive societal actions on climate change. These highly commoditized, corporate-led, consumer-focused climate branding activities have used a sustainability agenda reflecting corporate climate politics. Some companies have used affective and emotional appeals to encourage acts of branded food consumption as a form of "climate care," invoked normative discourses about gender or family, and used climate branding to depoliticize climate change actions and actors (Doyle et al., 2020).

"Through marketing" helps food vendors communicate their brand message to end customers "through" various media channel partners. The World Health Organization asserts that advertising for unhealthy food items is detrimental to health. In many countries, regulation restricts such marketing to younger children, but these regulations rarely address food marketing to adolescents. Adolescents often respond more positively to unhealthy food ads, as compared with healthy food or nonfood ads, and they react more positively to ads shared by peers or celebrities than to ads shared by a brand. They are also more likely to share unhealthy food posts, rate peers more positively when they have unhealthy posts in their feeds, recall and recognize a greater number of unhealthy food brands, and view unhealthy food advertising posts longer. Although interactions with peers, celebrities, and companies on social media are more complex, they favor unhealthy food advertising (Murphy et al., 2020).

Conclusion

Media framing affects food trends, individual food choices, policy changes, and risk perceptions. Scientists face growing challenges in delivering balanced, accurate, and credible food information that can compete with misinformation. Food misinformation spreads faster when convincing claims promote hope or fear or when a single food or factor is mischaracterized. When it spreads across social media, it crowds out accurate scientific information and provokes unhealthy choices. However, experts that can flag misinformation, promote food literacy, or express skepticism about misleading claims can help consumers evaluate food claims.

Only after people gain carbon footprint knowledge can they change their daily behavior. This knowledge also influences beliefs about raw materials, transportation, and manufacturing processes. Although media literacy is a necessary skill, news framing can lead people to overestimate metrics such as carbon footprint, calories, animal welfare impacts, and food safety. It also can promote false assumptions about the sustainability of food choices. Although solutions journalism in controversial food stories can promote policy deliberation, it also can promote perceptions of biased coverage even when stories are neutral. Coverage that ignores or conceals speciesism has led to widespread animal cruelty and suffering within the industrial farming complex.

Online images shape food cultures, and people often post food images to promote their identity and to interact with online communities. These images can provoke perceptions of appearance, flavor, texture, and hedonic value, and illuminate barriers to healthy eating. Provocative or stylized food images on social media known as "food porn" popularize the pleasure of cooking and eating and activate desire-based triggers to engage viewers. Influencers with perceived expertise about food products, places to eat or shop, and other facets of food markets encourage broad food image sharing across audiences.

Food and drink brands create distinct personalities on social media that leverage humor and references to hot topics, memes, and current events. However, ubiquitous junk food marketing is a major barrier to healthy eating and behavior change. The less nutritious a food product is, the more positive emotions people feel when exposed to hedonistic ads about it. Even more troubling, junk food ads often encourage social comparisons, emotional engagement, and attaching symbolic meanings to foods.

Media amplification and sensationalism of food safety incidents shape long-term public attitudes about food risks. Consumer risk perceptions not only include fears about ag biotech but also concerns about food-borne illnesses, food allergies, additives, chemical residues, hormones, and pesticides, and catastrophic incidents such as food contamination or bioterrorism. When consumers see food risks as involuntary, industrial, or unfair, they are more likely to react forcefully than if they believe a food risk is voluntary, known, natural, or fair and conveyed by trusted sources. The more negative statements a message or story contains, the more widely it is transmitted. And the more someone is exposed to media content about a particular food risk, the more concerned they are about it. Even when news stories accurately and thoroughly reveal the causes of and solutions to potential food safety problems, audiences still blame the government for food health risks.

Mobile apps that take advantage of existing social media habits and motivations have great potential in promoting healthy and sustainable eating. Emerging communication technologies are creating innovative opportunities for consumer access to real-time food information. These include the use of Augmented Reality, which involves overlaying virtual objects on a real-world environment, Virtual Reality simulations, sensors, cloud-based platforms, extensive knowledge bases, deep learning systems, mixed-reality headsets, games, neural networks, smart e-health systems, and Internet of People peer-to-peer networks. The Unified Medical Language System provides a platform for linking biomedical vocabularies about foods to various food concepts embedded in recipes. Sensor-based apps monitor food intake, emotional states, nutritional content, chemical additives, and stress eating.

Industry applications provide functions such as nutritional product monitoring, food process efficiency tracking, marketing, training, and food safety. Consumer apps provide features such as diet tracking, shopping simulations, automated food and drink recognition, calorie management, meal planning, reverse food image searches, text mining, and food item detection in real-world environments. For farmers, emerging technologies include Information and Communication Technology automation tools, livestock sensors, crop management and forecasting tools, machine learning algorithms, time series analysis of big data, IoT applications, AI, and blockchain technology, and robotics. Fourth Industrial Revolution digitization solutions promote smart automation, robust cybersecurity, extended product shelf life, and more effective connectivity and communication.

While media in developing countries tend to emphasize the affordability of sustainable food, news media in richer, developed countries instead highlight its long-term health and environmental benefits. For many people, sustainability messages are too abstract unless stories can help them understand complex issues such as food sustainability, nutritional challenges, practical solutions, trade-offs within food systems, agricultural system sustainability, food waste, and food loss.

Sometimes documentaries frame sustainable food production in terms of "good food" for capitalist consumers while blaming individuals rather than policies for systemic problems. Diet-related health problems are often framed as personal, lifestyle issues rather than problems requiring policy or structural change—or in terms of specific foods or ingredients rather than an overall diet. Stories about unhealthy diets tend to ignore societal factors including the impact of food production on climate change.

Although media framing shapes public opinion about diet-related causes of illnesses and obesity, news coverage often blames the agri-food industry for these problems and promotes industry restrictions. Social media users often shift story discussion to broader food-related frames. "Superfood" stories that link diet to climate change sometimes highlight novel solutions such as edible insects. Media framing of sustainability initiatives sometimes leans toward simplified, "just-so" accounts that highlight agrarian and environmental successes amid grim realities of globalization. When stories romanticize farming, such as depicting small, traditional, multigenerational family farms, they encourage public support for agri-food industry interests.

Agricultural research in outer space has explored how food could be produced in extreme conditions, as well as genetic food engineering, digital agriculture, 3D bioprinting, and synthetically grown meat. Coverage of this research has ignored how these discoveries could be used to address systemic and structural inequalities in the food system on Earth. Similarly, coverage of 3D food printing often uncritically frames it as futuristic, creative, healthy, efficient, and sustainable.

Media framing shapes agri-food policymaking when it presents opposing decision frames about societal choices or that balance the complexity of conflicting needs, expert claims, and institutional knowledge. These stories often discuss food security, food supply chains, farm yield practices, scientific expertise, rural and urban planning, land use conflicts, and written policy. A story about regulating the food industry is more likely to prompt junk food avoidance if it frames personal food choices in terms of gains such as disease prevention, instead of losses such as health consequences, or if it highlights concerns about how food is produced and traded.

A contaminated information environment impedes efforts to provide informative messages or policy solutions that might otherwise promote sustainable food choices or producer responsibility. However, when stakeholders can embed their interests into stories, the coverage can influence policy outcomes. The strength of this influence depends on tone, hashtags, social media hype highlighting controversy or partisan outrage, story timing, issue framing, interaction among stakeholders, and the extent of coverage across media outlets.

As sustainability communication scholarship continues to explore news framing of food issues, it highlights emerging trends in food production, distribution, preparation, consumption, and disposal practices. Conflicting media frames can confuse consumers and cause distrust. To truly inform audiences, journalists must focus on the impacts of food controversies by contextualizing conflicting frames and by interviewing scientists to explore underlying causes. Without this content, food stories can spread misinformation, provide confusing, competing narratives, or oversimplify the complex factors that contribute to climate change, land rights, price volatility, and global food shortages.

References

Ajibade, I., & Boateng, G. O. (2021). Predicting why people engage in pro-sustainable behaviors in Portland Oregon: The role of environmental self-identity, personal norm, and socio-demographics. *Journal of Environmental Management, 289*, 112538.

Akkem, Y., Biswas, S. K., & Varanasi, A. (2023). Smart farming using artificial intelligence: A review. *Engineering Applications of Artificial Intelligence*, *120*, 105899.

Aldossari, N., & Al-Mahish, M. (2021). Social media and unhealthy food nexus: Evidence from Saudi Arabia. *Acta Alimentaria*, *50*(1), 102–111. Available from https://doi.org/10.1556/066.2020.00178.

Allievi, F., Dentoni, D., & Antonelli, M. (2018). The role of youth in increasing awareness of food security and sustainability. *Encyclopedia of food security and sustainability*, (pp. 39–44). Elsevier.

Asplund, T., Hjerpe, M., & Wibeck, V. (2013). Framings and coverage of climate change in Swedish specialized farming magazines. *Climatic Change*, *117*(1–2), 197–209.

Ateş, H. (2020). Pre-service science teachers' perceptual biases regarding sustainable food consumption: Negative footprint illusion. *International Journal of Research in Education and Science*, *6*(4), 599–612.

Baker, P., Machado, P., Santos, T., Friel, S., & Lawrence, M. (2020). Ultra-processed foods and the nutrition transition: Global, regional and national trends, food systems transformations and political economy drivers. *Obesity Reviews*, *21*(12), e13126.

Baker, P., Russ, K., Kang, M., Boatwright, M., & McCoy, D. (2021). Globalization, first-foods systems transformations and corporate power: A synthesis of literature and data on the market and political practices of the transnational baby food industry. *Globalization and Health*, *17*(1), 58.

Banovic, M., & Barone, A. M. (2021). The hybrid enigma: The importance of self-construal for the effectiveness of communication messages promoting sustainable behaviour. *Food Quality and Preference*, *94*. Available from https://doi.org/10.1016/j.foodqual.2021.104334.

Barklamb, A. M., Molenaar, A., Brennan, L., Evans, S., Choong, J., Herron, E., Reid, M., & McCaffrey, T. A. (2020). Learning the language of social media: A comparison of engagement metrics and social media strategies used by food and nutrition-related social media accounts. *Nutrients*, *12*(9), 1–23. Available from https://doi.org/10.3390/nu12092839.

Barry, C. L., Jarlenski, M., Grob, R., Schlesinger, M., & Gollust, S. E. (2011). News media framing of childhood obesity in the United States from 2000 to 2009. *Pediatrics*, *128*(1), 132–145. Available from https://doi.org/10.1542/peds.2010-3924.

Basáñez, T., Majmundar, A., Cruz, T. B., & Unger, J. B. (2018). Vaping associated with healthy food words: A content analysis of Twitter. *Addictive Behaviors Reports*, *8*, 147–153. Available from https://doi.org/10.1016/j.abrep.2018.09.007.

Belaud, J. P., Adoue, C., Vialle, C., Chorro, A., & Sablayrolles, C. (2019). A circular economy and industrial ecology toolbox for developing an eco-industrial park: Perspectives from French policy. *Clean Technologies and Environmental Policy*, *21*(5), 967–985. Available from https://doi.org/10.1007/s10098-019-01677-1.

Bellotti, E., & Panzone, L. (2016). Media effects on sustainable food consumption. How newspaper coverage relates to supermarket expenditures. *International Journal of Consumer Studies*, *40*(2), 186–200. Available from https://doi.org/10.1111/ijcs.12242.

Blasco, M. M., & Jiménez-Morales, M. (2021). Breakfast food advertising and prevention of obesity: Analysis of the nutritional value of the products and discursive strategies used in the breakfast ads from 2015 to 2019. *Nutrients*, *13*(1), 231.

Bode, L., Vraga, E. K., & Tully, M. (2021). Correcting misperceptions about genetically modified food on social media: Examining the impact of experts, social media heuristics, and the gateway belief model. *Science Communication*, *43*(2), 225–251.

Boehm, J., Kayser, M., & Spiller, A. (2010). Two sides of the same coin? Analysis of the web-based social media with regard to the image of the agri-food sector in Gemany. *International Journal on Food System Dynamics*, *1*(3), 264–278.

Boeuf, B. (2019). Political ideology and health risk perceptions of food. *Social Science and Medicine*, *236*. Available from https://doi.org/10.1016/j.socscimed.2019.112405.

Borghini, A., Piras, N., & Serini, B. (2020). Ontological frameworks for food utopias. *Rivista di Estetica*, *75*, 120–142.

Boulos, M. N. K., Yassine, A., Shirmohammadi, S., Namahoot, C. S., & Brückner, M. (2015). Towards an "Internet of Food": Food ontologies for the internet of things. *Future Internet*, *7*(4), 372–392.

Bray, J., Hartwell, H., Appleton, K., & Price, S. (2020). Consumer communication when eating out of home: The role of technology. *British Food Journal*, *123*(1), 373–386. Available from https://doi.org/10.1108/bfj-12-2019-0932.

Briseno, M. V., Soto, O. N. G., Tellez, A. M., Hipolito, J. I. N., Prieto, S. I., & Lopez, J. D. D. S. (2019). Enhancing nutrition learning using interactive tools. *IEEE Latin America Transactions*, *17*(5), 751–758. Available from https://doi.org/10.1109/tla.2019.8891943.

Brito, R. P., Miguel, P. L. S., & Pereira, S. C. F. (2020). Climate risk perception and media framing. *RAUSP Management Journal, 55*(2), 247–262.

Butt, A. (2019). *Food security and planning. The routledge companion to rural planning* (pp. 581–592). Taylor and Francis Inc. Available from https://doi.org/10.4324/9781315102375.

Cankul, D., Ari, O. P., & Okumus, B. (2021). The current practices of food and beverage photography and styling in food business. *Journal of Hospitality and Tourism Technology, 12*(1).

Capasso, M., Oliano, D., & Caso, D. (2020). Promoting healthy eating through Instagram: The experience of the Teen Food page. *Psicologia della Salute* (3), 129–144.

Caracciolo, C., Stellato, A., Morshed, A., Johannsen, G., Rajbhandari, S., Jaques, Y., & Keizer, J. (2013). The AGROVOC Linked Dataset. *Semantic Web, 4*(3), 341–348. Available from https://doi.org/10.3233/sw-130106.

Carfora, V., Morandi, M., & Catellani, P. (2022). The influence of message framing on consumers' selection of local food. *Foods, 11*(9), 1268.

Carpio, D. A., Fernandez, A., & Urbano, B. (2020). How to gain image and positioning on social media: Spanish agribusiness firm image and position on social media. *Applied Economics, 52*(21), 2280–2291.

Castellini, G., Savarese, M., & Graffigna, G. (2021). Online fake news about food: Self-evaluation, social influence and the stages of change moderation. *International Journal of Environmental Research and Public Health, 18*(6), 1–14, 2934.

Cavazza, N., Graziani, A. R., & Guidetti, M. (2020). Impression formation via #foodporn: Effects of posting gender-stereotyped food pictures on Instagram profiles. *Appetite, 147*, 104565.

Çelik, E. D. (2016). FoodWiki: A mobile app examines side effects of food additives via semantic Web. *Journal of Medical Systems, 40*, 41.

Chang, W. J., Chen, L. B., Ou, Y. K., Lin, I. C., & Tsai, Y. H. (2021). iBuffet: A deep learning-based intelligent calories management system for eating buffet meals. In: *Digest of Technical Papers – IEEE International Conference on Consumer Electronics*, 9427718.

Choudhary, S., Nayak, R., Kumari, S., & Choudhury, H. (2019). Analysing acculturation to sustainable food consumption behaviour in the social media through the lens of information diffusion. *Technological Forecasting and Social Change, 145*, 481–492. Available from https://doi.org/10.1016/j.techfore.2018.10.009.

Chung, A., Vieira, D., Donley, T., Gouley, K. K., & Seixas, A. (2021). Adolescent peer influence on eating behaviors via social media: Scoping review. *Journal of Medical Internet Research, 23*(6), e19697.

Chung, Y. C. Y., Chang, H. H., & Kitamura, Y. (2021). Digital and traditional media advertising and business performance of agribusiness firms: Empirical evidence in Japan. *Agricultural Economics, 67*(2), 51–59.

Clapp, J., & Scrinis, G. (2017). Big food, nutritionism, and corporate power. *Globalizations, 14*(4), 578–595. Available from https://doi.org/10.1080/14747731.2016.1239806.

Clark, M., Macdiarmid, J., Jones, A. D., Ranganathan, J., Herrero, M., & Fanzo, J. (2020). The role of healthy diets in environmentally sustainable food systems. *Food and Nutrition Bulletin, 41*(2_suppl), 31S–58S. Available from https://doi.org/10.1177/0379572120953734.

Corvello, V., & Felicetti, A. M. (2014). Factors affecting the utilization of knowledge acquired by researchers from scientific social networks: An empirical analysis. *Knowledge Management, 13*(3). Available from https://doi.org/10.18848/2327-7998/cgp/v13i03/50821.

Craig, G. (2019). Media, sustainability and everyday life. *Palgrave studies in media and environmental communication* (pp. 1–25). Palgrave Macmillan. Available from https://www.springer.com/series/14612.

Cuevas, S., Patel, N., Thompson, C., Petticrew, M., Cummins, S., Smith, R., & Cornelsen, L. (2021). Escaping the Red Queen: Health as a corporate food marketing strategy. *SSM – Population Health, 16*. Available from https://doi.org/10.1016/j.ssmph.2021.100953.

Cupar, D., & Juric, M. (2019). Developing food and nutrition literacy with the Croatian Facebook group "Homemade Food for Babies. *In Communications in computer and information science* (989, pp. 3–13). Springer Verlag. Available from https://doi.org/10.1007/978-3-030-13472-3_1.

D'Ambrosi, L. (2018). Pilot study on food sharing and social media in Italy. *British Food Journal, 120*(5), 1046–1058. Available from https://doi.org/10.1108/bfj-06-2017-0341.

Danner, H., Hagerer, G., Pan, Y., & Groh, G. (2022). The news media and its audience: Agenda setting on organic food in the United States and Germany. *Journal of Cleaner Production, 354*. Available from https://doi.org/10.1016/j.jclepro.2022.131503.

Darnhofer, I., D'Amico, S., & Fouilleux, E. (2019). A relational perspective on the dynamics of the organic sector in Austria, Italy, and France. *Journal of Rural Studies, 68*, 200−212.

del Moral, R. G. (2020). Gastronomic paradigms in contemporary western cuisine: From French haute cuisine to mass media gastronomy. *Frontiers in Nutrition, 6*, 192.

de Olde, E. M., & Valentinov, V. (2019). The moral complexity of agriculture: A challenge for corporate social responsibility. *Journal of Agricultural and Environmental Ethics, 32*(3), 413−430.

Demestichas, K., Remoundou, K., & Adamopoulou, E. (2020). Food for thought: Fighting fake news and online disinformation. *IT Professional, 22*(2), 28−34, 9049290.

Demke, A., & Höhler, J. (2019). Agenda-setting in the agricultural and food industries of the case of green genetic engineering. *Journal of the Austrian Society of Agricultural Economics, 29*(25), 215−223. Available from https://oega.boku.ac.at/fileadmin/user_upload/Tagung/2019/AJARS29/28_Kasparek_DOI29_26.pdf.

Diaconeasa, M. C., Popescu, G., Maehle, N., Nelgen, S., & Capitello, R. (2022). Media discourse on sustainable consumption in Europe. *Environmental Communication, 16*(3), 352−370.

Dondokova, A., Aich, S., Kim, H. C., & Huh, G. H. (2019). A text mining approach to study individuals' food choices and eating behavior using Twitter feeds. *In Lecture Notes in Electrical Engineering* (542, pp. 520−527). Springer Verlag. Available from https://doi.org/10.1007/978-981-13-3648-5_60.

Doyle, J., Farrell, N., & Goodman, M. K. (2020). The cultural politics of climate branding: Project Sunlight, the bio-politics of climate care and the socialisation of the everyday sustainable consumption practices of citizens-consumers. *Climatic Change, 163*, 117−133.

Duffy, A., & Ashley, Y. Y. (2012). Bread and circuses: Food meets politics in the Singapore media. *Journalism Practice, 6*(1), 59−74. Available from https://doi.org/10.1080/17512786.2011.622892.

Ekebas-Turedi, C., Cilingir Uk, Z., Basfirinci, C., & Pinar, M. (2021). A cross-cultural analysis of gender-based food stereotypes and consumption intentions among millennial consumers. *Journal of International Consumer Marketing, 33*(2), 209−225. Available from https://doi.org/10.1080/08961530.2020.1771644.

Entman, R. M. (1993). Framing: Toward clarification of a fractured paradigm. *Journal of Communication, 43*(4), 51−58. Available from https://doi.org/10.1111/j.1460-2466.1993.tb01304.x.

Ertuğrul, C. D. (2016). FoodWiki: A mobile app examines side effects of food additives via semantic Web. *Journal of Medical Systems, 40*, Original work published 2016.

Felicetti, A. M., Volpentesta, A. P., & Ammirato, S. (2020). Analyzing app-based food information services: The case of olive oil sector. *VINE Journal of Information and Knowledge Management Systems, 50*(3), 427−453.

Fernandez, C. M., Alves, J., Gaspar, P. D., & Lima, T. M. (2021). Fostering awareness on environmentally sustainable technological solutions for the post-harvest food supply chain. *Processes, 9*(9). Available from https://doi.org/10.3390/pr9091611.

Folwarczny, M., Christensen, J. D., Li, N. P., Sigurdsson, V., & Otterbring, T. (2021). Crisis communication, anticipated food scarcity, and food preferences: Preregistered evidence of the insurance hypothesis. *Food Quality and Preference, 91*. Available from https://doi.org/10.1016/j.foodqual.2021.104213.

Fountaine, S., & Bulmer, S. (2020). Telling stories about farming: Mediated authenticity and New Zealand's Country Calendar. *Television & New Media, 23*(1), 81−99.

Fuchs, K., Haldimann, M., Grundmann, T., & Fleisch, E. (2020). Supporting food choices in the Internet of People: Automatic detection of diet-related activities and display of real-time interventions via mixed reality headsets. *Future Generation Computer Systems, 113*, 343−362. Available from https://doi.org/10.1016/j.future.2020.07.014.

Fusté-Forné, F., & Masip, P. (2018). Deciphering specialist journalism: Food in Spanish newspapers. *Observatorio, 12*(2), 108−121.

Gamborg, C., Millar, K., Shortall, O., & Sandøe, P. (2012). Bioenergy and land use: Framing the ethical debate. *Journal of Agricultural and Environmental Ethics, 25*(6), 909−925. Available from https://doi.org/10.1007/s10806-011-9351-1.

Ganeshkumar, C., Jena, S. K., Sivakumar, A., & Nambirajan, T. (2023). Artificial intelligence in agricultural value chain: Review and future directions. *Journal of Agribusiness in Developing and Emerging Economies, 13*(3), 379−398.

Gaspar, R., Domingos, S., & Demétrio, P. (2018). Serving science to the public: Deliberations by a sample of older adults upon exposure to a serving size recommendation for meat. *Food Quality and Preference, 66*, 85−94.

Gerritsen, S., Sing, F., Lin, K., Martino, F., Backholer, K., Culpin, A., & Mackay, S. (2021). The timing, nature and extent of social media marketing by unhealthy food and drinks brands during the COVID-19 pandemic in New Zealand. *Frontiers in Nutrition, 8*. Available from https://doi.org/10.3389/fnut.2021.645349.

Gess, H. (2012). Climate change and the possibility of 'slow journalism. *Ecquid Novi: African Journalism Studies, 33* (1), 54–65. Available from https://doi.org/10.1080/02560054.2011.636828.

Giampietri, E., Bugin, G., & Trestini, S. (2020). Exploring the interplay of risk attitude and organic food consumption. *Journal on Food System Dynamics, 11*(3).

Gigandet, S. (2012). Open food facts. Available from https://world.openfoodfacts.org/.

Giousmpasoglou, C., Brown, L., & Cooper, J. (2020). The role of the celebrity chef. *International Journal of Hospitality Management, 85*, 102358.

Goodman, M. K., & Jaworska, S. (2020). Mapping digital foodscapes: Digital food influencers and the grammars of good food. *Geoforum; Journal of Physical, Human, and Regional Geosciences, 117*, 183–193. Available from https://doi.org/10.1016/j.geoforum.2020.09.020.

Goody, J. (1982). *Cooking, cuisine and class: A study in comparative sociology.* Cambridge, UK: Cambridge University Press.

Gottwald, F., & Weder, F. (2021). *Food and morality: Deconstruction of value judgments in sustainability documentary films* (pp. 177–196). Springer Science & Business Media. Available from https://doi.org/10.1007/978-3-658-31883-3_11.

Gordon, C., & Hunt, K. (2019). Reform, justice, and sovereignty: A food systems agenda for environmental communication. *Environmental Communication, 13*(1), 9–22.

Greene, T., Seet, C., Barrio, A. R., Kelly, B., & Bragg, M. A. (2022). Brands with personalities – good for businesses, but bad for public health? A content analysis of how food and beverage brands personify themselves on Twitter. *Public Health Nutrition, 25*(1), 51–60.

Griffin, R. J., Dunwoody, S., & Neuwirth, K. (1999). Proposed model of the relationship of risk information seeking and processing to the development of preventive behaviors. *Environmental Research, 80*(2), S230–S245.

Hamm, M. W. (2009). Principles for framing a healthy food system. *Journal of Hunger and Environmental Nutrition, 4*(3–4), 241–250. Available from https://doi.org/10.1080/19320240903321219.

Handayani, B. (2021). Foodstragramming of solo dining experience scape: The fear of missing out. *International Journal on Food System Dynamics, 12*(1), 83–94. Available from https://doi.org/10.18461/ijfsd.v11i5.77.

Harris, J. L., Yokum, S., & Fleming-Milici, F. (2021). Hooked on junk: Emerging evidence on how food marketing affects adolescents' diets and long-term health. *Current Addiction Reports, 8*(1), 19–27.

Harun, A. F., Ruslan, N., Adnan, W. A. W., Suliman, S. I., Ismail, J., & Baharin, H. (2020). Digitizing food experience: Food taste perception on digital image and true form using hashtags. *Bulletin of Electrical Engineering and Informatics, 9*(5), 2141–2153. Available from https://doi.org/10.11591/eei.v9i5.2252.

Hawkins, L., Farrow, C., & Thomas, J. M. (2021). Does exposure to socially endorsed food images on social media influence food intake? *Appetite, 165*, 105424.

Hepworth, A. D., Kropczynski, J., Walden, J., & Smith, R. A. (2019). Exploring patterns of social relationships among food bloggers on Twitter using a social network analysis approach. *Journal of Social Structure, 20*(4), 1–21. Available from https://doi.org/10.21307/joss-2019-038.

Horta, P. M., Rodrigues, F. T., & Dos Santos, L. C. (2018). Ultra-processed food product brands on Facebook pages: Highly accessed by Brazilians through their marketing techniques. *Public Health Nutrition, 21*(8), 1515–1519.

Huamaní-Cahuana, J., & Cabanillas-Carbonell, M. (2021). Analysis of mobile applications reporting on nutritional recipes: A review of the scientific literature. *E3S Web of Conferences, 229*, 01060.

Huang, Q. (2020). How does news media exposure amplify publics' perceived health risks about air pollution in China? A conditional media effect approach. *International Journal of Communication, 14*, 1705–1724. Available from https://ijoc.org/index.php/ijoc.

Hunt, D. (2021). How food companies use social media to influence policy debates: A framework of Australian ultra-processed food industry Twitter data. *Public Health Nutrition, 24*, 3124–3135.

Huynh, T. T. G., Luu, T. D., & Phung, T. T. (2021). A fuzzy-set approach for multiple criteria: Decision making in sustainable consumption of organic food. *Decision Science Letters, 10*(3), 291–300.

IPCC. (2022). *Climate Change 2022: Mitigation of Climate Change. Working Group III contribution to the Sixth Assessment Report.* Intergovernmental Panel on Climate Change: Geneva, Switzerland. Retrieved from: https://www.ipcc.ch/report/ar6/wg3/.

Ireland, R., Muc, M., Bunn, C., & Boyland, E. (2021). Marketing of unhealthy brands during the 2018 Fédération Internationale de Football Association (FIFA) World Cup UK broadcasts: A frequency analysis. *Journal of Strategic Marketing*, 1–16.

Javed, M., Malik, F. A., Awan, T. M., & Khan, R. (2021). Food photo posting on social media while dining: An evidence using embedded correlational mixed methods approach. *Journal of Food Products Marketing*, 27(1), 10–26. Available from https://doi.org/10.1080/10454446.2021.1881861.

Jiang, H., Starkman, J., Liu, M., & Huang, M. C. (2018). Food nutrition visualization on Google Glass: Design tradeoff and field evaluation. *IEEE Consumer Electronics Magazine*, 7(3), 21–31. Available from https://doi.org/10.1109/MCE.2018.2797740.

Junqueira, A. H. (2019). Food porn: Images, social senses and virtualization of the pleasure of eating. *Discursos Fotograficos*, 15(26), 171–197. Available from https://doi.org/10.5433/1984-7939.2019v15n26p171.

Kāle, M., & Agbozo, E. (2020). *Healthy food depiction on social media: The case of kale on Twitter*. In CEUR Workshop Proceedings 2865, pp. 51–62. CEUR-WS. http://ceur-ws.org/.

Kanay, A., Hilton, D., Charalambides, L., Corrégé, J., Inaudi, E., Waroquier, L., & Cézéra, S. (2021). Making the carbon basket count: Goal setting promotes sustainable consumption in a simulated online supermarket. *Journal of Economic Psychology*, 83. Available from https://doi.org/10.1016/j.joep.2020.102348.

Keller, E., & Wyles, K. J. (2021). Straws, seals, and supermarkets: Topics in the newspaper coverage of marine plastic pollution. *Marine Pollution Bulletin*, 166. Available from https://doi.org/10.1016/j.marpolbul.2021.112211.

Khazaal, N., & Almiron, N. (2016). An angry cow is not a good eating experience": How U.S. and Spanish media are shifting from crude to camouflaged speciesism in concealing nonhuman perspectives. *Journalism Studies*, 17(3), 374–391. Available from https://doi.org/10.1080/1461670X.2014.982966.

Kidd, B., Mackay, S., Swinburn, B., Lutteroth, C., & Vandevijvere, S. (2021). AdHealth: A feasibility study to measure digital food marketing to adolescents through Facebook. *Public Health Nutrition*, 24(2), 215–222. Available from https://doi.org/10.1017/s1368980020001561.

Klassen, K. M., Borleis, E. S., Brennan, L., McCaffrey, T. A., & Lim, M. S. C. (2018). What people "like": Analysis of social media strategies used by food industry brands, lifestyle brands, and health promotion organizations on Facebook and Instagram. *Journal of Medical Internet Research*, 20(6), e10227.

Kley, S., Kleinen-von Königslöw, K., & Dunker, A. (2023). Media diets of vegetarians. How news consumption, social media use and communicating with one's social environment are associated with a vegetarian diet. *Environmental Communication*, 17(8), 875–890.

Knez, S., & Šajn, L. (2020). Food object recognition using a mobile device: Evaluation of currently implemented systems. *Trends in Food Science & Technology*, 99, 460–471. Available from https://doi.org/10.1016/j.tifs.2020.03.017.

Kogen, L. (2019). News you can use or news that moves?: Journalists' rationales for coverage of distant suffering. *Journalism Practice*, 13(1), 1–15. Available from https://doi.org/10.1080/17512786.2017.1400395.

Korthals, M. (2017). Ethics of dietary guidelines: Nutrients, processes and meals. *Journal of Agricultural and Environmental Ethics*, 30(3), 413–421. Available from https://doi.org/10.1007/s10806-017-9674-7.

Kramer, M. P., Bitsch, L., & Hanf, J. (2021). Blockchain and its impacts on agri-food supply chain network management. *Sustainability*, 13(4), 2168.

Kusá, A., Urmínová, M., Darázs, T., & Šalgovičová, J. (2021). Testing of standardized advertising slogans within the marketing communication of sustainable and local foods in order to reveal consumer preferences. *Frontiers in Sustainable Food Systems*, 5. Available from https://doi.org/10.3389/fsufs.2021.703223.

Kusumasondjaja, S., & Tjiptono, F. (2019). Endorsement and visual complexity in food advertising on Instagram. *Internet Research*, 29(4), 659–687. Available from https://doi.org/10.1108/intr-11-2017-0459.

Laakso, S., Aro, R., Heiskanen, E., & Kaljonen, M. (2021). Reconfigurations in sustainability transitions: A systematic and critical review. *Sustainability: Science, Practice and Policy*, 17(1), 15–31. Available from https://doi.org/10.1080/15487733.2020.1836921.

Ladhari, R., Rioux, M. C., Souiden, N., & Chiadmi, N. E. (2019). Consumers' motives for visiting a food retailer's Facebook page. *Journal of Retailing and Consumer Services*, 50, 379–385. Available from https://doi.org/10.1016/j.jretconser.2018.07.013.

Lee, H. M., Kang, J. W., & Namkung, Y. (2021). Instagram users' information acceptance process for food-content. *Sustainability*, 13(5), 1–15, 2638.

Lee, K. C. L., Newell, J. P., Wolch, J., Schneider, N., & Joassart-Marcelli, P. (2014). "Story-networks" of livestock and climate change: Actors, their artifacts, and the shaping of urban print media. *Society and Natural Resources*, 27(9), 948–963.

Lezaun, J. (2017). Afterword: The public's two bodies – food activism in digital media. *Digital Food Activism*, 220–227. Available from https://doi.org/10.4324/9781315109930.

Lupton, D. (2017). Download to delicious': Promissory themes and sociotechnical imaginaries in coverage of 3D printed food in online news sources. *Futures*, *93*, 44–53. Available from https://doi.org/10.1016/j.futures.2017.08.001.

Lutfeali, S., Ward, T., Greene, T., Arshonsky, J., Seixas, A., Dalton, M., & Bragg, M. A. (2020). Understanding the extent of adolescents' willingness to engage with food and beverage companies' Instagram accounts. *JMIR Public Health and Surveillance*, *6*(4). Available from https://doi.org/10.2196/20336.

Mahfuz, S., Mun, H. S., Dilawar, M. A., & Yang, C. J. (2022). Applications of smart technology as a sustainable strategy in modern swine farming. *Sustainability*, *14*(5), 2607.

Marinagi, C., Trivellas, P., & Sakas, D. P. (2014). The impact of information technology on the development of supply chain competitive advantage. *Procedia-Social and Behavioral Sciences*, *147*, 586–591.

Marinescu, V., & Rodat, S. (2020). Food sustainability, healthy eating, on-trend diets: Media representations of nutrition in Romania and Germany. *Social Change Review*, *18*(1), 105–143. Available from https://doi.org/10.2478/scr-2020-0004.

Marinescu, V., Fox, B., Cristea, D., Roventa-Frumusani, D., Marinache, R., & Branea, S. (2021). Talking about sustainability: How the media construct the public's understanding of sustainable food in Romania. *Sustainability*, *13*(9). Available from https://doi.org/10.3390/su13094609.

Massari, S., Allievi, F., & Recanati, F. (2021). Fostering empathy towards effective sustainability teaching: From the Food Sustainability Index educational toolkit to a new pedagogical model. *In Handbook on teaching and learning for sustainable development*. Edward Elgar Publishing.

Maye, D., Fellenor, J., Potter, C., Urquhart, J., & Barnett, J. (2021). What's the beef?: Debating meat, matters of concern, and the emergence of online issue publics. *Journal of Rural Studies*, *84*, 134–146. Available from https://doi.org/10.1016/j.jrurstud.2021.03.008.

McCarthy, J. F., & Obidzinski, K. (2017). Framing the food poverty question: Policy choices and livelihood consequences in Indonesia. *Journal of Rural Studies*, *54*, 344–354. Available from https://doi.org/10.1016/j.jrurstud.2017.06.004.

McCluskey, J. J., Kalaitzandonakes, N., & Swinnen, J. (2016). Media coverage, public perceptions, and consumer behavior: Insights from new food technologies. *Annual Review of Resource Economics*, *8*(1), 467–486.

Meijers, M. H., Smit, E. S., de Wildt, K., Karvonen, S. G., van der Plas, D., & van der Laan, L. N. (2022). Stimulating sustainable food choices using virtual reality: Taking an environmental vs health communication perspective on enhancing response efficacy beliefs. *Environmental Communication*, *16*(1), 1–22.

Meixner, O., Haas, R., Moosbrugger, H., & Magdits, P. (2013). Interaction with customers: The application of social media within the Austrian supply chain for food and beverages. *International Journal on Food System Dynamics*, *4*(1), 26–37. Available from https://doi.org/10.18461/ijfsd.v4i1.413.

Middha, B. (2018). Everyday digital engagements: Using food selfies on Facebook to explore eating practices. *Communication Research and Practice*, *4*(3), 291–306. Available from https://doi.org/10.1080/22041451.2018.1476796.

Millet, M., Keast, V., Gonano, S., & Casabianca, F. (2020). Product qualification as a means of identifying sustainability pathways for place-based agri-food systems: The case of the GI Corsican grapefruit (France). *Sustainability*, *12*(17), 7148.

Miller, G., Serjeant, S., & Oakden, L. (2021). Engaging citizens in sustainability research: Comparing survey recruitment and responses between Facebook, Twitter and Qualtrics. *British Food Journal*, *123*(9), 3116–3132.

Moberg, E., Allison, E. H., Harl, H. K., Lei, X. G., & Halpern, B. S. (2021). Combined innovations in public policy, the private sector and culture can drive sustainability transitions in food systems. *Nature Food*, *2*(4), 282–290.

Molenaar, A., Saw, W. Y., Brennan, L., Lim, M. S. C., & McCaffrey, T. A. (2021). Effects of advertising: A qualitative analysis of young adults' engagement with social media about food. *Nutrients*, *13*(6), 1934.

Monnier, L., Colette, C., El Azrak, A., Bauduceau, B., Bordier, L., Essekat, N., & Schlienger, J. L. (2020). The two faces of nutritional studies: From "fake" to "real" news. *Medecine Des Maladies Metaboliques*, *14*(2), 101–113. Available from https://doi.org/10.1016/j.mmm.2019.11.001.

Murphy, G., Corcoran, C., Tatlow-Golden, M., Boyland, E., & Rooney, B. (2020). See, like, share, remember: Adolescents' responses to unhealthy-, healthy- and non-food advertising in social media. *International Journal of Environmental Research and Public Health*, *17*(7). Available from https://doi.org/10.3390/ijerph17072181.

Myrick, J. G., & Evans Comfort, S. (2020). The pope may not be enough: How emotions, populist beliefs, and perceptions of an elite messenger interact to influence responses to climate change messaging. *Mass Communication and Society*, *23*(1), 1–21. Available from https://doi.org/10.1080/15205436.2019.1639758.

Neff, R. A., Chan, I. L., & Smith, K. C. (2009). Yesterday's dinner, tomorrow's weather, today's news? U.S. newspaper coverage of food system contributions to climate change. *Public Health Nutrition, 12*(7), 1006—1014.

Ngqangashe, Y., Maldoy, K., De Backer, C. J., & Vandebosch, H. (2021). *Exploring adolescents' motives for food media consumption using the theory of uses and gratifications.* Communications.

Nixon, L., Mejia, P., Dorfman, L., Cheyne, A., Young, S., Friedman, L. C., Gottlieb, M. A., & Wooten, H. (2015). Fast-food fights: News coverage of local efforts to improve food environments through land-use regulations, 2000-2013. *American Journal of Public Health, 105*(3), 490—496. Available from https://doi.org/10.2105/AJPH.2014.302368.

Okamoto, K., & Yanai, K. (2019). Analyzing regional food trends with geo-tagged Twitter food photos. In *Proceedings — International Workshop on Content-Based Multimedia Indexing (Vols. 2019).* IEEE Computer Society. https://doi.org/10.1109/CBMI.2019.8877473.

Open Food Facts. (2018). *List of NOVA groups — World.* Available from https://world.openfoodfacts.org/nova-groups.

Orste, L., Krumina, A., Kilis, E., Adamsone-Fiskovica, A., & Grivins, M. (2021). Individual responsibilities, collective issues: The framing of dietary practices in Latvian media. *Appetite, 164.* Available from https://doi.org/10.1016/j.appet.2021.105219.

Otterbach, S., Oskorouchi, H. R., Rogan, M., & Qaim, M. (2021). Using Google data to measure the role of Big Food and fast food in South Africa's obesity epidemic. *World Development, 140.* Available from https://doi.org/10.1016/j.worlddev.2020.105368.

Özkaya, F. T., Durak, M. G., Doğan, O., Bulut, Z. A., & Haas, R. (2021). Sustainable consumption of food: Framing the concept through Turkish expert opinions. *Sustainability (Switzerland), 13*(7). Available from https://doi.org/10.3390/su13073946.

Panzone, L. A., Lemke, F., & Petersen, H. L. (2016). Biases in consumers' assessment of environmental damage in food chains and how investments in reputation can help. *Technological Forecasting and Social Change, 111,* 327—337.

Panzone, L. A., Sniehotta, F. F., Comber, R., & Lemke, F. (2020). The effect of traffic-light labels and time pressure on estimating kilocalories and carbon footprint of food. *Appetite, 155.* Available from https://doi.org/10.1016/j.appet.2020.104794.

Pappa, I. C., Iliopoulos, C., & Massouras, T. (2018). What determines the acceptance and use of electronic traceability systems in agri-food supply chains? *Journal of Rural Studies, 58,* 123—135.

Park, S. (2020). How celebrities' green messages on Twitter influence public attitudes and behavioral intentions to mitigate climate change. *Sustainability (Switzerland), 12*(19). Available from https://doi.org/10.3390/SU12197948.

Pascucci, S. (2010). Governance structure, perception and innovation in credence food transactions: The role of food community networks. *International Journal on Food System Dynamics, 1*(3), 224—236. Available from https://doi.org/10.18461/ijfsd.v1i3.136.

Pérez Perales, D., Verdecho, M. J., & Alarcón-Valero, F. (2019). Enhancing the sustainability performance of agri-food supply chains by implementing Industry 4.0. In *Collaborative Networks and Digital Transformation: 20th IFIP WG 5.5 Working Conference on Virtual Enterprises,* PRO-VE 2019, Turin, Italy, Proceedings 20 (pp. 496-503). Springer International Publishing.

Perl, A., Howlett, M., & Ramesh, M. (2018). Policymaking and truthiness: Can existing policy models cope with politicized evidence and willful ignorance in a "post-fact" world? *Policy Sciences, 51*(4), 581—600.

Pham, G. V., Shancer, M., & Nelson, M. R. (2019). Only other people post food photos on Facebook: Third-person perception of social media behavior and effects. *Computers in Human Behavior, 93,* 129—140.

Phillipov, M. (2016). The new politics of food: Television and the media/food industries. *Media International Australia, 158*(1), 90—98.

Phillipov, M. (2017). *Media and food industries: The new politics of food* (pp. 1—259). London: Palgrave MacMillan.

Phillipov, M., & Loyer, J. (2019). In the wake of the supermarket 'milk wars': Media, farmers and the power of pastoral sentimentality. *Discourse, Context and Media* (32, p. 100346).

Pilař, L., Stanislavská, L. K., & Kvasnička, R. (2021). Healthy food on the Twitter social network: Vegan, homemade, and organic food. *International Journal of Environmental Research and Public Health, 18*(7). Available from https://doi.org/10.3390/ijerph18073815.

Pilař, L., Stanislavská, L. K., Kvasnička, R., Hartman, R., & Tichá, I. (2021). Healthy food on Instagram social network: Vegan, homemade and clean eating. *Nutrients, 13*(6), 1991.

Pilgeram, R., & Meeuf, R. (2015). Good food, good intentions: Where pro-sustainability arguments get stale in U. S. food documentaries. *Environmental Communication*, 9(1), 100−117. Available from https://doi.org/10.1080/17524032.2014.967706.

Pindado, E., & Barrena, R. (2021). Using Twitter to explore consumers' sentiments and their social representations towards new food trends. *British Food Journal*, 123(3), 1060−1082. Available from https://doi.org/10.1108/BFJ-03-2020-0192.

Pinto, P. A., Antunes, M. J. L., & Almeida, A. M. P. (2021). Public health on Instagram: An analysis of health promotion strategies of Portugal and Brazil. *Procedia Computer Science*, 181, 231−238.

Pitigraisorn, P. (2021). *Disinformation, science communication, and trust: Food rumours in Thailand.* Doctoral dissertation, University of Essex.

Plamondon, G., Labonté, M. È., Pomerleau, S., Vézina, S., Mikhaylin, S., Laberee, L., & Provencher, V. (2022). The influence of information about nutritional quality, environmental impact and eco-efficiency of menu items on consumer perceptions and behaviors. *Food Quality and Preference*, 102, 104683.

Platania, M., & Spadoni, R. (2018). How people share information about food: Insights from tweets regarding two Italian Regions. *International Journal on Food System Dynamics*, 9(2), 149−165. Available from https://doi.org/10.18461/ijfsd.v9i2.924.

Pollack, C. C., Gilbert-Diamond, D., Emond, J. A., Eschholz, A., Evans, R. K., Boyland, E. J., & Masterson, T. D. (2021). Twitch user perceptions, attitudes, and behaviours in relation to food and beverage marketing on Twitch compared with YouTube. *Journal of Nutritional Science*, 10, e32. Available from https://doi.org/10.1017/jns.2021.22.

Popovski, G., Korousic-Seljak, B., & Eftimov, T. (2019). FoodOntoMap: Linking food concepts across different food ontologies. *In Knowledge discovery, knowledge engineering, and knowledge management* (pp. 195−202). Springer.

Prabowo, H., & Alversia, Y. (2020). The influence of value and browsing activity on impulse buying behavior moderated by source of online review case study: Food and beverage products on Instagram. In *ACM International Conference Proceeding Series* (pp. 103−108). Association for Computing Machinery. https://doi.org/10.1145/3409929.3414745.

Rachakonda, L., Mohanty, S. P., & Kougianos, E. (2020). ILog: An intelligent device for automatic food intake monitoring and stress detection in the IoMT. *IEEE Transactions on Consumer Electronics*, 66(2), 115−124, 9011599.

Ragelienė, T., & Grønhøj, A. (2021). The role of peers, siblings, and social media for children's healthy eating socialization: A mixed methods study. *Food Quality and Preference*, 93. Available from https://doi.org/10.1016/j.foodqual.2021.104255.

Rejeb, A., Rejeb, K., & Keogh, J. G. (2021). Enablers of augmented reality in the food supply chain: A systematic literature review. *Journal of Foodservice Business Research*.

Rewane, R. & Chouragade, P. M. (2019). Food nutritional detection, visualization and recommendation for health monitoring using image processing. In: *Proceedings of the International Conference on Trends in Electronics and Informatics*, 556−559.

Ricci, E. C., & Banterle, A. (2020). Do major climate change-related public events have an impact on consumer choices? *Renewable and Sustainable Energy Reviews*, 126. Available from https://doi.org/10.1016/j.rser.2020.109793.

Rintyarna, B. S. (2021). Mapping acceptance of Indonesian organic food consumption under COVID-19 pandemic using sentiment analysis of Twitter dataset. *Journal of Theoretical and Applied Information Technology*, 99(5), 1009−1019. Available from http://www.jatit.org/volumes/Vol99No5/1Vol99No5.pdf.

Rose, J. T., James, A. K., & Patel, Z. (2016). Contesting the edge: Analyzing environmental decision-making as it is represented in the media. *South African Geographical Journal*, 98(2), 235−253.

Rowe, S. B., & Alexander, N. (2017b). Communicating health and nutrition information after the death of expertise. *Nutrition Today*, 52(6), 285−288.

Rowe, S. B., & Alexander, N. (2017a). On post-truth, fake news, and trust. *Nutrition Today*, 52(4), 179−182.

Rowe, S., & Alexander, N. (2018). Fake science/nutrition news: Part 2 − trust. *Nutrition Today*, 53(4), 166−168.

Rowe, S., & Alexander, N. (2019). Fake nutrition/health news, part 3: How (and why) did we get here? *Nutrition Today*, 54(4), 170−173.

Ruggeri, A., & Samoggia, A. (2018). Twitter communication of agri-food chain actors on palm oil environmental, socio-economic, and health sustainability. *Journal of Consumer Behaviour*, 17(1), 75−93. Available from https://doi.org/10.1002/cb.1699.

Rummo, P. E., Cassidy, O., Wells, I., Coffino, J. A., & Bragg, M. A. (2020). Examining the relationship between youth-targeted food marketing expenditures and the demographics of social media followers. *International Journal of Environmental Research and Public Health, 17*(5). Available from https://doi.org/10.3390/ijerph17051631.

Russell, G. (2015). Improving the understanding of climate change factors with images. *Impact of meat consumption on health and environmental sustainability* (pp. 43–63). IGI Global. Available from https://doi.org/10.4018/978-1-4666-9553-5.ch003.

Rutsaert, P., Regan, Á., Pieniak, Z., McConnon, Á., Moss, A., Wall, P., & Verbeke, W. (2013). The use of social media in food risk and benefit communication. *Trends in Food Science & Technology, 30*(1), 84–91. Available from https://doi.org/10.1016/j.tifs.2012.10.006.

Ryan, C. D., Schaul, A. J., Butner, R., & Swarthout, J. T. (2020). Monetizing disinformation in the attention economy: The case of genetically modified organisms (GMOs). *European Management Journal, 38*(1), 7–18. Available from https://doi.org/10.1016/j.emj.2019.11.002.

Sacks, G., & Looi, E. S. Y. (2020). The advertising policies of major social media platforms overlook the imperative to restrict the exposure of children and adolescents to the promotion of unhealthy foods and beverages. *International Journal of Environmental Research and Public Health, 17*(11), 1–11. Available from https://doi.org/10.3390/ijerph17114172.

Samoggia, A., Riedel, B., & Ruggeri, A. (2020). Social media exploration for understanding food product attributes perception: The case of coffee and health with Twitter data. *British Food Journal, 122*(12), 3815–3835.

Santeramo, F. G., Bevilacqua, A., Caroprese, M., Speranza, B., Ciliberti, M. G., Tappi, M., & Lamonaca, E. (2021). Assessed versus perceived risks: Innovative communications in agri-food supply chains. *Foods, 10*(5), 1001.

Saurabh, S., & Dey, K. (2021). Blockchain technology adoption, architecture, and sustainable agri-food supply chains. *Journal of Cleaner Production, 284.* Available from https://doi.org/10.1016/j.jclepro.2020.124731.

Schiemer, C., Halloran, A., Jespersen, K., & Kaua, P. (2018). *Marketing insects: Superfood or solution-food? In Edible insects in sustainable food systems* (pp. 213–236). Springer International Publishing. Available from https://doi.org/10.1007/978-3-319-74011-9_14.

Schneider, T., Eli, K., McLennan, A., Dolan, C., Lezaun, J., & Ulijaszek, S. (2019). Governance by campaign: The co-constitution of food issues, publics, and expertise through new information and communication technologies. *Information Communication and Society, 22*(2), 172–192. Available from https://doi.org/10.1080/1369118X.2017.1363264.

Segovia-Villarreal, M., & Rosa-Díaz, I. M. (2022). Promoting sustainable lifestyle habits: "Real food" and social media in Spain. *Foods, 11*(2). Available from https://doi.org/10.3390/foods11020224.

Shaw, R., & Soma, T. (2022). To the farm, Mars, and beyond: Technologies for growing food in space, the future of long-duration space missions, and earth implications in English news media coverage. *Frontiers in Communication, 7.* Available from https://doi.org/10.3389/fcomm.2022.1007567.

Shimul, A. S., Cheah, I., & Lou, A. J. (2021). Regulatory focus and junk food avoidance: The influence of health consciousness, perceived risk and message framing. *Appetite, 166*, 105428.

Siegrist, M., & Hartmann, C. (2020). Consumer acceptance of novel food technologies. *Nature Food, 1*(6), 343–350. Available from https://doi.org/10.1038/s43016-020-0094-x.

Šikić, F. (2021). Using Instagram as a communication channel in green marketing digital mix: A case study of bio&bio organic food chain in Croatia. *Critical studies on corporate responsibility, governance and sustainability 14* (pp. 221–236). Emerald Group Holdings. Available from https://doi.org/10.1108/S2043-905920210000015013.

Stefanik-Sidener, K. (2013). Nature, nurture, or that fast food hamburger: Media framing of diabetes in the New York Times from 2000 to 2010. *Health Communication, 28*(4), 351–358. Available from https://doi.org/10.1080/10410236.2012.688187.

Steils, N., & Obaidalahe, Z. (2020). Social food": Food literacy co-construction and distortion on social media. *Food Policy, 95.*

Stevens, T. M., Aarts, N., Termeer, C. J. A. M., & Dewulf, A. (2018). Social media hypes about agro-food issues: Activism, scandals and conflicts. *Food Policy, 79*, 23–34. Available from https://doi.org/10.1016/j.foodpol.2018.04.009.

Stirling, E., Willcox, J., Ong, K. L., & Forsyth, A. (2021). Social media analytics in nutrition research: A rapid review of current usage in investigation of dietary behaviours. *Public Health Nutrition, 24*(6), 1193–1209. Available from https://doi.org/10.1017/S1368980020005248.

Tankard, J. W. (2001). The empirical approach to the study of media framing. In S. D. Reese, O. H. Gandy, & A. E. Grant (Eds.), *Framing public life: Perspectives on media and our understanding of the social world* (pp. 111–121). Oxfordshire, UK: Routledge.

Taylor, N., & Keating, M. (2018). Contemporary food imagery: Food porn and other visual trends. *Communication Research and Practice, 4*(3), 307–323. Available from https://doi.org/10.1080/22041451.2018.1482190.

Tepper, S., Kaufman-Shriqui, V., & Shahar, D. R. (2020). Mapping young adults' concerns and attitudes toward food-related sustainability issues in Israel: Implications for food policy. *Nutrients, 12*(10), 1–13, 3190.

Terlau, W., & Hirsch, D. (2015). Sustainable consumption and the attitude-behaviour-gap phenomenon — causes and measurements towards a sustainable development. *International Journal on Food System Dynamics, 6*(3), 159–174. Available from https://doi.org/10.18461/ijfsd.v6i3.634.

Thakurta, P. G., & Chaturvedi, S. (2012). Food and nutrition justice: How to make it more newsworthy? *IDS Bulletin, 43*(1), 58–64. Available from https://doi.org/10.1111/j.1759-5436.2012.00347.x.

Turner-McGrievy, G., Karami, A., Monroe, C., & Brandt, H. M. (2020). Dietary pattern recognition on Twitter: A case example of before, during, and after four natural disasters. *Natural Hazards, 103*(1), 1035–1049. Available from https://doi.org/10.1007/s11069-020-04024-6.

van der Bend, D. L. M., Jakstas, T., van Kleef, E., Shrewsbury, V. A., & Bucher, T. (2022). Making sense of adolescent-targeted social media food marketing: A qualitative study of expert views on key definitions, priorities and challenges. *Appetite, 168.* Available from https://doi.org/10.1016/j.appet.2021.105691.

Vasconcelos, C., Da Costa, R. L., Dias, Á. L., Pereira, L., & Santos, J. P. (2021). Online influencers: Healthy food or fake news. *International Journal of Internet Marketing and Advertising, 15*(2), 149–175. Available from https://doi.org/10.1504/IJIMA.2021.114334.

Vassallo, A. J., Kelly, B., Zhang, L., Wang, Z., Young, S., & Freeman, B. (2018). Junk food marketing on instagram: Content analysis. *JMIR Public Health and Surveillance, 4*(6). Available from https://doi.org/10.2196/publichealth.9594.

Ventura, V., Cavaliere, A., & Iannò, B. (2021). #Socialfood: Virtuous or vicious? A systematic review. *Trends in Food Science and Technology, 110*, 674–686.

Verdouw, C. N., Robbemond, R. M., Verwaart, T., Wolfert, J., & Beulens, A. J. M. (2018). A reference architecture for IoT-based logistic information systems in agri-food supply chains. *Enterprise Information Systems, 12*(7), 755–779. Available from https://doi.org/10.1080/17517575.2015.1072643.

Vermeir, I., & Roose, G. (2020). Visual design cues impacting food choice: A review and future research agenda. *Foods, 9*(10). Available from https://doi.org/10.3390/foods9101495.

von Zazie, D., Heyder, M., & Theuvsen, L. (2010). Media analysis on volatile markets' dynamics and adaptive behavior for the agri-food system. *Journal on Food System Dynamics, 1*(3).

Vraga, E. K., Bode, L., & Tully, M. (2020). Creating news literacy messages to enhance expert corrections of misinformation on Twitter. *Communication Research*, 0093650219898094.

Vydiswaran, V. G. V., Romero, D. M., Zhao, X., Yu, D., Gomez-Lopez, I., Lu, J. X., Iott, B. E., Baylin, A., Jansen, E. C., Clarke, P., Berrocal, V. J., Goodspeed, R., & Veinot, T. C. (2020). Uncovering the relationship between food-related discussion on Twitter and neighborhood characteristics. *Journal of the American Medical Informatics Association, 27*(2), 254–264. Available from https://doi.org/10.1093/jamia/ocz181.

Walsh, M. J., & Baker, S. A. (2020). Clean eating and Instagram: Purity, defilement, and the idealization of food. *Food, Culture and Society, 23*(5), 570–588. Available from https://doi.org/10.1080/15528014.2020.1806636.

Wang, Y., McKee, M., Torbica, A., & Stuckler, D. (2019). Systematic literature review on the spread of health-related misinformation on social media. *Social Science and Medicine, 240.* Available from https://doi.org/10.1016/j.socscimed.2019.112552.

Weinstein, L. C., Chilton, M., Turchi, R., Klassen, A. C., LaNoue, M., Silvero, A., & Cabassa, L. J. (2021). 'It's common sense that an individual must eat': Advocating for food justice with people with psychiatric disabilities through photovoice. *In Health Expectations, 24*(1), 161–173, Blackwell Publishing. Available from https://doi.org/10.1111/hex.13101.

Weitkamp, E., Milani, E., Ridgway, A., & Wilkinson, C. (2021). Exploring the digital media ecology: Insights from a study of healthy diets and climate change communication on digital and social media. *Journal of Science Communication, 20*(3), 1–22.

Wells, R. (2017). Mediating the spaces of diet and health: A critical analysis of reporting on nutrition and colorectal cancer in the UK. *Geoforum; Journal of Physical, Human, and Regional Geosciences, 84*, 228–238. Available from https://doi.org/10.1016/j.geoforum.2016.05.001.

Whiting, A., & Williams, D. (2013). Why people use social media: A uses and gratifications approach. *Qualitative Market Research*, 16(4), 362–369. Available from https://doi.org/10.1108/qmr-06-2013-0041.

Wojciechowska-Solis, J., Kowalska, A., Bieniek, M., Ratajczyk, M., & Manning, L. (2022). Comparison of the purchasing behaviour of Polish and United Kingdom consumers in the organic food market during the COVID-19 pandemic. *International Journal of Environmental Research and Public Health*, 19(3), 1137.

Wong, I. A., Liu, D., Li, N., Wu, S., Lu, L., & Law, R. (2019). Foodstagramming in the travel encounter. *Tourism Management*, 71, 99–115. Available from https://doi.org/10.1016/j.tourman.2018.08.020.

Woodruff, S. J., Coyne, P., Fulcher, J., Reagan, R., Rowdon, L., Santarossa, S., & Pegoraro, A. (2021). Reaction on social media to online news headlines following the release of Canada's food guide. *Canadian Journal of Dietetic Practice and Research*, 82(1), 16–20. Available from https://doi.org/10.3148/cjdpr-2020-022.

Wu, T., Hu, P., Zou, M., Zhang, F., Zeng, H., Sharma, M., & Shi, Z. (2022). An exploratory social media intervention for perception and behavior on eating away from home: A cluster randomized trial in Chongqing, China. *Journal of Cleaner Production*, 347, 131206.

Yakubu, B. N., Salamzadeh, A., Bouzari, P., Ebrahimi, P., & Fekete-Farkas, M. (2022). Identifying the key factors of sustainable entrepreneurship in the Nigerian food industry: The role of media availability. *Entrepreneurial Business and Economics Review*, 10(2), 147–162. Available from https://doi.org/10.15678/EBER.2022.100209.

Yamashita, L., Hayes, K., & Trexler, C. J. (2017). How pre-service teachers navigate trade-offs of food systems across time scales: A lens for exploring understandings of sustainability. *Environmental Education Research*, 23(3), 365–397.

Yan, Z., Huang, Z. H., Wang, Y., & Zhou, J. H. (2019). Are social embeddedness associated with food risk perception under media coverage? *Journal of Integrative Agriculture*, 18(8), 1804–1819. Available from https://doi.org/10.1016/S2095-3119(19)62586-4.

Yang, Y., & Hobbs, J. E. (2020). The power of stories: Narratives and information framing effects in science communication. *American Journal of Agricultural Economics*, 102(4), 1271–1296. Available from https://doi.org/10.1002/ajae.12078.

Yi-Frazier, J. P., Cochrane, K., Mitrovich, C., Pascual, M., Buscaino, E., Eaton, L., Panlasigui, N., Clopp, B., & Malik, F. (2015). Using Instagram as a modified application of photovoice for storytelling and sharing in adolescents with type 1 diabetes. *Qualitative Health Research*, 25(10), 1372–1382. Available from https://doi.org/10.1177/1049732315583282.

Yildirim, S. (2021). Do green women influencers spur sustainable consumption patterns? Descriptive evidences from social media influencers. *Ecofeminism and Climate Change*, 2(4).

Zhang, H., Zhang, D., Wei, Z., Li, Y., Wu, S., Mao, Z., He, C., Ma, H., Zeng, X., Xie, X., Kou, X., & Zhang, B. (2023). Analysis of public opinion on food safety in Greater China with big data and machine learning. *Current Research in Food Science*, 6. Available from https://doi.org/10.1016/j.crfs.2023.100468.

Zimmerer, K. S. (2007). Agriculture, livelihoods, and globalization: The analysis of new trajectories (and avoidance of just-so stories) of human-environment change and conservation. *Agriculture and Human Values*, 24(1), 9–16. Available from https://doi.org/10.1007/s10460-006-9028-y.

Food culture and food sustainability on social media

Yelena Mejova

ISI Foundation, Torino, Italy

Introduction

Topics regarding sustainability, environment, wellness, and food justice have become an important part of public discourse in the past century. The 2002 World Summit on Sustainable Development in Johannesburg established a strategic priority in "transforming (unsustainable) patterns of consumption" (Evans, 2011). Sustainable Consumption and Production (SCP) concerns both the sustainable production and sustainable consumption, the latter including the consumption of organic and local food, as they use fewer resources (Reisch et al., 2013). Concurrently, the nutrition-related illnesses including obesity and diabetes have contributed to millions of deaths worldwide. In the European Union, 22.8% of the population in 2016 was estimated to be obese (Damiani, 2021), while in the United States 42.4% in 2017–2018 (Hales et al., 2020); obesity is a major contributor to the incidence of heart disease, stroke, diabetes, and some types of cancer (Damiani, 2021). The European Commission has been battling these developments with initiatives such as *Farm to Fork*, which aims to increase the availability and affordability of healthy and sustainable food choices, and the *HealthyLifestyle4All* campaign that seeks to promote a healthy lifestyle (Damiani, 2021). The struggle to reform the food system thus concerns both systemic and individual changes.

Despite the best efforts of governments, the culture of food is complex, and it takes substantial shifts in the populace to impact the its food preferences. At the time of this writing, the Twitter account of McDonalds, the world's largest restaurant chain by revenue (The World's Biggest Restaurants In 2017, 2017), has 4,198,992 followers. Every day, they are presented with an assortment of brightly colored animated graphics, celebrity endorsements, and meal photoshoots with internet-savvy, curt catch phrases. Their tweets are regularly reshared ("retweeted") over a thousand times. It is estimated that children and adolescents see food marketing 30 and 189 times on average per week on social media apps, respectively

Food Sustainability and the Media
DOI: https://doi.org/10.1016/B978-0-323-91227-3.00010-X

(Potvin Kent et al., 2019). However, the social, fast-paced nature of social media democratizes online speech beyond traditional big players in media and food industry. When the company used #McDStories for a campaign, the hashtag quickly took a life of its own, subverted by other Twitter users making humorous posts about not-so-perfect experiences, and more serious ones expressing concerns about low labor standards and animal welfare (Thomases, 2012). The hashtag became a warning story of how digital marketing can go horribly wrong.

The recent advent of internet and social media brought a slew of new venues for deliberation and reform, potentially democratizing the conversation, connecting different actors in new ways, and even capturing relevant data. Insofar as sustainability is a matter concerning the business, governmental, and individual actions, social media has been used by all of these actors to further their agendas and empower action. In this chapter, we will examine the uses of social media by both individual and organizational actors, as a one-way communication platform, as a community deliberation space and knowledge resource, and as a data source for monitoring issues related to dietary health and food sustainability.

Social media analysis methods

Social media data can be analyzed within many research frameworks, some of which we outline here (for more details on the methods mentioned here, we refer the reader to the references). The study of social media often begins with the identification of relevant keywords, places, or people, such that the social media websites can be queried only for the relevant posts. For instance, Twitter Application Programming Interface (API)[1] allows us to ask for tweets mentioning a particular word (such as "environment") or for tweets recently posted by a particular user (such as the account of the US Environmental Protection Agency, @EPA). Others, such as Reddit, allow bulk download of its data,[2] which then can be filtered to select the relevant communities, such as r/Green, which focuses on environmental issues.

The sheer volume and semistructured nature of the data lend itself to quantitative analysis using methodologies from data science. To tackle the large quantities of text on social media, the textual elements can be processed using Natural Language Processing (NLP) techniques, which can produce, for instance, words and topics most discussed (via topic modeling (Ramage et al., 2009)), list places, people, and events most mentioned (via named entity extraction (Al-Moslmi et al., 2020)), and finally, produce a summary as a piece of text (Torres Moreno, 2014) or a knowledge base of concepts (Shin et al., 1310). Other elements of the posts, such as information about the user, geolocation of the post, time and date of the posting, as well as social engagement (such as "liking," commenting, and reposting), can also be summarized and used for analysis (Toivonen et al., 2019) for an overview of methods in the context of conservation science. For instance, information users choose to reveal about themselves in their social media profiles may point to

---

[1] Available from: https://developer.twitter.com/en/docs

[2] Available from: https://files.pushshift.io/reddit/

whether they are an organizational or private accounts or where they originate geographically. Time and date allow for time series analysis, making possible to detect events that spur bursts of online activity and discussion. Finally, social engagement signals can point us to the particularly popular (or even "viral") pieces of information and reveal social structures and communities.

Once the large amounts of data have been processed, they can be examined to reveal qualitative insights. Both grounded theory and theory-driven approaches to data coding have been applied to samples of social media to reveal social trends, opinions, and behaviors related to food and sustainability. Social media is often included as one of the channels in a broader "communicative ecology" (Tacchi et al., 2003) consisting of a technological layer (the devices and software), the discursive layer (ideas and symbols expressed), and social layer (concerning the groups and institutions involved) (Hearn et al., 2014). The social campaigns and communities are often examined as case studies in an effort to develop a theory from a collection of rich data sources (Eisenhardt & Graebner, 2007) and develop a rich holistic knowledge of a phenomenon (Silverman, 2015). Finally, surveys concerning the use and attitudes toward social media have been employed to reveal cognitive preferences toward this communication medium (Choudhary et al., 2019). Although the quantitative and qualitative approaches often come from different academic communities, the efforts in multidisciplinarity are creating new research fields including computational social science (Zhang et al., 2020) and digital epidemiology (Salathé, 2018), some of the insights from which we share below.

Social media and the individual

Social media has become an integral part of food experience for many people, from finding a restaurant or local market, discovering new recipes, seeing friends post their meals, to being accosted by explicit and more insidious advertising. These new ecosystems of information, social influence, and marketing allow for new insights not only into individual behaviors around nutrition and health but also into the beliefs and attitudes toward the questions of health, sustainability, and other relevant values. Below, we examine the latest efforts in tracking the health aspects of food choices that are revealed on social media and the effects of it on consumer choice within the sustainability context.

Food consumption on social media and health

The combination of social media and mobile phone technology has allowed people to document their daily lives using not only text but also high-resolution photography and GPS-enabled geolocation. Computational social scientists and epidemiologists have used the digital traces people leave on social media to monitor health-related behaviors at an unprecedented scale. The individual records of meals, photos, and accompanying metadata have been aggregated to quantitative signals, which correlate with noncommunicable diseases such as obesity and diabetes. Text provides the easiest way to gage the content and context of food consumption. For instance, the text accompanying Twitter posts has been annotated for food-related keywords and enriched with their caloric values (Abbar et al., 2015). When aggregated per state (in the United States), these caloric values have a Pearson correlation with the state-wide obesity rates of $r = 0.77$ (Abbar et al., 2015). The mentioning of particular foods, emotions, and activities has also been linked to obesity (Gore et al., 2015). Text analysis has further been used to successfully track teen birth rates, smoking, limited access to healthy food (Culotta, 2014), heart disease (Eichstaedt et al., 2015), and life satisfaction (Schwartz et al., 2013). Images provide yet another source of rich information about the healthiness of the meals posted to social media. First attempts at extracting nutritional information from images involved augmenting the laborious and expensive expert annotation with crowdsourcing—having many untrained workers label an image—which may attain nearly as accurate results (Noronha et al., 2011). Recent developments in deep learning have resulted in several automated architectures for image classification (He et al., 2020; Jiang et al., 2020; Metwalli et al., 2020), which are trained on annotated datasets including Food-101 (Bossard et al., 2014) or images found on the internet (Yunus et al., 2019). It has been shown that automatically extracted image tags may benefit modeling of specific behaviors, such as excessive drinking (Kiran et al., 2016). Further, both text and images have been used to detect alcohol use, for instance, using Instagram posts by extracting demographic information from the images and contextual information from the text (Pang et al., 2015).

In aggregate, these statistics can tell us not only about the individual consumption patterns but also relate them to environmental factors. Food deserts have been defined as neighborhoods characterized by poor access to health and affordable food, and it is estimated that about 19 million people, or roughly 6% of the U.S. population, lived in a food desert in 2015, according to the USDA (U.S. Department of Agriculture, n.d.). The impact of this situation on the daily diets of U.S. residents can be monitored through social media. For instance, using Instagram posts, De Choudhury et al. (2016) find that people living in food deserts post about foods that are higher in fat, cholesterol, and sugar, so much so that a model can be built to predict whether people live in a food desert from the content that they post (at an accuracy of over 80%). A study of tweets also shows a lower proportion of tweets about healthy foods with a positive sentiment, and a higher proportion of unhealthy tweets in general (Widener & Li, 2014).

Such studies, although reaching large numbers of people, are limited by the platform API limitations and privacy concerns, making it difficult to obtain a truly global view of the platforms' users. An alternative source of information—one preserving the privacy of the users—is to use the advertising platforms of these websites that may reveal some attributes of their users. For instance, Facebook Advertising platform[3] allows potential advertisers to refine the audience of their ad by gender, age, language, location (down to a circle of 1 km in radius), and many other interests and demographics. By querying the proportion of Facebook users who are interested in relevant activities, such as sports and outdoor activities, Chunara et al. (2013) were able to show that an increase of users with activity-related interests was associated with a lower prevalence of obese and/or overweight people, while a similar increase in interest in television was associated with a

---

[3] Available from: https://developers.facebook.com/docs/marketing-apis/

higher prevalence. However, later study has shown that such correlations may be achieved by tracking nonrelevant interests such as music or technology (Mejova et al., 2018), necessitating "placebo" baselines for understanding the strengths of the relationships between online and offline health measures, much like in medical research.

Beyond purely observational study, attempts have been made to influence healthiness of social media users' diets through the platforms they use. For instance, Takeuchi et al. (2014) propose a system Yumlog where a user can upload a picture of their meal, which can then be evaluated by other users, as well as the posting user. Thus, they attempt to employ what in behavior science is called "expectation assimilation" to change the imagined palatability of the meal and unconsciously enhance satisfaction. A similar application called Tagliatelle (Linehan et al., 2010) was also proposed. However, motivating users to engage as posters and taggers is a constant battle with such specialized tools. General-purpose social media platforms may serve the same purpose by allowing the social circle of the user to "like" or "upvote" posts, with the additional benefit that these users are more likely to continuously engage with the platform. A study of Instagram posts labeled with the #*foodporn* hashtag has shown that posts that also have health-related hashtags such as #*lowcarb*, #*gym*, and #*healthy* have higher levels of social engagement than those mentioning unhealthy tags or those concerning sentiment or social interactions (Mejova et al., 2016). Although when rigorously tested, the use of mobile technology and social media to quantitatively improve diet and physical activity has been shown to work only some of the time (for a systematic review of social media-based interventions, see (Chau et al., 2018)). For instance, Chung et al. (2017) use a combination of Fitbit wearable device, Twitter posting, and gamification to help participants improve their lifestyle and show a decrease in fruit/vegetable intake and decrease in sugar-sweetened beverage intake. On the other hand, Svetkey et al. (2015) find that self-monitoring and a "buddy system" on the cellphone did not result in behavioral improvements. Beyond general-purpose platforms, communities focused on a theme or goal allow users with a common interest to express themselves, seek social support, and trade useful information. For instance, Reddit platform consists of small communities centered on themes. One such community is *r/loseit*, self-described as "A place for people of all sizes to discuss healthy and sustainable methods of weight loss." Studies show that users who are more active, and who engage in discussions that might provide social support, achieve greater weight loss, as well as those who self-declare a higher body mass index (BMI) level (who possibly have a greater dissatisfaction with their weight) (Pappa et al., 2017). In fact, the users who receive positive feedback at the beginning of their participation are more likely to remain active in the community (Cunha et al., 2016), pointing to easy interventions that could be done by the moderators to make sure new members are welcomed and retained.

The above efforts testify to the extent to which social media has permeated our daily lives, such that it can be used to monitor food-related activities and related health outcomes. Beyond being a data source, it provides an avenue of interpersonal interaction, community deliberation, and outside advertisement, which may affect these activities. Next, we focus on the sustainable food choices with which these communication platforms may interact.

Sustainable consumer choice

Social media and applications have numerous ways of swaying consumer choice and individual behavior. As an advertising platform, these websites have a massive audience for advertisement. As was mentioned earlier, it is estimated that children and adolescents see food marketing 30 and 189 times on average per week on social media apps, respectively (Potvin Kent et al., 2019). However, social media also provides a platform for engagement among the users, creating new avenues of consumer education and opinion formation. Environmentalist movement has been at the forefront of using new technologies to engage the public, from conservation photography (Farnsworth, 2011), to blogs (Schuenemann & Wagner, 2014), to online messaging (Adams & Gynnild, 2013) (for an overview, see (Pearson et al., 2016)). Food-related communication in the sustainability framework has gathered a lesser attention from researchers; however, notable insights show the potential of social media to impact consumer behavior.

Social media "influencers" or popular peers have become a new important player in the mass communication about food. For instance, König et al. (2017) show that popular peers are perceived to eat more healthily than unpopular peers, and that own eating behavior such as snacking is more likely to be ascribed to the popular, rather than unpopular, peers. Through such influences, social media may also help cultures intersect and behaviors spread. Using semistructured interviews, Choudhary et al. (2019) find that social media can be seen as an "enabler" and not an "imposer" when cross-cultural information is shared, due to a lesser adaptation stress for those not familiar with the new culture may experience. Specifically, the new cultural information may include techniques that may lead to a more sustainable consumption behavior, including information related to best before date, cooking from scratch, deep-freezing, and advanced storage techniques. In fact, the study found that most of the respondents interviewed said they "either regularly browsed or invariably came across content related to food waste and sustainable consumption in the form of videos, photographs, blogs, live videos, stories, and cooking classes" (Choudhary et al., 2019). Much like in previous studies, these respondents largely agreed that content shared by their role models such as celebrities, religious leaders, celebrity chefs, and global organizations influenced them the most. They have also revealed that social media was the primary source of information on environmental pollution, sustainability, and food waste—information that may change their behavior in the long term. A survey of Chinese consumers has also shown a sensitivity to food safety information on social media, with those encountering negative food safety information being less willing to purchase the product (Cui et al., 2019). Note that, again, they show that the influence of friends and opinion leaders remains important in these decisions.

However, another consumer survey has shown that the effect of social media on the sustainability of food-related behaviors is not always positive (Simeone & Scarpato, 2020). They find that consumers who are mainly staying informed through social media are less likely to consume local varieties of fish and opt for more well-known species. This behavior contributes to the overconsumption (and increased farming) of certain kinds of fish, while local fish stocks remain unsold. The mixed picture continues, as respondents who report consuming "biological products" are less likely to be getting most of their information from social media, potentially pointing to the use of alternative sources of information by those interested in health and environmental safety. On the other hand, social media use is positively related to the "purchase of zero miles products," possibly showing usefulness of the platforms in local food distribution (see next section for more on the subject). In all, they do find that users of social media are also more likely to reach out to people they may not know, possibly supporting cross-cultural information sharing revealed in Choudhary et al. (2019).

Further, social media seems to be associated with the environmental awareness, especially for the younger people. A structured questionnaire of wine consumers (Sogari et al., 2017) has found that millennials are more likely to pay attention to the energy type and its use, as well as soil/land preservation when choosing their wine. The authors speculate that "younger people, who have grown up in a period in which the topic of limited natural resources is highly debated by the media, give more value in energy issues and biodiversity preservation." This then translates into willingness to pay higher prices for the products, especially by the younger generation. The exposure to new information and new emphasis on the aspect of environmental sustainability issues may have gained in importance when social media users choose a product (Pomarici & Vecchio, 2014).

As in previous section, it is also possible to measure attitudes toward elements of sustainable food production and consumption through social media. For instance, an analysis of Instagram posts at farmers' markets revealed emphasis on "healthy" in the association with positive emotions (Pilař et al., 2016). This concept was often accompanied with "organic," "vegan," "cleaneating," "fresh," and "local," along side "foodporn" and "good," further revealing value judgments of the experience. Although the absence of posts with words reflecting negative feelings (only 4.7% were such posts) may suggest that the purpose of these posts may be ostensibly to share only positive experiences, with negative ones possibly self-censored. Others have studied emotions and opinions around social posting about veganism and plant-based diets (Aleixo et al., 2021), organic produce (Martina et al., 2021), and even urban parks (Wan et al., 2021). Such marketing research may enable parties to be able to better speak to the consumers, though not necessarily address their concerns, as those do not seem to be expressed as often as positive emotions.

Food waste

Food waste in particular has been the focus of many governmental and business initiatives. For example, Love Food Hate Waste campaign[4] based in the United Kingdom comes with an app that helps shoppers reduce food waste by cooking, organizing their shopping lists, and keeping track of their produce. Such initiatives depend on successful communication campaigns to engage with their target audience and spread their message.

An analysis of 26 case studies of communication campaigns has shown that managing attention and achieving a large scale were especially important key success factors (Aschemann-Witzel et al., 2017), making social media an important outlet for mass communication. Continuous long-term initiatives were needed to establish the messenger as the authoritative source around food waste. The aim of the campaigns was often to put a positive spin on the sustainable practices, such as calling fruits and vegetables "inglorious,"

[4] Available from: https://www.lovefoodhatewaste.com/

giving them their own personalities, and employing humor. Another case study of From Waste to Delicacy campaign lead by bloggers identified "creativity" as a path to reducing food waste: "innovating new recipes based on ingredients, adapting recipes, and creatively using ingredients as well as creatively handling and storing food" (Närvänen et al., 2018). The bloggers were acting as "artists" who would create a beautiful dish out of "waste" that was not conventionally seen as such and encouraging their readers to experiment and create. The Waste Week campaign in Finland encouraged people to commit to pledges to use up all of the ingredients they bought or make new dishes from leftovers (Närvänen et al., 2018). All of these campaigns were accompanied with beautiful and styled photographs of the dishes, esthetically similar to the #foodporn trend, turning the desirability of food to the service of sustainability (Kozinets et al., 2017).

The effectiveness of these campaigns remains to be established. An analysis of self-reported food waste around the time of a social influence intervention by a large retailer has shown that both respondents who engaged with the social media and e-newsletter interventions and the control group all showed significant reductions in self-reported food waste over the study period (Young et al., 2017). The messaging was consistent with the Love Food Hate Waste campaign. Researchers conclude that "social media such as Facebook cannot replicate enough of the interaction shown by face to face social influence interventions to change reported behaviour more than the control group" and that this should dampen the enthusiasm expressed by previous studies. In fact, the authors suggest that social media tool should be classified as an information intervention, as it does not have the affordances or impact of the other physical face-to-face interventions (Young et al., 2017). This conclusion has been disputed, however, by Grainger and Stewart (2017), who say more studies need to be evaluated, as there is a large heterogeneity that has been attributed to the type and number of behavior change techniques employed in the past. To this critique, the authors of the original study point out that more field experiments such as their study are necessary to establish the efficacy of the behavior change efforts (Young et al., 2017). Thus, quantitative evaluation of social media as a consumer behavior intervention platform remains an active area of research.

Social media and organizations

Corporate communication

Social media provides unique benefits and dangers to businesses aiming to reach their target audience with sustainability messages. Unlike traditional media, it allows two-way interaction, allowing not only the businesses and organizations to promote their message and their brand but also the consumers to express their opinions on that message and the business behind it (Reilly & Hynan, 2014). Thus, it allows for instantaneous feedback about consumer preferences, lifestyle, and opinions about sustainable behaviors (Lynn and Valette-Florence, 2012).

A survey of consumers in the United States, Germany, and South Korea has found that there are significant differences in social media motives for sustainability among countries and the social media website (Minton et al., 2012). For both Facebook and Twitter, the responsibility level of attitude commitment was associated with contributions to sustainable charities and organic food purchasing, but compared with heavy Twitter users, heavy Facebook users were significantly less likely to participate in all sustainable behaviors. Beyond social media, the countries display a marked difference in sustainable behaviors such as recycling and taking green transportation, leading the authors to recommend cross-cultural analysis in the future. Further, an analysis of Polish customers on Facebook (Bojanowska & Kulisz, 2020) has shown that women are more likely to be aware of zero-waste and proecological activities of companies on social media, suggesting gender should be another variable researchers should strive to control.

The social nature of social media allows the individuals on these platforms to share their experiences, to review and recommend brands and businesses. Such electronic word-of-mouth (WOM) support can be invaluable to a business, effectively providing free advertising for their product (Strähle & Gräff, 2016). Those who have had positive experiences with online WOM tend to also return and contribute reviews (Ng et al., 2015). This may benefit especially more niche products, as the insights about them would provide more impact in building a business image (Phang et al., 2013). Unfortunately, the increased reliance of consumers on online WOM has led to a clandestine industry of generating fake reviews on consumer-facing websites—automatically finding such reviews is an open area of natural language processing research (for a survey on such methods, see (Mohawesh et al., 2021)).

Perhaps one of the most controversial companies engaged in social media is Monsanto, an American agrochemical and agricultural biotechnology corporation, which often appears at the center of genetically modified food (or "organism" making for abbreviation GMO) debate. Despite, or perhaps because, of its standing as one of the largest US corporations by revenue (Forture 500 Companies, 2018), the company's products attract the ire of health, environmental, and social justice activists—recently much of it on social media. In 2013, a U.S. bill that may benefit the company caught the attention of social media users and was dubbed as the "Monsanto Protection Act" (Sprinkle, 2013), forcing politicians to make explicit statements on the subject. A 2015 analysis of mentions of GMO on different social media sites showed an overall positive sentiment, the GMO-related posts on Twitter displayed a markedly negative attitude toward the topic, leading the authors to conclude that "Twitter users do not represent the broader difference in opinion found in other media" (Munro et al., 2015). However, not all social media users are private individuals—news agencies are known to use it to disseminate their articles. Indeed, within this negative discourse around GMOs, Russian news agencies Russia Today (RT) and Sputnik have been found to present GMOs in a negative light and include related keywords in irrelevant articles (as "click bait") (Dorius & Lawrence-Dill, 2018). Although these outlets have been banned from advertising on Twitter (Russel, 2017), the content generated by them can still be propagated via retweeting by other accounts within the platform.

Another more direct example of the interaction between agriculture advertising and activism response was the #Februdairy campaign launched to promote the British dairy industry in February 2018 (Rodak, 2020). In response, it led to a campaign #Veganuary, which encouraged people to go vegan for the month of January. Further, the hashtag #Februdairy was "hijacked" by animal protection activists and used in posts critical of the dairy industry. Another example is the McDonalds' #McDStories promotion meant to provide a venue for suppliers to share their stories that was mentioned at the beginning of this chapter (Thomases, 2012). Such hashtag hijacking has been encountered in many controversial discussions, and attempts have been made to automatically detect it (Hadgu et al., 2013). Further, Veil et al. (2014) describe how a hoax video was used as an excuse by activists to bring public attention to Kraft Foods' use of food dyes. Although the original hoax was quickly dismissed, the attention encouraged activists to post on the company's Facebook page, gaining a platform for their message.

Although the communication of corporate social responsibility (CSR) on social media has been linked to the increased eWOM (Fatma et al., 2020), these examples illustrate the larger movement of social media users to prevent companies from "greenwashing" their products and product chains. A possible side effect of CSR messaging, Lyon and Maxwell define greenwashing as "selective disclosure of positive information about a company's environmental or social performance, while withholding negative information on these dimensions" (Lyon & Maxwell, 2011). In 2013, Lyon and Montgomery hypothesized that the increased public voice and scrutiny online should diminish corporate greenwash (Lyon & Montgomery, 2013). Despite this, "green marketing" has been thriving in the past decade, with possible dangers of stretching the truth and employing "greenspeak" to the point of greenwashing (Hjorth & Kato, 2016; Lewis, 2017). Besides social media crowd efforts, attempts have been made to pursue dishonest statements by the governmental authorities, for instance, in the United States by Federal Trade Commission (FTC), which has the power to prosecute false and misleading advertisement claims outlines in the Guides for the Use of Environmental Marketing Claims (Guides for the Use of Environmental Marketing Claims, 2008).

The double-edged sword of social media attention thus must be wielded by companies with caution—a potentially cheap source of advertisement audience, and in the best case organic WOM, it is also a source of increased scrutiny by consumers and activist groups, and in the worst case, a firestorm of bad publicity. The advice of Reilly and Hynan in 2014 for companies to have a robust social media strategy with dedicated resources to quickly handle adverse reactions is likely to be relevant in the near future (Reilly & Hynan, 2014).

Grassroots organization

Digital connectivity has been hailed as a new era of "political relationships and power relations" (Lupton, 2014), which support the continual rise of consumer movements toward ethical consumption and environmental awareness (Lewis, 2018). In order to reconnect with the food production chain and establish local communities, both generic social media websites and specialized applications have been employed to connect its users under a cause or locale (which Tania Lewis dubs "apptivism") (Lewis, 2018).

Urban farming has benefited greatly from the new communication technologies of the social media age. Paradoxically, despite being global in potential reach, social media and other new communication platforms are often used to promote "local, community-level action" (Hearn et al., 2014). Initiatives such as Landshare Australia[5], Kokoza in Czech Republic[6], and numerous local community gardens provide a place for community to connect and for the

[5] Available from: https://www.facebook.com/Landshare-Australia-1826931607535154

[6] Available from: https://www.mapko.cz

sharing of resources on management, soil and water considerations, and final distribution of the produce. Social media accounts such as the Kansas City Community Gardens Twitter account[7] and Community Gardens Ireland Facebook page[8] allow the social media users to find them on the platforms they already use and to get involved in the local gardening efforts. Online marketplaces for local produce then allow for the consumers to connect to the local producers on platforms such as Local Food Loop in Australia[9] and Central Oregon Locavore[10]. The app Olio[11] allows neighbors to share food and other household items in order to decrease food waste and connect local communities. A study of eight Alternative Food Networks and 21 online spaces around England finds that not only is social media used for advertising and promotions, they also capitalize "on customers and members as promoters who assist in the development of organisational reputation and ultimately, income or sustainability" (Bos & Owen, 2016). Interestingly, even though civic food networks often focus on social inequality and injustices, they find that the people found engaged in these networks are often highly educated, and women, pointing to a potential barriers to engagement. Traditional farming, on the other hand, as an established commercial enterprise, has been only slowly getting on the social media bandwagon. A series of interviews with Welsh farmers in 2017 showed the lack of fast internet infrastructure and limited technical knowledge as obstacles to using social media (Morris & James, 2017). The farmers have also indicated that they saw no need in changing their sales and marketing strategies, preferring to "pick up the phone" and speak with their business partners. However, one farmer did acknowledge that "there is a massive power in social media and a massive opportunity for many, many farmers to sell their produce, and not by selling it directly to them but selling us as an industry." One respondent also pointed to a successful local law change in response to his tweet to a local minister after a disaster.

Alternatively, social media can serve as a knowledge base for people having interests in sustainability. For instance, FallingFruit.org is a collaborative map where "foragers, freegans, and foresters everywhere" point to food sources around the world. Reddit community *r/foraging* also provides resources, tips, and recipes for those seeking to benefit from "wild edable food."[12] Those who are interested more generally in learning about their natural surroundings can contribute to applications such as Leafsnap[13] that aims to identify plants from photographs. Further, the #Agchat hashtag has been used on Twitter by farmers to exchange information about soil management (Mills et al., 2019) and contributed to a "blended learning" (Cullen et al., 2016) wherein the hands-on interaction with the subject matter is supplemented by online information.

The above examples are only some of the efforts in food sustainability area, which is a part of the global sphere of environmentalism and food justice. There are many other

[7] Available from: https://twitter.com/KC_Gardens

[8] Available from: https://www.facebook.com/cgireland.org

[9] Available from: https://localfoodloop.com/

[10] Available from: https://centraloregonlocavore.org/

[11] Available from: https://olioex.com/

[12] Available from: https://www.reddit.com/r/foraging/

[13] Available from: https://plantidentifier.info/

relevant movements, which include advocates for the reduction in meat production and consumption, fair labor practices throughout food production systems, and the reduction of agriculture-related deforestation. From global campaigns, such as the World Environment Day (Pang & Law, 2017) to local collective action (such as that in Malaysia (Leong et al., 2019), China (Lee & Ho, 2014), or Tasmania (Wells, 2018)), social media allows for diversification of participation, introduction of heterogeneous capabilities of the resulting community, and organization of mass action (Leong et al., 2019). For further reading on the use of new communication technologies by environmental movements, we direct the reader to recent books aimed both at researchers and environmentalists looking to use the technology for furthering their cause (Fow, 2021; Narula et al., 2018).

Case study

In order to illustrate the span of topics and communities social media foster, we examine the main players in the conversation around sustainability on Twitter during the period from June 25, 2020 to June 1, 2021. Twitter Streaming API was queried with keywords and phrases relevant to food and sustainability,[14] acquiring a sample of publicly available tweets mentioning these for nearly 1 year. The dataset contains 97,675,155 tweets and is available upon request by the chapter's author.

To glimpse the content of this conversation, we applied Natural Language Processing to the text of these tweets by removing URLs and special characters and splitting up the words ("tokenizing" the text). Fig. 3.1 shows the 30 most frequent words, with common words such as "the" and "of" removed. We can see that *environment* and *climate change* are dominating the conversation, followed by *farming* and *agriculture*. We can also find *government* mentioned, as well as *Biden* and *Trump*—the two contenders for the U.S. Presidency in 2020. The ongoing *covid* pandemic is also mentioned. Interestingly, large companies are not present here, whereas political actors are, pointing to the focus of those engaged in the Twitter discussion on sustainability.

Alternatively, we can examine the retweet network in this data (due to the size of the dataset, we select only tweets from May 2021). Sometimes called an "endorsement" network, it consists of Twitter users as nodes and retweets as links (Garimella et al., 2018). Since when users retweet each other (verbatim), it is often considered in agreement with the original message, those who are close to each other in such a network tend to circulate similar content and often agree on a particular stance on it. Fig. 3.1 shows such a network for May 2021 (layout is determined by ForceAtlas2 algorithm), with automatically detected communities in colors (using Louvian algorithm[15]) and the most popular (by number of

---

[14] The full keyword list is as follows: food sustainability, sustainable food, sustainable eating, sustainable meat, sustainable diet, local sustainable, local food, eco friendly food, eco friendly restaurant, famine, food scarcity, food system, food security, food waste, agriculture, farming, environment, environmentally friendly, environmental sustainability, sustainable food, sustainable meals, smartfarming, ImprovingFoodTogether, foodsecurity, climate change, global warming, carbon dioxide, carbon footprint, carbon tax, climate model, emission, greenhouse gas, renewable, sea level.

[15] Network layout and community detection done with Gephi https://gephi.org/

links, or degree) user's name shown by the community. The largest community is one with Greta Thunberg as the most retweeted person. Greta Thunberg is a Swedish environmental activist who has found visibility after speaking at the United Nations Climate Change Conference in 2018 and UN Climate Action Summit in 2019. Other central figures include other advocates, including Ben See (@ClimateBen), Svein T Veitdal (@tveitdal), and Michael E. Mann (@MichaelEMann), as well as more topic-specific accounts such as @ECOWARRIORSS and @climatecouncil. Another community is centered on (Fig. 3.2).

the Former U.S. Secretary of State Mike Pompeo (@mikepompeo), as well as BreitbartNews (a right-leaning news source). Third largest community centers on @POTUS, or the account of the President of the United States, and includes accounts of U.S. Senators Bernie Sanders (@SenSanders) and Chuck Schumer (@SenSchumer) and Governor of California Gavin Newsom (@GavinNewsom). The network also includes non-U.S.-centric communities, including around Ro Nay San Lwin (@nslwin), a cofounder of the Free Rohingya Coalition & Boycott Myanmar[16] and Kisan Ekta Morcha (@Kisanektamorcha), which posts updates about an ongoing protest by farmers in India. Further, a community exists around the entrepreneur Elon Musk, although the other accounts in this community concern cryptocurrency.

In that month of May, some of the most retweeted (reshared) posts revolved around a mixture of ongoing topics (such as the COVID pandemic) and campaigns by individuals and organizations around environmental and trade issues. For instance, a tweet simply stating "It is high time there was a carbon tax!" by then-billionaire and entrepreneur Elon Musk gathered 16.8 K retweets (the number of Twitter users who saw the tweet is likely to be at least an order of magnitude larger). On the other hand, Greenpeace Thailand posted a tweet in opposition of the Comprehensive and Progressive Agreement for Trans-Pacific Partnership (CPTPP), a trade agreement signed in 2018, saying "Because the collection and transmission of plant species is not a crime. access to seeds It's our basic right!" (translated from Thai by Google Translate). Yet other personalities were mentioning the rise of the cryptocurrencies and the environmental impact they create: "Yesterday I was pleased to host a meeting between @elonmusk & the leading Bitcoin miners in North America. The miners have agreed to form the Bitcoin Mining Council to promote energy usage transparency & accelerate sustainability initiatives worldwide." However, among these popular tweets are those posted by

FIGURE 3.1   Top 30 words in Twitter dataset around food and sustainability on June 25, 2020 − June 1, 2021.

[16] Available from: https://www.linkedin.com/in/ronslwin/

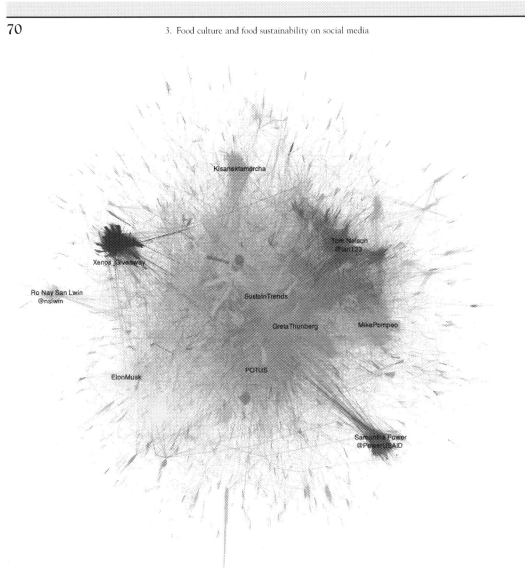

FIGURE 3.2 Retweet network around food and sustainability in May 2021.

accounts that were later suspended by Twitter, suggesting that some activities in this sphere of discourse may break the platform's Terms of Service and may even be harmful to their audience. Future research on social media must take into consideration potentially negative actors coopting the conversation or creating unwelcome online environment.

From this brief exercise, we can find the voices with most reach to belong to politicians and business people; however, some prominent environmentalists can be seen at the center of the conversation. The network also contains many small communities, which may be more localized both geographically and in terms of interest when it comes to sustainability.

Looking forward

There is no doubt social media will continue to be an important platform for communication—by organizations, and especially by the individuals around the world. As the acceptance of social media as a legitimate communication platform grows, new norms are beginning to develop in the message design, audience interaction, and brand management. The increased importance of these platforms brings new questions of data ownership, privacy, and its use for research both by governments and for market research by businesses. Censorship by the governments, and especially by the platforms themselves, is yet another consideration for engaging in discourse on platforms that are owned by for-profit companies. Nonetheless, the advent of new technologies, including 5 G, virtual reality, and the increased use of wearables and connected devices will likely result in even more digital data and platforms, which will move forward both the food culture and the activism around sustainability.

References

Abbar, S., Mejova, Y., & Weber, I. (2015). You tweet what you eat: Studying food consumption through twitter. *In Conference on Human Factors in Computing Systems - Proceedings* (2015, pp. 3197–3206). Association for Computing Machinery. Available from https://doi.org/10.1145/2702123.2702153.

Adams, P. C., & Gynnild, A. (2013). Environmental messages in online media: The role of place. *Environmental Communication, 7*(1), 113–130. Available from https://doi.org/10.1080/17524032.2012.754777.

Aleixo, Marina G. B., Sass, Carla A. B., Leal, Rafael M., Dantas, Tiago M., Pagani, M. ônica M., Pimentel, Tatiana C., Freitas, M. ônica Q., Cruz, Adriano G., Azeredo, Denise R. P., & Esmerino, Erick A. (2021). Using Twitter ® as source of information for dietary market research: A study on veganism and plant-based diets. *International Journal of Food Science & Technology, 56*(1), 61–68. Available from https://doi.org/10.1111/ijfs.14743.

Al-Moslmi, Tareq, Gallofre Ocana, Marc, L. Opdahl, Andreas, & Veres, Csaba (2020). Named entity extraction for knowledge graphs: A literature overview. *IEEE Access, 8*, 32862–32881. Available from https://doi.org/10.1109/access.2020.2973928.

Aschemann-Witzel, Jessica, de Hooge, Ilona E., Rohm, Harald, Normann, Anne, Bossle, Marilia Bonzanini, Grønhøj, Alice, & Oostindjer, Marije (2017). Key characteristics and success factors of supply chain initiatives tackling consumer-related food waste – A multiple case study. *Journal of Cleaner Production, 155*, 33–45. Available from https://doi.org/10.1016/j.jclepro.2016.11.173.

Bojanowska, Agnieszka, & Kulisz, Monika (2020). Polish consumers' response to social media eco-marketing techniques. *Sustainability, 12*(21). Available from https://doi.org/10.3390/su12218925.

Bos, Elizabeth, & Owen, Luke (2016). Virtual reconnection: The online spaces of alternative food networks in England. *Journal of Rural Studies, 45*, 1–14. Available from https://doi.org/10.1016/j.jrurstud.2016.02.016.

Bossard, Lukas, Guillaumin, Matthieu, & Van Gool, Luc (2014). *Food-101 – Mining discriminative components with random forests* (Vol. 8694, pp. 446–461). Springer Science and Business Media LLC Issue 6. Available from https://doi.org/10.1007/978-3-319-10599-4_29.

Chau, Michelle M., Burgermaster, Marissa, & Mamykina, Lena (2018). The use of social media in nutrition interventions for adolescents and young adults—A systematic review. *International Journal of Medical Informatics, 120*, 77–91. Available from https://doi.org/10.1016/j.ijmedinf.2018.10.001.

Choudhary, Sonal, Nayak, Rakesh, Kumari, Sushma, & Choudhury, Homagni (2019). Analysing acculturation to sustainable food consumption behaviour in the social media through the lens of information diffusion. *Technological Forecasting and Social Change, 145*, 481–492. Available from https://doi.org/10.1016/j.techfore.2018.10.009.

Chunara, Rumi, Bouton, Lindsay, Ayers, John W., Brownstein., John, S., & Pappalardo, Francesco (2013). Assessing the online social environment for surveillance of obesity prevalence. *PLoS One, 8*(4). Available from https://doi.org/10.1371/journal.pone.0061373.

Chung, A. E., Skinner, A. C., Hasty, S. E., & Perrin, E. M. (2017). Tweeting to health: A novel mHealth intervention using fitbits and twitter to foster healthy lifestyles. *Clinical Pediatrics*, *56*(1), 26−32. Available from https://doi.org/10.1177/0009922816653385.

Cui, Li, Jiang, Haiyang, Deng, Hepu, & Zhang, Tao (2019). The influence of the diffusion of food safety information through social media on consumers' purchase intentions. *Data Technologies and Applications*, *53*(2), 230−248. Available from https://doi.org/10.1108/dta-05-2018-0046.

Cullen, B., Amos, D., & Padel, S. (2016). Organic Knowledge Network Arable-D2. 1 Description of farmer innovation groups. (Original work published 2016).

Culotta, A. (2014). Estimating county health statistics with twitter. *In Conference on Human Factors in Computing Systems - Proceedings* (pp. 1335−1344). Association for Computing Machinery. Available from https://doi.org/10.1145/2556288.2557139.

Cunha, T. O., Weber, I., Haddadi, H., & Pappa, G. L. (2016). The effect of social feedback in a Reddit weight loss community. *In DH 2016 - Proceedings of the 2016 Digital Health Conference* (pp. 99−103). Association for Computing Machinery, Inc. Available from https://doi.org/10.1145/2896338.2896353.

Damiani, A. (2021). Europe faces 'obesity epidemic' as figure almost tripled in 40 years. Centers for Disease Control and Prevention. https://www.euractiv.com/section/healthconsumers/news/europe-faces-obesity-epidemic-as-figure-almost-tripled-in-40-years/ (Original work published 2011).

De Choudhury, M., Sharma, S., & Kiciman, E. (2016). Characterizing dietary choices, nutrition, and language in food deserts via social media. *In Proceedings of the ACM Conference on Computer Supported Cooperative Work, CSCW* (Vol. 27, pp. 1157−1170). Association for Computing Machinery. Available from https://doi.org/10.1145/2818048.2819956.

Dorius, S. F., & Lawrence-Dill, C. J. (2018). Sowing the seeds of skepticism: Russian state news and anti-GMO sentiment. *GM Crops and Food*, *9*(2), 53−58. Available from https://doi.org/10.1080/21645698.2018.1454192.

Eichstaedt, J. C., Schwartz, H. A., Kern, M. L., Park, G., Labarthe, D. R., Merchant, R. M., Jha, S., Agrawal, M., Dziurzynski, L. A., Sap, M., Weeg, C., Larson, E. E., Ungar, L. H., & Seligman, M. E. P. (2015). Psychological language on twitter predicts county-level heart disease mortality. *Psychological Science*, *26*(2), 159−169. Available from https://doi.org/10.1177/0956797614557867.

Eisenhardt, K. M., & Graebner, M. E. (2007). Theory building from cases: Opportunities and challenges. *Academy of Management Journal*, *50*(1), 25−32. Available from https://doi.org/10.5465/amj.2007.24160888.

Evans, D. (2011). Thrifty, green or frugal: Reflections on sustainable consumption in a changing economic climate. *Geoforum; Journal of Physical, Human, and Regional Geosciences*, *42*(5), 550−557. Available from https://doi.org/10.1016/j.geoforum.2011.03.008.

Farnsworth, B. E. (2011). Conservation photography as environmental education: Focus on the pedagogues. *Environmental Education Research*, *17*(6), 769−787. Available from https://doi.org/10.1080/13504622.2011.618627.

Fatma, M., Ruiz, A. P., Khan, I., & Rahman, Z. (2020). The effect of CSR engagement on eWOM on social media. *International Journal of Organizational Analysis*, *28*(4), 941−956. Available from https://doi.org/10.1108/IJOA-10-2019-1895.

Forture 500 Companies. (2018). available from: https://fortune.com/fortune500/2018.

Fow, S. J. (2021). Be Decent: Environmental Activism 2.0. Decent Publications Ltd.

Garimella, K., Morales, G. D. F., Gionis, A., & Mathioudakis, M. (2018). Quantifying controversy on social media. *ACM Transactions on Social Computing*, *1*(1), 1−27. Available from https://doi.org/10.1145/3140565.

Gore, R. J., Diallo, S., & Padilla, J. (2015). You are what you tweet: Connecting the geographic variation in America's obesity rate to twitter content. *PLoS One*, *10*(9). Available from https://doi.org/10.1371/journal.pone.0133505.

Grainger, M. J., & Stewart, G. B. (2017). The jury is still out on social media as a tool for reducing food waste a response to Young et al. (2017). *Resources, Conservation and Recycling*, *122*, 407−410. Available from https://doi.org/10.1016/j.resconrec.2017.04.001.

Guides for the Use of Environmental Marketing Claims. (2008). Available from: https://web.archive.org/web/20081117145015/ http://www.ftc.gov/bcp/grnrule/guides980427.htm.

Hadgu, A. T., Garimella, K., & Weber, I. (2013). Political hashtag hijacking in the U.S. *WWW 2013 Companion - Proceedings of the 22nd International Conference on World Wide Web* (pp. 55−56). Association for Computing Machinery. Available from https://doi.org/10.1145/2487788.2487809.

Hales, C. M., Carroll, M. D., Fryar, C. D., & Ogden, C. L. (2020). Prevalence of obesity and severe obesity among adults: United States, 2017–2018. *NCHS Data Brief, 360*, 1–8.

He, J., Shao, Z., Wright, J., Kerr, D., Boushey, C., & Zhu, F. (2020). Multi-task image-based dietary assessment for food recognition and portion size estimation. *Proceedings - 3rd International Conference on Multimedia Information Processing and Retrieval, MIPR 2020* (pp. 49–54). Institute of Electrical and Electronics Engineers Inc. Available from https://doi.org/10.1109/MIPR49039.2020.00018.

Hearn, G., Collie, N., Lyle, P., Choi, J. H. J., & Foth, M. (2014). Using communicative ecology theory to scope the emerging role of social media in the evolution of urban food systems. *Futures, 62*, 202–212. Available from https://doi.org/10.1016/j.futures.2014.04.010.

Hjorth, L., & Kato, F. (2016). Keitai mizu: A mobile game reflection in a post-3/11 Tokyo, Japan. *Green Asia: Ecocultures, Sustainable Lifestyles, and Ethical Consumption (pp. 129–141)*. Taylor and Francis. Available from https://doi.org/10.4324/9781315722351.

Jiang, L., Qiu, B., Liu, X., Huang, C., & Lin, K. (2020). DeepFood: Food image analysis and dietary assessment via deep model. *IEEE Access, 8*, 47477–47489. Available from https://doi.org/10.1109/ACCESS.2020.2973625.

Kiran, V. R., Alfayad, A., & Weber, I. (2016). Social media image analysis for public health. *Conference on Human Factors in Computing Systems - Proceedings* (pp. 5543–5547). Association for Computing Machinery. Available from https://doi.org/10.1145/2858036.2858234.

König, L. M., Giese, H., Stok, F. M., & Renner, B. (2017). The social image of food: Associations between popularity and eating behavior. *Appetite, 114*, 248–258. Available from https://doi.org/10.1016/j.appet.2017.03.039.

Kozinets, R., Patterson, A., & Ashman, R. (2017). Networks of desire: How technology increases our passion to consume. *Journal of Consumer Research, 43*(5), 659–682. Available from https://doi.org/10.1093/jcr/ucw061.

Lee, K., & Ho, M. S. (2014). The maoming anti-PX protest of 2014: An environmental movement in contemporary China. *China Perspectives, 3*, 33–39. Available from http://www.cefc.com.hk/rubrique.php?id = 17.

Leong, C., Pan, S. L., Bahri, S., & Fauzi, A. (2019). Social media empowerment in social movements: Power activation and power accrual in digital activism. *European Journal of Information Systems, 28*(2), 173–204. Available from https://doi.org/10.1080/0960085X.2018.1512944.

Lewis, T. (2017). *Sustainability, lifestyle and consumption in Asia. Green Asia: Ecocultures, sustainable lifestyles, and ethical consumption* (pp. 1–19). New York: Routledge.

Lewis, T. (2018). Digital food: From paddock to platform. *Communication Research and Practice, 4*(3), 212–228. Available from https://doi.org/10.1080/22041451.2018.1476795.

Linehan, C., Doughty, M., Lawson, S., Kirman, B., Olivier, P., & Moynihan, P. (2010). Tagliatelle: Social tagging to encourage healthier eating. *Conference on Human Factors in Computing Systems - Proceedings, 3331–3336*. Available from https://doi.org/10.1145/1753846.1753980.

Lupton, D. (2014). *Digital sociology* (pp. 1–230). Taylor and Francis. Available from https://doi.org/10.4324/9781315776880.

Lynn, K., & Valette-Florence, P. (2012). Marketplace lifestyles in an age of social media: Theory and methods.

Lyon, T. P., & Maxwell, J. W. (2011). Greenwash: Corporate environmental disclosure under threat of audit. *Journal of Economics and Management Strategy, 20*(1), 3–41. Available from https://doi.org/10.1111/j.1530-9134.2010.00282.x.

Lyon, T. P., & Montgomery, A. W. (2013). Tweetjacked: The impact of social media on corporate greenwash. *Journal of Business Ethics, 118*(4), 747–757. Available from https://doi.org/10.1007/s10551-013-1958-x.

Martina, C., Ladislav, P., & Stanislav, R. (2021). Organic wine as an Instagram star using a design thinking approach. *In Transdisciplinary case studies on design for food and sustainability* (pp. 149–163). Elsevier. Available from https://doi.org/10.1016/B978-0-12-817821-8.00011-4.

Mejova, Y., Abbar, S., & Haddadi, H. (2016). Fetishizing Food in Digital Age: #foodporn Around the World. *Proceedings of the International AAAI Conference on Web and Social Media, 10*(1), 250–258. Available from https://doi.org/10.1609/icwsm.v10i1.14710.

Mejova, Y., Weber, I., & Fernandez-Luque, L. (2018). Online health monitoring using facebook advertisement audience estimates in the United States: Evaluation study. *JMIR Public Health and Surveillance, 4*(3). Available from https://doi.org/10.2196/publichealth.7217.

Metwalli, A. S., Shen, W., & Wu, C. Q. (2020). Food image recognition based on Densely connected convolutional neural networks. *2020 International Conference on Artificial Intelligence in Information and Communication, ICAIIC*

2020 (pp. 027–032). Institute of Electrical and Electronics Engineers Inc. Available from https://doi.org/10.1109/ICAIIC48513.2020.9065281.

Mills, J., Reed, M., Skaalsveen, K., & Ingram, J. (2019). The use of Twitter for knowledge exchange on sustainable soil management. *Soil use and management* (Vol. 35, pp. 195–203). Blackwell Publishing Ltd Issue 1. Available from https://doi.org/10.1111/sum.12485.

Minton, E., Lee, C., Orth, U., Kim, C.-H., & Kahle, L. (2012). Sustainable marketing and social media. *Journal of Advertising, 41*(4), 69–84. Available from https://doi.org/10.1080/00913367.2012.10672458.

Mohawesh, R., Xu, S., Tran, S. N., Ollington, R., Springer, M., Jararweh, Y., & Maqsood, S. (2021). Fake reviews detection: A survey. *IEEE Access, 9,* 65771–65802. Available from https://doi.org/10.1109/ACCESS.2021.3075573.

Morris, W., & James, P. (2017). Social media, an entrepreneurial opportunity for agriculture-based enterprises. *Journal of Small Business and Enterprise Development, 24*(4), 1028–1045. Available from https://doi.org/10.1108/jsbed-01-2017-0018.

Munro, K., Christopher, M., Hartt, G., & Pohlkamp. (2015). Social media discourse and genetically modified organisms. *The Journal of Social Media in Society, 4.*

Narula, S., Rai, S., & Sharma, A. (2018). *Environmental awareness and the role of social media.* IGI Global.

Närvänen, E., Mesiranta, N., Sutinen, U. M., & Mattila, M. (2018). Creativity, aesthetics and ethics of food waste in social media campaigns. *Journal of Cleaner Production, 195,* 102–110. Available from https://doi.org/10.1016/j.jclepro.2018.05.202.

Ng, N. K. Y., Chow, P. S., & Choi, T. M. (2015). Impacts of social media mediated electronic words of mouth on young consumers' disposal of fashion apparel: A review and proposed model. *Sustainable fashion supply chain management: From sourcing to retailing* (pp. 47–58). Springer International Publishing. Available from https://doi.org/10.1007/978-3-319-12703-3_3.

Noronha, J., Hysen, E., Zhang, H., & Gajos, K. Z. (2011). PlateMate: Crowdsourcing nutrition analysis from food photographs. *UIST'11 - Proceedings of the 24th Annual ACM Symposium on User Interface Software and Technology,* 1–11. Available from https://doi.org/10.1145/2047196.2047198.

Pang, N., & Law, P. W. (2017). Retweeting #WorldEnvironmentDay: A study of content features and visual rhetoric in an environmental movement. *Computers in Human Behavior, 69,* 54–61. Available from https://doi.org/10.1016/j.chb.2016.12.003.

Pang, R., Baretto, A., Kautz, H., & Luo, J. (2015). Monitoring adolescent alcohol use via multimodal analysis in social multimedia. *Proceedings - 2015 IEEE International Conference on Big Data, IEEE Big Data 2015* (pp. 1509–1518). Institute of Electrical and Electronics Engineers Inc. Available from https://doi.org/10.1109/BigData.2015.7363914.

Pappa, G. L., Cunha, T. O., Bicalho, P. V., Ribeiro, A., Couto Silva, A. P., Meira, W., & Beleigoli, A. M. R. (2017). Factors associated with weight change in online weight management communities: A case study in the loseit reddit community. *Journal of Medical Internet Research, 19*(1). Available from https://doi.org/10.2196/jmir.5816.

Pearson, E., Tindle, H., Ferguson, M., Ryan, J., & Litchfield, C. (2016). Can we tweet, post, and share our way to a more sustainable society? A review of the current contributions and future potential of #socialmediaforsustainability. *Annual Review of Environment and Resources, 41,* 363–397. Available from https://doi.org/10.1146/annurev-environ-110615-090000.

Phang, C. W., Zhang, C., & Sutanto, J. (2013). The influence of user interaction and participation in social media on the consumption intention of niche products. *Information and Management, 50*(8), 661–672. Available from https://doi.org/10.1016/j.im.2013.07.001.

Pilař, L., Balcarová, T., & Rojík, S. (2016). Farmers' markets: Positive feelings of instagram posts. *Acta Universitatis Agriculturae et Silviculturae Mendelianae Brunensis, 64*(6), 2095–2100. Available from https://doi.org/10.11118/actaun201664062095.

Pomarici, E., & Vecchio, R. (2014). Millennial generation attitudes to sustainable wine: An exploratory study on Italian consumers. *Journal of Cleaner Production, 66,* 537–545. Available from https://doi.org/10.1016/j.jclepro.2013.10.058.

Potvin Kent, M., Pauzé, E., Roy, E. A., de Billy, N., & Czoli, C. (2019). Children and adolescents' exposure to food and beverage marketing in social media apps. *Pediatric Obesity, 14*(6). Available from https://doi.org/10.1111/ijpo.12508.

Ramage, D., Rosen, E., Chuang, J., Christopher, d, Manning, D.A., & McFarland. (2009). Topic modeling for the social sciences. In NIPS 2009 workshop on applications for topic models: Text and beyond (Vol. 5).

Reilly, A. H., & Hynan, K. A. (2014). Corporate communication, sustainability, and social media: It's not easy (really) being green. *Business Horizons, 57*(6), 747−758. Available from https://doi.org/10.1016/j. bushor.2014.07.008.

Reisch, L., Eberle, U., & Lorek, S. (2013). Sustainable food consumption: An overview of contemporary issues and policies. *Sustainability: Science, Practice, and Policy, 9*(2), 7−25. Available from https://doi.org/10.1080/15487733.2013.11908111.

Rodak, O. (2020). Hashtag hijacking and crowdsourcing transparency: Social media affordances and the governance of farm animal protection. *Agriculture and Human Values, 37*(2), 281−294. Available from https://doi.org/10.1007/s10460-019-09984-5.

Russel, J. (2017). Twtter Bans Russia Today and Sputnik from advertising on its service. Available from: https://techcrunch.com/2017/10/26/twitter-bans-russiatoday-and-sputnik-from-advertising-on-its-service.

Salathé, M. (2018). Digital epidemiology: What is it, and where is it going? *Life Sciences, Society and Policy, 14*(1). Available from https://doi.org/10.1186/s40504-017-0065-7.

Schuenemann, K., & Wagner, R. (2014). Using student-generated blogs to create a global perspective on climate change. *Journal of Geoscience Education, 62*(3), 364−373. Available from https://doi.org/10.5408/13-065.1.

Schwartz, H., Eichstaedt, J., Kern, M., Dziurzynski, L., Lucas, R., Agrawal, M., Park, G., Lakshmikanth, S., Jha, S., Seligman, M., & Ungar, L. (2013). Characterizing geographic variation in well-being using tweets. *Proceedings of the International AAAI Conference on Web and Social Media, 7*(1), 583−591. Available from https://doi.org/10.1609/icwsm.v7i1.14442.

Shin, J., Wu, S., Wang, F., Sa, C.D., Zhang, C., & Ré, C. (1310). Incremental knowledge base construction using deepdive. In Proceedings of the VLDB Endowment International Conference on Very Large Data Bases (Vol. 8). NIH Public Access.

Silverman, D. (2015). *Interpreting qualitative data*. Sage.

Simeone, M., & Scarpato, D. (2020). Sustainable consumption: How does social media affect food choices? *Journal of Cleaner Production, 277*. Available from https://doi.org/10.1016/j.jclepro.2020.124036.

Sogari, G., Pucci, T., Aquilani, B., & Zanni, L. (2017). Millennial generation and environmental sustainability: The role of social media in the consumer purchasing behavior for wine. *Sustainability (Switzerland), 9*(10). Available from https://doi.org/10.3390/su9101911.

Sprinkle, T. (2013). Social Media Uproar of the Moment: The 'Monsanto Protection Act.' Available from: https://finance.yahoo.com/blogs/the-exchange/socialmedia-uproar-week-monsanto-protection-act-170146036.html.

Strähle, J., & Gräff, C. (2016). The role of social media for a sustainable consumption (pp. 225−247). Springer Science and Business Media LLC. https://doi.org/10.1007/978-981-10-2440-5_12.

Svetkey, L. P., Batch, B. C., Lin, P. H., Intille, S. S., Corsino, L., Tyson, C. C., Bosworth, H. B., Grambow, S. C., Voils, C., Loria, C., Gallis, J. A., Schwager, J., & Bennett, G. B. (2015). Cell phone intervention for you (CITY): A randomized, controlled trial of behavioral weight loss intervention for young adults using mobile technology. *In Obesity, Vol. 23*(Issue 11), 2133−2141. Available from https://doi.org/10.1002/oby.21226.

Tacchi, J., Slater, D., & Hearn, G. (2003). Ethnographic action research.

Takeuchi, T., Narumi, T., Fujii, T., Tanikawa, T., Ogawa, K., & Hirose, M. (2014). *Using social media to change eating habits without conscious effort. In UbiComp 2014 - Adjunct Proceedings of the 2014 ACM International Joint Conference on Pervasive and Ubiquitous Computing* (pp. 527−535). Association for Computing Machinery, Inc. Available from https://doi.org/10.1145/2638728.2641330.

The World's Biggest Restaurants In 2017. (2017). Available from: https://www.forbes.com/pictures/591c79084bbe6f1b730a5811/2017-global-2000-restaura.

Thomases, H. (2012). McDonald's twitter mess: What went wrong. Available from: https://www.inc.com/hollis-thomases/mcdonalds-mcdstoriestwitter-mess.html.

Toivonen, T., Heikinheimo, V., Fink, C., Hausmann, A., Hiippala, T., Järv, O., Tenkanen, H., & Di Minin, E. (2019). Social media data for conservation science: A methodological overview. *Biological Conservation, 233*, 298−315. Available from https://doi.org/10.1016/j.biocon.2019.01.023.

Torres Moreno, J. M. (2014). *Automatic Text Summarization* (Vol. 9781848216686, pp. 1−348). Wiley Blackwell. Available from https://doi.org/10.1002/9781119004752.

U.S. Department of Agriculture. (n.d.). Food Access Research Atlas. Retrieved June 1, 2021, Available from: https://www.ers.usda.gov/data-products/food-access-researchatlas/documentation/

Veil, S. R., Reno, J., Freihaut, R., & Oldham, J. (2014). Online activists vs. Kraft foods: A case of social media hijacking. *Public Relations Review*, *41*(1), 103−108. Available from https://doi.org/10.1016/j.pubrev.2014.11.017.

Wan, C., Shen, G. Q., & Choi, S. (2021). Eliciting users' preferences and values in urban parks: Evidence from analyzing social media data from Hong Kong. *Urban Forestry and Urban Greening*, *62*. Available from https://doi.org/10.1016/j.ufug.2021.127172.

Wells, P.E. (2018). Social Movement Leadership and the Tasmanian Environmental Movement: A Case Study. Available from: https://doi.org/10.25959/100.00028572.

Widener, M. J., & Li, W. (2014). Using geolocated Twitter data to monitor the prevalence of healthy and unhealthy food references across the US. *Applied Geography*, *54*, 189−197. Available from https://doi.org/10.1016/j.apgeog.2014.07.017.

Young, C. W., Russell, S. V., & Barkemeyer, R. (2017). Social media is not the 'silver bullet' to reducing household food waste, a response to Grainger and Stewart (2017). *Resources, Conservation and Recycling*, *122*, 405−406. Available from https://doi.org/10.1016/j.resconrec.2017.04.002.

Young, W., Russell, S. V., Robinson, C. A., & Barkemeyer, R. (2017). Can social media be a tool for reducing consumers' food waste? A behaviour change experiment by a UK retailer. *Resources, Conservation and Recycling*, *117*, 195−203. Available from https://doi.org/10.1016/j.resconrec.2016.10.016.

Yunus, R., Arif, O., Afzal, H., Amjad, M. F., Abbas, H., Bokhari, H. N., Haider, S. T., Zafar, N., & Nawaz, R. (2019). A framework to estimate the nutritional value of food in real time using deep learning techniques. *IEEE Access*, *7*, 2643−2652. Available from https://doi.org/10.1109/ACCESS.2018.2879117.

Zhang, J., Wang, W., Xia, F., Lin, Y. R., & Tong, H. (2020). Data-driven computational social science: A survey. *Big Data Research*, *21*. Available from https://doi.org/10.1016/j.bdr.2020.100145.

Food security in the Eurobarometer opinion trends

Arianna Marcolin[1] and Elena Cadel[2]

[1]Department of Economics, Management and Quantitative Methods University of Milano Via Conservatorio, Milano, Italy [2]"Riccardo Massa" Department of Human Sciences for Education, Piazza dell'Ateneo lavoro Nuovo, Milano, Italy

Introduction

The Eurobarometer[1] consists of a series of surveys (called waves) administered to representative samples of citizens from all the EU countries since 1973. They are used by the European Commission, the European Parliament, and other EU institutions and agencies to monitor twice a year the state of public opinion in Europe on issues regarding the European Union and attitudes on political or social nature topics (Schmitt, 2003). For example, the key Eurobarometer questions address citizens' attitudes toward European institutions, European policies, and the integration process in general.

The Eurobarometer is characterized by a set of questions that recur over a long period and specific questions created ad hoc for each wave. Moreover, the Standard Eurobarometer is supplemented by ad hoc surveys on specific topics, called Special Eurobarometer and the Flash Eurobarometer. All the Eurobarometer Surveys are an important open source of transnational data, which have a broad geographical coverage and regularly deal with an extensive range of topics (Gatto & Panarello, 2022).

Food security has always been a critical issue for humankind. Despite the efforts of the United Nations to achieve the Sustainable Development Goal to end hunger and all form of malnutrition by 2030 (e.g., SDG 2 Zero hunger), recent data are more likely to suggest that such goals will not be reached if current trends remain unchanged. In fact, according to the latest report of the Food and Agriculture Organization (FAO et al., 2023), between 691 and 783 million people faced hunger in 2022, with a mid-range of 735 million. That represents an increase of 122 million people compared to 2019, presumably

[1] https://europa.eu/eurobarometer/about/eurobarometer

Food Sustainability and the Media
DOI: https://doi.org/10.1016/B978-0-323-91227-3.00003-2

due to the pandemic and repeated weather shocks and conflicts, including the war in Ukraine.

In general, at present, food availability is currently not at stake in the EU, but FAO data FAO et al. (2023) highlights that Europe is facing food insecurity, especially in urban areas. In particular, moderate or severe levels of food insecurity are increasing again in all European areas since 2020, and in 2022, they affected roughly 10% of the European population.

Within these premises, the study presented in this chapter investigates the attitudes and beliefs of European citizens regarding food security, using data retrieved from several Eurobarometer surveys. In particular, this study aims at replying to the following questions: *what do EU citizens think about food security? Has their opinion changed over time? Is the Eurobarometer survey a reliable tool for this topic?*

Getting data on food security using the Eurobarometer surveys

Since food security has many facets, the classic definition of food security provided by FAO (2006) has been used as an investigation framework to collect the questions—and corresponding answers. It states: *"Food security exists when all people, at all times, have physical and economic access to sufficient, safe and nutritious food that meets their dietary needs and food preferences for an active and healthy life"* (World Food Summit, 1996—FAO 2006, p.1). This definition points to the following dimensions of food security: "availability", "access", "use/utilization", "and stability". In addition, in recent years, two more dimensions, "agency" and "sustainability", have also been recognized as important dimensions of food security (HLPE, 2020). In this study, all these six pillars have been used as keys for the investigation research.

According to FAO (2006), **food availability** refers to the availability of a sufficient amount of (quality) food, supplied through domestic production or imports (including food aid). **Food access**, instead, represents the chances that individuals have to access adequate resources (entitlements) to acquire appropriate foods for eating a nutritious diet. These entitlements are defined as the legal, political, economic, and social arrangements of the community in which people live (including traditional rights, such as access to common resources). **Utilization** refers to the use of food through adequate diet, clean water, sanitation, and health care to achieve a state of nutritional well-being in which all physiological needs are met. **Stability** highlights the importance of constant access to adequate food, despite sudden shocks, such as climate crises or economic downturns. This pillar is strongly connected to both the availability and access dimensions of food security. **Agency** refers to individuals or groups of individuals that can change the food system through individual actions, such as producing their own food and participating in the policy process. Finally, **sustainability** encompasses all the practices that shape the long-term disposition of natural, social, and economic resources, which ensure the food needs of today are met without compromising the food needs of future generations. Last but not least, since

safety is a key aspect of security, as it aims to have food that is safe to eat[2], this word has also entered the database repository of Gesis[3].

Data have been collected from the Eurobarometer and Special Eurobarometer from 2010 to 2022, using a qualitative approach. In particular, waves have been retrieved from the Gesis database[4], which is a repository website of all the Eurobarometer waves, including Traditional waves, Special and Flash Eurobarometer. The following keywords: "Food Security", "Food Availability", "Food Access", "Food Use/Utilization", "Food Stability", "Food Agency", "Food Sustainability", and "Food Safety" have been entered into the wave. When the perfect wording was not matched in the repository, different proxies have been used to explore in depth each concept. In total, 161 questions have been collected: 30 Availability, 5 Access, 1 Utilization, 2 Stability, 4 Agency, 113 Sustainability, 6 Food Safety. The questions were then analyzed through descriptive statistics. In the next section, the results of the search described in the previous paragraph are reported following the six food security areas plus food safety.

Availability

For this area, 30 questions have been found.

Impact of humans on food availability. In 2014, 43% of the respondents stated that humans would have a positive impact on food availability, which means human actions will increase food availability in the future. This question has not been repeated over time; hence, it is not possible to compare it with subsequent questions (Table 4.1).

Characteristics of food. In 2020, only 3.4% of the respondents identified availability as the most important characteristic of sustainable food. As the second most important characteristic, this percentage slightly increased (12%). However, other issues, such as "being nutritious and healthy" (21%) and "being affordable" (12%), were deemed more relevant. This question has not been repeated over time; hence, it is not possible to compare it with subsequent questions (Table 4.2).

Food aid. For three waves—from 2016 to 2019—the Eurobarometer repeated the same question regarding the most pressing challenges for the future of Developing Countries. The answers have been quite stable over time, with 25% of individuals considering food agriculture as a challenge and a source of food insecurity for Developing Countries (Table 4.3).

Interestingly, in the Eurobarometer surveys, no questions have been found regarding food aid or similar, such as Food Bank (i.e., a nonprofit organization that distributes food to help people in need), whose workload increased in Europe after the COVID-19 pandemic (Capodistrias et al., 2022).

Domestic production. The pillar of food availability encompasses both food production and imports. Regarding food production, in 2012, 71% of the total respondents were concerned about international food production. Instead, national and European food

---

[2] https://www.fao.org/food-safety/background/qa-on-food-safety/en/

[3] The GESIS Leibniz Institute for the Social Sciences is a European data archive and research infrastructure (Gesellschaft Sozialwissenschaftlicher Infrastruktureinrichtungen). The GESIS ZACAT catalogue can be used to locate data series, including Eurobarometer surveys (standard and special topics)

[4] https://www.gesis.org/home

TABLE 4.1   Impact of humans on food availability.

	Percentage 2014
A positive impact	43
No impact	22.3
A negative impact	19.5
Do not know	15.1
Total (N)	27.910

Note: *81.5 ZA5929 (2014). QB1: 15 years from now, what impact do you think people's actions and behaviour will have on the following areas …?*

TABLE 4.2   Characteristics of food.

	Percentage 2014
A positive impact	43
No impact	22.3
A negative impact	19.5
Do not know	15.1
Total (N)	27.910

Note: *GB excluded. 93.2 (ZA7739) (2020). QB2a: Which of the following do you consider to be the most important characteristic of "sustainable" food? Firstly?*

TABLE 4.3   Food security as an international threat.

	Percentage 2016	Percentage 2018	Percentage 2019
Not mentioned	74.37	77.55	74.4
Food security and agriculture	25.63	22.45	25.6
Total (N)	27.929	27.732	32.524

Note: *86.3 ZA6791 (2016); 89.3 (2018) ZA7483; 91.5 (2019) ZA 7576. QC2: Which of the following challenges do you consider the most pressing for the future of Developing Countries?*

TABLE 4.4   Concern about the sufficiency of food production.

	Percentage 2012		
	Country	European Union	World
Concerned	44.9	41.2	71.9
Not concerned	54	55.9	25.2
Do not know	1.1	2.9	2.9
Total (N)	26.593	26.593	26.593

Note: *77.2 (ZA5598) (2012). QD1: To what extent are you concerned that sufficient food is produced to meet the needs of the population in…?*

TABLE 4.5 Perception on food imports.

	No import barriers			Trade barriers Developing Countries			EU standards compliance		
	Percentage 2017	Percentage 2020	Percentage 2022	Percentage 2017	Percentage 2020	Percentage 2022	Percentage 2017	Percentage 2020	Percentage 2022
Agree	37.2	37.7	42.1	52.4	54.5	55.4	85.8	92.5	87.2
Disagree	51.3	56.2	51.1	33.1	34.1	35.7	6.3	8.6	8.1
Do not know	11.5	6.2	6.8	14.5	7.7	8.9	7.9	4.1	4.7
Total (N)	28.031	28.030	26.502	28.031	27.237	26.502	28.031	27.237	26.502

Note: In 93.2 and 97.1 waves in the year 2020, GB excluded.
88.4 (2017) ZA6939; 93.2 ZA7739 (2020); 97.1 ZA7886 (2022). QA16: Do you agree or disagree with each of the following statements about the European Union and trade barriers to imports of agricultural products? The EU should have no trade barriers to imports of agricultural products, regardless of their origin; the EU should have trade barriers to imports of agricultural products with the exception of imports from Developing Countries; agricultural imports from any origin should only enter the EU if their production has complied with the EU's environmental and animal welfare standards?

production concerns only around 40% of the respondents. This question was asked in 2012 and has not been repeated in other waves (Table 4.4).

Regarding *food imports*, the Eurobarometer surveys include three questions repeated from 2018 to 2022. Half of the EU respondents were more likely to maintain high import barriers in the food sector. At the same time, almost 90% of the respondents agreed that agricultural imports from any origin should only enter the EU if their production complied with the EU's environmental and animal welfare standards (Table 4.5).

In 2017, the Eurobarometer included questions on *organic food production* specifically. This topic has been added here because it represents one feature of domestic production that has become more relevant in the last few years. In 2020, organic food production represented 9.1% of the total European agricultural territory, and from 2015 to 2020, the consumption of organic products doubled (EC, 2023).

From 2017 to 2022, the percentage of EU respondents agreed that organic food production improved environmental practices, compliance with the rules on pesticides, and respect for animal welfare increased. The feature that has increased the most (by 5% in the 4 years) is the respect of animal welfare. Morever, the aspect considered most important in 2022 was the compliance with EU law on the use of pesticides (82%) (Table 4.6-4.8). Each question was asked three times.

TABLE 4.6 Perception of organic food production.

	Better environmental practices		
	Percentage 2017	Percentage 2020	Percentage 2022
Agree	77.5	79.9	80.2
Disagree	12.8	14	13.5
Do not know	9.7	6.1	6.4
Total (N)	28.301	28.300	26.502

Note: In 93.2 and 97.1 waves in the year 2020, GB excluded.
88.4 (2017) ZA6939; 93.2 ZA7739 (2020); 97.1 ZA7886 (2022). QC16_5: Do you agree or not with the following statements related to food products coming from "organic" agriculture...?

TABLE 4.7 Perception of organic food production.

	Comply with use rules of pesticides		
	Percentage 2017	Percentage 2020	Percentage 2022
Totally agree	78.8	81.8	82.3
Totally disagree	11.7	12.1	10.9
Do not know	9.4	6.1	6.8
Total (N)	28.301	28.300	26.502

Note: In 93.2 and 97.1 waves in the year 2020, GB excluded.
88.4 (2017) ZA6939; 93.2 ZA7739 (2020); 97.1 ZA7886 (2022). QC16_6: Do you agree or not with the following statements related to food products coming from "organic" agriculture...?

TABLE 4.8 Perception of organic food production.

	Higher animal welfare		
	Percentage 2017	Percentage 2020	Percentage 2022
Totally agree	75.4	79.2	80.2
Totally disagree	13.1	14.5	13.5
Do not know	11.5	6.3	6.3
Total (N)	28.301	28.300	26.502

Note: In 93.2 and 97.1 waves in the year 2020, GB excluded.
88.4 (2017) ZA6939; 93.2 ZA7739 (2020); 97.1 ZA7886 (2022). QC16_7: Do you agree or not with the following statements related to food products coming from "organic" agriculture...?

TABLE 4.9 Preferences when purchasing food.

	Percentage 2010	Percentage 2012	Percentage 2017
Quality	70	96	76
Not mentioned	30	4	24
Total (N)	26.691	26.593	28.031

Note: *Eurobarometer 73.5 ZA5235 (2010); 77.2 ZA5598(2012); 88.4 ZA6939 (2017). QB1T: When you buy food, which of the following are the most important to you? Firstly?*

Quality. An important feature of the food availability is the quality of food. The question, which covers this topic, was asked three times. From 2010 to 2017, the trend in quality food did not show a linear direction. According to the given answers, in 2021, quality was considered very important when purchasing food for 96% of the EU respondents (with an increase of almost 30% in 2 years). Then that percentage decreased, again, by 20 percentage points in 2017 (Table 4.9).

In the Eurobarometer waves, information related to the quality of food can be retrieved from food *labeling perception*. From 2017 to 2022, a stable percentage of 80% of respondents answered that specific "quality label" is a factor that drives food purchases. The two other important factors -when purchasing food- were "local tradition", with an increase of 4%, and "knowing geographical area", with an increase of 5% (Table 4.10).

Moreover, regarding the quality of food, in 2020, the Eurobarometer measured the *factors that are considered sustainable and healthy in a diet*. From this question, which has not repeated over times, it is possible to identify two main blocks of aspects. "Eating a variety of different foods, having a balanced diet", "Eating more fruit and vegetables" and "Eating seasonally" are identified as the three most important aspects of a sustainable and healthy diet (more than 50% of the respondents mentioned them). These aspects are followed by "eating more home-cooked meals", "avoiding waste" and "avoiding or do not eating too much food high in fat, sugars and/or salt" (more than 40% of the respondents mentioned them).

This question has not been repeated over time. Hence, it is not possible to identify a trend of features to conduct a sustainable and healthy diet (Table 4.11).

TABLE 4.10 Preferences when purchasing food.

	Knowing geographical area			Local tradition			Specific quality label			Part of short supply chain		
	2017	2020	2022	2017	2020	2022	2017	2020	2022	2017	2020	2022
Important	74.5	81.2	79.4	75.7	80.8	79.9	75.7	82.3	79.5	–	85.7	85.9
Not important	22.7	17.9	19.4	20.4	17.2	18.1	21.2	16.4	19	–	12.5	12.6
Do not buy food products	0.4	0.2	0.3	0.5	0.3	0.5	0.4	0.2	0.4	–	0.3	0.3
Do not know	2.3	0.8	0.9	3.4	1.6	1.5	2.7	1.0	1.1	–	1.4	1.2
Total (N)	28.031	28.300	26.502	28.031	28.300	26.502	28.031	28.300	26.502	–	28.300	26.502

Note: In 93.2 and 97.1 waves in the year 2020, GB excluded. Percentage values in the cells.
88.4 (2017) ZA6939; 93.2 ZA7739 (2020); 97.1 ZA7886 (2022). QA12: How important or not are the following factors in your decision to buy food products?

TABLE 4.11 Aspects for a sustainable and healthy diet.

	Percentage 2020
Eating a variety of different foods, having a balanced diet	57.5
Avoiding or do not eating too much food high in fat, sugars and/or salt	37.6
Eating more fruit and vegetables	56.6
Eating more home-cooked meals	42.3
Eating more wholegrain, high-fiber foods	29.8
Eating meat less often	30
Eating fish more often	33.2
Eating vegetarian or vegan	10.6
Eating seasonally, local	44.9
Not eating too many calories	26.2
Eating organic food	24.8
Little or no pesticides	40
Eating foods with a low carbon footprint	21.7
Eating foods produced by companies that protect workers' social rights	17.4
Avoiding wasting food	39.7
Total (N)	28.300

Note: GB excluded. Each of these questions has been singularly asked, requiring the respondents to choose between two options: to mention the feature or to not mention it. Here, only the mentioned features have been reported (and not the "not mentioned" category). Hence, the total is not 100%.
93.2 ZA7739 (2020). QB4: We often hear people talking about the importance of eating a healthy and sustainable diet. What do you think about "eating a healthy and sustainable diet"?

Access

For this area, five questions have been found. In the Eurobarometer waves, access to food has been investigated through *price*. Price has always been an important driver when buying food. Usually, purchasing behaviors do not seem to be necessarily driven by the lowest price, but price considerations deem among the main determinants of purchasing decisions (Nicolau et al., 2021). Questions about price were asked two times: one in 2010 and the other in 2012. From 2010 to 2012, its importance increased by four points percentage (Table 4.12).

One question, repeated for three years, regards *access to organic food* in the supermarket. The EU respondents have polarized on this topic. Half of them thought it is difficult to find organic food, while the others stated it is easy to access organic food in the supermarket. That probably depends on the fact that the distribution of organic products differs considerably among EU Member States (USDA, 2023) (Table 4.13).

Utilization

For this area, one question has been found. In this document, "Utilization" has been associated with diet. However, in the Eurobarometer, the questions about diet are not numerous. In 2020, a set of questions about a sustainable diet has been added. Almost 52% of the respondents identified the healthy aspects of a sustainable diet as crucial. This

TABLE 4.12 Importance of price when purchasing food.

	Percentage 2010	Percentage 2012
Yes	86	90.6
Not	13	8.3
Do not know	1	1.1
Total (N)	26.691	26.593

Note: 73.5 ZA5235 (2010); 77.2 ZA5598 (2012). QF1.4: To what extent do you associate food and eating with each of the following? Affordable Price?. QD3: When buying food, how important are the following for you personally…? Price?

TABLE 4.13 Access to organic agriculture via supermarkets.

	Access to organic agriculture via supermarkets		
	Percentage 2017	Percentage 2020	Percentage 2022
Agree	44.3	46.7	44.7
Disagree	49	49.9	51.6
Do not know	6.7	3.4	3.7
Total (N)	28.301	28.300	26.502

Note: In 93.2 and 97.1 waves in the year 2020, GB excluded.
88.4 (2017) ZA6939; 93.2 ZA7739 (2020); 97.1 ZA7886 (2022). QC16_8: Do you agree or not with the following statements related to food products coming from "organic" agriculture…?

factor was followed by the importance given to the local economy (9.4% of respondents) and to companies that pay fair wages (9.2%) (Table 4.14). This question has not been repeated over time; hence it is not possible to compare it with subsequent questions.

Stability

For this area, two questions have been found (two different questions asked only once each). In 2022, national and European *perception of food security* has been constant over the last decade for around 40% of the respondents. Unfortunately, no further questions are available for comparing data (Table 4.15).

During the same year, the most important perceived features that affect food security have been investigated. The feature that has been considered as the most critical for food security is "Extreme weather events" with 53.8% of the total, followed by "Natural Resource Scarcity/Degradation" with 41.2% (Table 4.16).

TABLE 4.14 Most important features of a sustainable diet.

	Percentage 2020
What you eat is good for the planet	6.8
What you eat is healthy for you	51.5
What you eat supports local economy	9.4
What you eat has been produced in a way that minimizes waste	5.9
What you eat is organic	8.2
The food you eat has minimal packaging	3.9
The food packaging is recyclable	4.1
What you eat is produced by a company that pays fair wages and respects workers'	9.2
Do not know	1
Total(N)	28.300

Note: In 93.2 wave in the year 2020, GB excluded.
93.2 ZA7739 (2020). QB3a: What aspects of a sustainable diet are important to you? Firstly?

TABLE 4.15 The perception of national and international food security.

	Country	European Union
Increased	40.4	38.6
Stayed the same	40.9	41.3
Decreased	13.6	11.1
Do not know	5.1	8.9
Total (N)	26.502	26.502

Note: *97.1ZA7886 (2022). QA20: Compared with 10 years ago, do you think that the level of food security in the EU has increased, stayed the same, or decreased?*

TABLE 4.16 The most important perceived features that affect food security.

	Percentage 2022
Diminishing Number of Farm Holdings	31.1
Stagnating EU agricultural production and increasing dependency on imports	25.6
Extreme weather events (severe droughts, floods, etc...) and climate change	53.8
Natural Resource Scarcity/Degradation	41.2
Human health events, such as COVID-19	17
Economic downturns and poverty	27.7
Geopolitical events, e.g., large-scale trade disputes	22.7
Technological events, such as cyber threats	7
Total (N)	26.502

Note: *97.1 ZA7886 (2022). QA21: What do you think are the most important risks to food security in the EU?*

Agency

For this area, four question has been found (the same question has been replicated for four years). The pillar of "Agency" has been explored through the *actions of farmers*, because, in the agriculture sector, they represent the individuals who can exercise voice and decide what to produce and how, impacting the whole community (Clapp et al., 2022). In the Eurobarometer, from 2015 to 2022, farmers' responsibility has been tracked. Results suggest that half of the EU citizens believe farmers should focus on providing "Safe and Healthy High-Quality Food." However, a growing number of respondents stated that farmers are responsible for "Securing Food Supply in the EU." The rate of respondents that gives farmers the responsibility to diversify quality products decreased by 20% (Table 4.17).

Sustainability

For this area, 113 questions have been found from 2013 to 2022. Sustainability is the last pillar that contributes to describing food security. In the previous ones, the feature of sustainability has also been introduced because some aspects deem overlapping with previous pillars. Following the definition of sustainability as *"all the practices that shape long-term disposition of natural and economic resources, ensuring future requests of food"* (FAO, 2006, HPLE, 2022), this area has been associated with the policies implemented in the agricultural sector, focusing on the *CAP* (the European Common Agricultural Policy)[5].

Over the years—from 2013 to 2022—the awareness of the CAP has increased by 30 percentage points (Table 4.18). The same question was repeated four times in nine years.

[5] https://agriculture.ec.europa.eu/common-agricultural-policy/cap-overview/cap-glance_en

TABLE 4.17 Farmers' responsibility regarding food security.

	Percentage 2015	Percentage 2017	Percentage 2020	Percentage 2022
Environment Protection	25.9	21.9	23.7	21.4
Growth/Jobs in Rural Areas	32.5	20.4	17.5	19.9
Securing Food Supply in the EU	24.4	17.2	21.6	27.2
Provide Safe and Healthy High-Quality Food	–	54.7	56.6	51
Diversity of Quality Products	44.6	24.1	23.1	24.2
Farmed Animals Welfare	31.2	25.2	25.9	25.3
Life in Countryside	26.7	20.3	19.6	20.5
Total (N)	27.822	28.031	28.300	28.300

Note: Each of these questions has been singularly asked, requiring the respondents to choose between two options: to mention the feature or to not mention it. Here, only the mentioned features have been reported (and not the "not mentioned" category). Hence, the total is not 100%. In 93.2 and 97.1 waves in the year 2020, GB excluded.

84.2 ZA6642 (2015); 88.4 (2017) ZA6939; 93.2 ZA7739 (2020); 97.1 ZA7886 (2022). QA6: What do you think should be the two main responsibilities of farmers in our society?

TABLE 4.18 Awareness of the CAP within European Member States.

	Percentage 2013	Percentage 2017	Percentage 2020	Percentage 2022
Yes, and you know the details	8	10	9.9	10.1
Yes, but you don't really know the details	56	57	69.5	65.7
No, you have never heard of it	34	32	20.2	24
Do not Know	2	1	0.4	0.2
Total (N)	26.917	28.031	28.030	26.502

Note: In 93.2 and 97.1 waves in the year 2020, GB excluded.

368 (2013), 88.4 (2017) ZA6939; 93.2 ZA7739 (2020); 97.1 ZA7886 (2022). QC2: Have you ever heard of the support that the EU gives farmers through its Common Agricultural Policy (CAP)?

From 2015 to 2022, the fluctuations in the perception of the CAP benefits have been minor, and the percentage of people who considered the CAP beneficial for EU citizens was stable at 80% (Table 4.19). From 2015 to 2022, the same question was asked four times.

Regarding the *CAP objective* (i.e., Fair Standard of Living for Farmers; Growth/Jobs in Rural Areas; Reasonable Food Prices for Consumers; Securing Food Supply in the EU; Sustainable Way to Produce Food; Protecting the Environment & Tackling Climate Change; Safe and Healthy High-Quality Food), respondents' beliefs remained stable over time. In particular, Safe and Healthy High-Quality Food was identified as the most important objective of the CAP (Table 4.20).

Instead, regarding *CAP performance* (i.e., Fair Standard of Living for Farmers; Growth/Jobs in Rural Areas; Reasonable Food Prices for Consumers; Securing Food Supply in the EU; Sustainable Way to Produce Food; Protecting the Environment & Tackling Climate Change; Safe and Healthy High-Quality Food), from 2013 to 2022, Securing Food Supply in the EU was considered the main role of the CAP, followed by Safe and Healthy High-Quality Food (Table 4.21).

TABLE 4.19 CAP benefits.

	Percentage 2015	Percentage 2017	Percentage 2020	Percentage 2022
Agree	68.6	67.6	79.2	79.9
Disagree	20.7	17.7	16.9	15
Do not know	10.7	14.7	3.9	3.8
Total (N)	27.822	28.031	28.030	26.502

Note: In 93.2 and 97.1 waves in the year 2020, GB excluded.
84.2 ZA6642 (2015); 88.4 (2017) ZA6939; 93.2 ZA7739 (2020); 97.1 ZA7886 (2022). QC3: To what extent do you agree or disagree with the following statement: the Common Agricultural Policy (CAP) benefits all European citizens and not only farmers?

Moreover, concerning *CAP contribution* (i.e., Boost Invest/Growth & Create Jobs; Help Tackle Climate Change; Securing Food Supply in the EU; Smooth Running of the EU Single Market; Improve Trade with the Rest of the World; Extending Internet Services in Rural Areas; Research and Digital Solutions; Strengthening Role of Farmers; Sustainable Management of Natural Resources; Reduce Regional Development Disparities) the majority of the respondents stated that CAP should contribute to helping tackle climate change, followed by securing food supply in the EU. The first three questions have been asked four times from 2015 to 2022; instead the last four questions have been asked only in 2020 and in 2022 (Table 4.22).

Lastly, one feature regarding the CAP has changed: the *perception of the budget*, the same question has been asked four times from 2015 to 2022. In fact, from 2015 to 2022, the number of people who believe that CAP financial resources should be devoted to "Environment/ Climate Change Rural Areas" and "Almost Entirely Financed" increased. Moreover, from 2013 to 2022, also the number of respondents stated that financial resources should be addressed to guarantee food supply and sustainable food production increased (Table 4.23).

Food safety

For this area, six questions have been found. Although it is an independent discipline, food safety is an issue closely related to food security, as the world will not achieve food and nutrition security without food safety. Health protection is the aim of all EU laws and standards in the agriculture, animal husbandry, and food production sectors. So it wasn't a surprise to find two Special Waves commissioned by the European Food Safety Authority (EFSA)[6] that have been conducted on the awareness and perception of food products.

In 2010, the majority of the respondents stated that European food products were safe. Unfortunately, this question has been included only in the 2010 EFSA Special wave and it is not possible to compare this result with previous or following responses (Table 4.24).

However, the opinion regarding food safety and subsequent concern has changed over the years- the same question has been asked in 2010 and 2019. In fact, the percentage of European people concerned about food safety has increased by 15 percentage points from 2015 to 2019 (Table 4.25).

[6] https://www.efsa.europa.eu/en

TABLE 4.20 CAP objectives.

	Fair standard of living for farmers				Growth/jobs in rural areas				Reasonable food prices for consumers				Securing food supply in the EU				Sustainable way to produce food				Protecting the environment & tackling climate change				Safe and healthy high quality food			
	2015	2017	2020	2022	2015	2017	2020	2022	2015	2017	2020	2022	2015	2017	2020	2022	2015	2017	2020	2022	2015	2017	2020	2022	2015	2017	2020	2022
Not mentioned	50.2	51.3	50.1	48.3	49.4	51.8	53.4	53.0	46	46.9	48.5	44.6	57.4	58.9	54.2	49.9	54.6	53	50.1	51.3	56.6	51.7	50.9	52.9	40.6	36.1	34.8	38.6
Yes	49.8	48.7	49.9	51.7	50.6	48.2	46.6	47.0	54	53.1	51.5	55.4	42.6	41.1	45.8	50.1	45.4	47	49.9	48.7	43.4	48.3	49.1	47.1		63.9	65.2	61.4
Total (N)	27.822	28.031	28.300	26.502	27.822	28.031	28.300	26.502	27.822	28.031	28.300	26.502	27.822	28.031	28.300	26.502	27.822	28.031	28.300	26.502	27.822	28.031	28.300	26.502	27.822	28.031	28.300	26.502

Note: In 93.2 and 97.1 waves in the year 2020, GB excluded. Percentage values in the cells.
84.2 ZA6642 (2015); 88.4 (2017) ZA6939; 93.2 ZA7739 (2020); 97.1 ZA7886 (2022). QA4: *In your opinion, which of the following should be the main objectives of the EU in terms of agriculture and rural development policy?*

TABLE 4.21 CAP performance.

	Fair standard of living for farmers				Growth/jobs in rural areas				Reasonable food prices for consumers				Securing food supply in the EU				Sustainable way to produce food				Protecting the environment & Tackling climate change				Safe and healthy high quality food			
	2015	2017	2019	2022	2015	2017	2019	2022	2015	2017	2019	2022	2015	2017	2019	2022	2015	2017	2019	2022	2015	2017	2019	2022	2015	2017	2019	2022
Agree	53.9	57.5	62.8	64	58.1	52.8	57.5	59.4	56.4	57.4	64.2	63.8	72.9	72.7	81.2	80.9	0	63.2	68.8	71.9	58	58.6	63.2	6.8	66.9	65.5	73.4	75.5
Disagree	34.7	29.4	31.5	29.5	28.9	33.4	36.7	33.8	34.2	31.7	31.4	31.7	15.6	14.7	13.7	13.7	0	22.5	24.3	20.9	29.4	28.6	31.1	26.6	23.7	24.3	22.3	19.6
Do not know	11.4	13.2	5.7	6.5	12.9	13.7	5.9	6.8	9.4	10.9	4.3	4.6	11.5	12.6	5.2	5.4	–	14.2	6.9	7.1	12.6	12.8	5.7	6.6	9.4	10.1	4.3	4.9
Total (N)	27.822	28.031	28.300	26.502	27.822	28.031	28.300	26.502	27.822	28.031	28.300	26.502	27.822	28.031	28.300	26.502	–	28.031	28.300	26.502	27.822	28.031	28.300	26.502	27.822	28.031	28.300	26.502

Note: In 93.2 and 97.1 waves in the year 2020, GB excluded. Percentage values in the cells.
84.2 ZA6642 (2015); 88.4 (2017) ZA6939; 93.2 ZA7739 (2020); 97.1 ZA7886 (2022).QA5: *To what extent do you think the EU through the Common Agricultural Policy (CAP) is fulfilling its role in…?*

TABLE 4.22 CAP contribution.

	Boost invest/growth & create jobs				Help tackle climate change				Securing food supply in the EU				Smooth running of EU single market		Improve trade with the rest of the World		Extend internet services in rural areas	
	2015	2017	2020	2022	2015	2017	2020	2022	2015	2017	2020	2022	2015	2017	2015	2017	2015	2017
Agree	65.9	65.6	67.8	68.1	58	60.8	68.2	70.6	58	60.8	80.4	83	65.4	67.1	66.3	67.9	52.9	61.9
Disagree	21.7	19.9	25.6	24.4	25.8	23.5	25.1	22	25.8	23.5	11	11.6	18.3	15	18.1	15	21.8	17.5
Do not know	12.4	14.5	6.6	7.4	16.3	15.8	6.6	7.4	16.3	15.8	4.9	5.4	16.3	17.9	15.7	17.1	25.4	20.6
Total (N)	27.822	28.031	28.300	26.502	27.822	28.031	28.031	26.502	27.822	28.031	28.300	26.502	27.822	28.031	27.822	28.031	27.822	28.031

	Research and digital solutions		Strengthening role of farmers		Sustainable management of natural resources		Reduce regional development disparities	
	2020	2022	2020	2022	2020	2022	2020	2022
Agree	69.9	70.7	70.9	70.9	70.4	72.7	64.1	64.7
Disagree	19.1	17.1	22.8	22.2	22.3	19.4	28	26.2
Do not know	11	12.3	6.2	6.9	7.3	8	7.9	9.1
Total (N)	28.031	26.502	28.300	26.502	28.300	26.502	28.300	26.502

Note: In 93.2 and 97.1 waves in the year 2020, GB excluded. Percentage values in the cells.
84.2 ZA6642 (2015); 88.4 (2017) ZA6939; 93.2 ZA7739 (2020); 97.1 ZA7886 (2022).QA8: *Do you agree or not that the Common Agricultural Policy (CAP) contributes to...?*

TABLE 4.23 CAP budget.

	Agri sector almost entirely financed				Guarantee food supply				Heavy investments required				Sustainable food production				Costs of strict standards				Environment/climate change rural areas			
	2015	2017	2020	2022	2015	2017	2020	2022	2015	2017	2020	2022	2015	2017	2020	2022	2015	2017	2020	2022	2015	2017	2020	2022
Not mentioned	82.2	84.7	80	80.7	60.6	70.4	64.4	62.9	67.5	70.9	69.1	69.1	63.3	67.3	66	65.7	75.6	75.8	69.9	69.8	–	82.2	80	78.9
Yes	17.8	15.3	20	19.3	39.4	29.6	35.6	37.1	32.5	29.1	30.9	30.9	36.7	32.7	34	34.3	24.4	24.2	30.1	30.2	–	17.8	20	21.1
Total (N)	27.822	28.031	28.300	26.502	27.822	28.031	28.300	26.502	27.822	28.031	28.300	26.502	27.822	28.031	28.300	26.502	27.822	28.031	28.300	26.502	–	28.031	28.300	26.502

Note: In 93.2 and 97.1 waves in the year 2020, GB excluded.
84.2 ZA6642 (2015); 88.4 (2017) ZA6939; 93.2 ZA7739 (2020); 97.1 ZA7886 (2022). QA10: What do you think are the main reasons why the EU spends a significant proportion of its budget (around 30% of the total EU budget) on the Common Agricultural Policy (CAP)?

TABLE 4.24 Perceived food safety of European food products.

	Percentage 2010
Agree	64.1
Disagree	25.7
Do not Know	10.2
Total (N)	26.691

Note: *73.5 ZA5235 (2010). QF6.2: To what extent do you agree or not with each of the following statements? Food produced in the EU is safer than food imported from outside the EU?*

TABLE 4.25 Concern about food safety.

	Percentage 2010	Percentage 2019
Yes	65	79
Not	33	20.5
Do not Know	2	0.5
Total (N)	26.691	27.655

Note: *73.5 ZA5235 (2010); 91.5 ZA7576 (2019). QF1.6 To what extent do you associate food and eating with each of the following? Being concerned about the safety of food Concern About Food Safety.*

TABLE 4.26 EU Level: Ensuring Food Safety.

	Percentage 2011	Percentage 2012
More EU decision-making	72.1	72.7
Less EU decision-making	21.8	21
No change needed	4.1	3.8
Do not Know	1.9	2.6
Total (N)	26.693	27.739

Note: *76.4 ZA5596 (2011); 78.2 ZA5686 (2012). QA16: For each of the following areas, please tell me if you believe that more decision-making should take place at a European level or on the contrary that less decision-making should take place at a European level?*

If the perception of food safety were included only in two Special Waves (the same question has been asked in 2011 and 2012), the role of the European Union in ensuring food safety has been further developed. The result of the analysis depicts that almost the total of the respondents agree that public authority should put more effort in assuring food safety. The percentage remained stable in the 2 years under analysis (Table 4.26).

Finally, the two waves in 2010 and 2019 measured the concern of European individuals regarding food. In fact, the same question has been asked twice: in 2010 and then in 2019. Both waves have been released in cooperation with EFSA. In the two waves, the questions have been changed and not all of them can be fully compared. In 2010, "Pesticide residues

TABLE 4.27 Concern on food.

	Percentage 2010	Percentage 2019
Genetically modified ingredients in food or drinks	8	7
Additives such as colors, preservatives, or flavorings used in food or drinks	9	10
Food poisoning from bacteria	12	9
Pesticide residues in food	19	10
Antibiotic, hormone, or steroid residues in meat	–	14
Environmental pollutants in fish, meat, or dairy	3	10
Traces of materials that come into contact with food, e.g., plastic or aluminum in packaging	9	3
Genome editing	–	1
Diseases found in animals	3	7
Plant diseases in crops	–	1
Nano particles found in food	1	2
Poisonous molds in food and feed crops	–	2
Food hygiene	5	11
Allergic reactions to food or drinks	4	5
Microplastics found in food	–	6
Bovine spongiform encephalopathy (BSE - mad-cow disease)	2	–
We do not know what we are eating/traceability of the products, origin of products	7	–
Food is not natural/industrial/artificial	6	–
Lack of freshness, expiry dates	9	–
Poor food quality	5	–
Obesity, overweight	9	–
Diet too high in fat, sugar or calories/Unbalanced diet	7	–
Diet-related diseases (high cholesterol, cardiovascular problems, Diabetes,…)	10	–
Anorexia/Bulimia	1	–
Digestive problems and discomforts (indigestions, ulcers, etc.)	3	–
Cancer	5	–
Prices (prices too high/food too expensive)	3	–
Problem of poverty/lack of food/hunger in the world	2	–
New technologies (e.g., animal cloning, nanotechnology, irradiation)	1	

(*Continued*)

TABLE 4.27   (Continued)

	Percentage 2010	Percentage 2019
No Problem	9	1
Other	15	—
Do not Know	8	1
Total (N)	26.691	25.893

Note: 73.5 ZA5235 (2010), 91.5 ZA7576 (2019). QF3: Could you tell me in your own words, what are all the things that come to your mind when thinking about possible problems or risks associated with food and eating? Just say out loud whatever comes to mind, and I will write it down. Anything else? (Open Question – Multiple Answers Possible). QD4a Please tell me which of these topics you have heard about concern you most when it comes to food? Firstly?

in food" represented one of the most important concerns of food, followed by "Food poisoning from bacteria." Instead, in 2019, the concern regarding these precedent issues was not mentioned, but they have been substituted by "Antibiotic, hormone or steroid residues in meat" followed by "Food hygiene" (Table 4.27).

Discussion

This study gathered 11 Eurobarometer waves that include information on food security plus food safety, from 2010 to 2022. Using a mixed-method approach and the food security definition as a framework (FAO, 2006; HLPE, 2020), 161 questions have been collected and analyzed.

The first result that emerges is the content discontinuity of the questions related to food security. Most of the questions, in fact, have been asked only once, making the comparison among the years very difficult or impossible. Only the two Special Eurobaromter surveys, commissioned by EFSA, were comparable because their questions on food safety have been repeated in 2010 and in 2019, allowing to draw a possible trend in European opinions and beliefs. However, in general, the data presented in this section do not reflect the evolution of specific attitudes and beliefs on a certain topic but rather, portray different thoughts in a variable period between 2010 and 2022.

As far as the pillar "Availability" of food security is concerned, according to the retrieved data, Europeans are more likely to believe that food availability does not represent a critical issue for Europeans (even if in 2022, food insecurity affected roughly 10% of the European population). Instead, agriculture and food insecurity are still perceived as a challenge for Developing Countries. That could be partially explained by the theory of psychological distance (Trope & Liberman, 2010) whereby, to protect one's health, threats are perceived far from one's self.

In the analyzed waves, more attention has been paid to domestic food production, with questions that investigated the perception of sufficiency and attitudes on import rules, testing individuals' opinion changes in 4 years. Again, respondents do not seem worried by the sufficiency of domestic European production and agree to maintain high import barriers for the Countries that do not respect European quality standards. Interestingly, in these Eurobarometer surveys, no questions have been found regarding food aid or similar, such as the Food Bank, even though their role has become more evident in fighting food insecurity after the COVID-19 emergency and the consequences of the war in Ukraine.

Food availability means disposing of a sufficient amount of food of a given quality. Questions on organic food provided interesting insights. From the retrieved data, it seems that respondents' preference for organic food production increased over the years because, in their opinion, organic food production better complies with the EU standard on pesticides. However, the geographic distribution of organic products differs considerably among EU Member States (USDA, 2023), making its access rather difficult for half of the European citizens.

Other food characteristics that respondents seek at the supermarket, as purchasing drivers, are geographical area, local traditions, short supply chain, and quality label. From 2018 to 2022, the perceived feature of quality and short supply chain kept constant (respectively around 80% and 85%), while local traditions, geographical area, and quality label increased by 5% from 2018 to 2022, from 75% to 80%. Interestingly, the quality of fresh fruit and vegetables was considered a key driver for conducting a sustainable and healthy diet in 2022. That suggests that some health messages have been internalized by people over time.

However, despite the growing—and recent—attention attributed to the issues such as sustainability and freshness, "Access" to food is still determined by price. Coherently with the literature (see, for example, Khoury et al., 2020), price determines 90% of the respondents' decision when purchasing food. In particular, the last Eurobarometer waves highlighted the trade-off between the desire to utilize more sustainable food and the price to pay, the so-called "attitude-behavior gap", which means that the food consumption practice relies on price more than the individual beliefs (Amilien et al., 2022).

Concerning the "Utilization" pillar, the single datum retrieved from the database regarded diet preferences. The respondents stated that they utilize sustainable food resources because they are helpful for their health. This result supports the idea that individuals seek sustainable diets mainly for their personal/health benefit (Hansen & Thomsen, 2018) and not necessarily for environmental purposes, which could be nudged at an institutional level (Latka et al., 2021).

Similarly, as far as Agency is concerned, respondents believe that the main role of EU farmers is to provide safe and high-quality food (51% in 2022), whereas securing food supply follows with less than 30% (however, it must be said that these questions have been singuralarily asked, hence the sum of the percentage is not up to 100).

Instead, the perception of food security, in terms of Stability, has maintained unaltered during the last decade (2012–2022), and coherently with the actual environmental context, climate change has been identified in 2022 as the feature that mostly affects the future of food security.

Finally, the sustainability pillar is the mostly covered one pillar by Eurobarometer waves from 2010 to 2022. The repetition of questions on the CAP might be referred to the institutions' interests to verify the perception of an implemented policy and for setting future agendas. On the contrary, the pillars that are more rarely included in Eurobarometer are "Utilization" (e.g., the times an individuals cook at home or their ability to cook) and "Agency" (that in this study has been associated with the role of farmers), and thus the Eurobarometer does not measure individual agency in shaping the Food System. Generally, the individuals' awareness of CAP has been stable over the years. However, the total percentage of respondents that think the CAP benefits European citizens has increased by 10%. Moreover, in 2022, coherently with the increased concern on climate change, the respondents agreed that CAP performance and contributions should be devoted to help tackle climate change.

Instead, if the perception of food security has remained stable, the concern for food safety increased. Interestingly, in 2019, when asked which are the features that mostly concern food safety, the percentage of individuals concerned about antibiotics, additives, and pesticides has risen by 7% from 2010. As Nisbet and Myers (2007) suggested, shifts in poll trends might be derived from external events. Hence, we are more likely to believe this increased concern might derive from the changed context and the increased concern about chemical additives, which is in line with the literature about "chemophobia" (Rollini et al., 2022).

In addition, the Eurobarometer, as a tool, has some intrinsic limitations (well-known in the literature) such as long and complicated questions, translation inaccuracies, and biases related to the self-reporting nature of the employed data collection techniques (see, for example, Bréchon, 2009; Félonneau & Becker, 2008; Gatto & Panarello, 2022). However, it is a useful instrument for testing the general perception of individuals over time (whenever possible) in particular regarding policies implemented by the European Institutions. It provides insight into the topics considered urgent by Europe, coherently with what Haverland et al. (2016) have already shown. However, as far as food security is concerned, this tool only provided some insights regarding a complex topic like food security, and a whole picture of the phenomenon was not possible to be taken. Instead, a more accurate idea about the perception of food safety over the years in Europe has been retrieved.

Conclusion

This qualitative study intended to contribute to a better understanding of the individual perception of food security and food safety through the cross-country survey Eurobarometer, to establish whether this tool is useful for drawing opinion trends and then setting the institutional agenda for the most important issues to tackle.

The present study has some limits. Such limitations include the intrinsic limits of the survey itself, and the lack of comparability across Eurobarometer waves because several questions have not been replicated over the years. Another limitation is represented by the single-case study approach: European cross-country survey is not compared to American or Asian cross-country surveys. As regards the former, it would be useful to replicate the same research with the next waves, to include the effects of the COVID-19 pandemic and the Ukrainian war, which have significantly impacted the food system, as well as the perception of Farm to Fork Strategy, embedded in the European Green Deal, which aims at making food system fair, healthy, and environmentally friendly.

In general, based on the assessments the 161 questions retrieved from all the Eurobarometer waves released from 2010 to 2022, the present study suggests that individuals are less concerned with Food Security than Food Safety. However, when analyzing each pillar of Food Security, it emerges a more nuanced picture emerges due also to the partial insights that the Eurobarometer provides. It results in an increasing percentage of respondents being worried about the Stability of Food Security due to the extreme natural events caused by climate change. This concern is not translated into a change in the Utilization of food, which is driven by diet and personal health issues, but into a higher request for actions from the European Institutions. In fact, during the studied decade, more individuals believe the role of European institutions should be strengthened in order to ensure Sustainability of Food Security within the Member States.

Instead, the opinion trend on food safety is more linear: the concern about Food Safety has increased. In particular, respondents are worried about additives, pesticides, and food hygiene.

In general, Eurobarometer changes questions over time, adapting them to the current most pressing issues regarding the Food System. As far as the questions of this study are concerned, it was a useful tool for understanding how opinions change over time in particular vis-à-vis food safety and to a less extent vis-à-vis food security, due to its complexity.

Acknowledgment

The authors would like to thank Dr. Antonelli and Professor Isernia for the support and constructive comments.

References

Amilien, V., Discetti, R., Lecoeur, J. L., Roos, G., Tocco, B., Gorton, M., Biasini, B., Menozzi, D., Duboys de Labarre, M., Filipović, J., Meyer, K., Áron, T., Veljković, S., Wavresky, P., Haugrønning, V., Csillag, P., Simons, J., & Ognjanov, G. (2022). European food quality schemes in everyday food consumption: An exploration of sayings and doings through pragmatic regimes of engagement. *Journal of Rural Studies*, *95*, 336–349. Available from https://doi.org/10.1016/j.jrurstud.2022.09.009.

Bréchon, P. (2009). 2 A breakthrough in comparative social research. The International Social Survey Programme 1984–2009: Charting the Globe, 28.

Capodistrias, P., Szulecka, J., Corciolani, M., & Strøm-Andersen, N. (2022). European food banks and COVID-19: Resilience and innovation in times of crisis. *Socio-Economic Planning Sciences*, *82*101187.

Clapp, J., Moseley, W. G., Burlingame, B., & Termine, P. (2022). Viewpoint: The case for a six-dimensional food security framework. *Food Policy*, *106*, 102–164. Available from https://doi.org/10.1016/j.foodpol.2021.102164.

European Commission. (2023). *Organic farming in the EU – A decade of organic growth*. Brussels: European Commission DG Agriculture and Rural Development. Retrived from. Available from https://agriculture.ec.europa.eu/system/files/2023-04/agri-market-brief-20-organic-farming-eu_en.pdf.

FAO. (2006). Policiy Brief. Food securityAvailable from https://www.fao.org/fileadmin/templates/faoitaly/documents/pdf/pdf_Food_Security_Cocept_Note.pdf. (2006).

FAO, IFAD, UNICEF, WFP and WHO. (2023). *In Brief to The State of Food Security and Nutrition in the World 2023. Urbanization, agrifood systems transformation and healthy diets across the rural–urban continuum*. Rome: FAO. Available from https://doi.org/10.4060/cc6550en.

Félonneau, M. L., & Becker, M. (2008). Pro-environmental attitudes and behavior: Revealing perceived social desirability. *Revue Internationale de Psychologie Sociale*, *21*(4), 25–53.

Gatto, A., & Panarello, D. (2022). Misleading intentions? Questioning the effectiveness and biases of Eurobarometer data for energy sustainability, development and transition research. *Energy Research & Social Science*, *93*102813.

Hansen, T., & Thomsen, T. U. (2018). The influence of consumers' interest in healthy eating, definitions of healthy eating, and personal values on perceived dietary quality. *Food Policy*, *80*(C), 55–67.

Haverland, M., de Ruiter, M., & Van de Walle, S. (2016). Agenda-setting by the European Commission. Seeking public opinion? *Journal of European Public Policy*, *25*, 1–19. Available from https://doi.org/10.1080/13501763.2016.1249014.

HLPE. (2020). Food security and nutrition: building a global narrative towards 2030. A report by the High Level Panel of Experts on Food Security and Nutrition of the Committee on World Food Security, RomeAvailable from https://www.fao.org/3/ca9731en/ca9731en.pdf. (2020).

Khoury, C. K., Jarvis, A., & Jones, A. D. (2020). Trade and its trade-offs in the food system. *Nature Food, 1*, 665–666. Available from https://doi.org/10.1038/s43016-020-00169-6.

Latka, C., Kuiper, M., Frank, S., Heckelei, T., Havlík, P., Witzke, H. P., Leip, A., Cui, H. D., Kuijsten, A., Geleijnse, J. M., & van Dijk, M. (2021). Paying the price for environmentally sustainable and healthy EU diets. *Global Food Security, 28*(100437). Available from https://doi.org/10.1016/j.gfs.2020.100437.

Nicolau, M., Esquivel, L., Schmidt, I., Fedato, C., Leimann, L., Samoggia, A., Monticone, F., Prete, D.M., Ghelfi, R., Saviolidis, M.N., Olafsdottir, G., Sigurdardottir, H., Aubert, P.M., Huber, E., Aditjandra, A., Hubbard, C., De, A., Gorton, M., Čechura, L., … Schamari, D. (2021). Food consumption behaviours in Europe. Mapping drivers, trends and pathways towards sustainability. VALUMICS project. Available from https://doi.org/10.5281/zenodo.5011691.

Nisbet, M. C., & Myers, T. (2007). Twenty years of public opinion about global warming. *Public Opinion Quarterly, 71*(3), 444, 70.

Rollini, R., Falciola, L., & Tortorella, S. (2022). Chemophobia: A systematic review. *Tetrahedron, 113132758.*

Schmitt, H. (2003). The Eurobarometers: Their evolution, obvious merits, and ways to add value to them. *European Union Politics, 4*(2), 243–251.

Trope, Y., & Liberman, N. (2010). Construal-level theory of psychological distance. *Psychological Review, 117*(2), 440.

USDA. (2023). EU Consumers save on food and buy less organic in 2022. Voluntary report. Retrieved from: https://apps.fas.usda.gov/newgainapi/api/Report/DownloadReportByFileName?fileName = EU% 20Consumers%20save%20on%20food%20and%20buy%20less%20organic%20in%202022_Berlin_European% 20Union_E42023-0005.pdf.

Can we slow climate change while feeding a hungry world? Media narratives about the food − water − energy nexus

Kristen Alley Swain

Department of Integrated Marketing Communications, University of Mississippi, Oxford, MS, United States

Moving beyond animal-based food systems to a sustainable food system has emerged as a societal grand challenge (Morris et al., 2021). What humans eat and how these foods are produced significantly affect climate change. There is a growing consensus among scientists that the production and consumption of meat and dairy are a major contributor to GHG emissions (Brummans et al., 2016; Tjärnemo & Södahl, 2015). Despite this strong evidence, meat consumption continues to increase worldwide (Olausson, 2018), and animal agriculture contributions to GHG emissions are represented in less than 1% of all U.S. and Australian newspaper articles about climate change (Mayes, 2016).

A global shift away from consumer diets dominated by meat, dairy, and eggs to mainly plant-based diets is needed for mitigating climate change, as much as a shift away from fossil fuels (Mayes, 2016). As the meat market continues to grow, increased global demand for protein sources has not only changed climate patterns but also has necessitated larger farms, promoted antimicrobial resistance and the spread of zoonotic diseases, threatened profit margins, and complicated efforts to monitor the health and behavior of farm animals (Mahfuz et al., 2022).

Industrialized animal food production is exacerbating global food supply problems, animal suffering, and diet-related public health maladies (Broad, 2020). Shifting the public toward plant-based diets is critical for achieving both environmental and public health

Food Sustainability and the Media
DOI: https://doi.org/10.1016/B978-0-323-91227-3.00006-8

outcomes (Niemiec et al., 2021). Increasing medical data indicate that excessive consumption of red and processed meats can result in serious, adverse health outcomes and lead to the conversion of natural habitats into crop land to produce the cereal plants needed for intensive livestock feeding (Monnier et al., 2020).

This chapter explores media framing of critical global climate change factors that impact food sustainability, including the food − water − energy nexus, agri-food corporate social responsibility, food safety, and impacts of the COVID-19 pandemic on the agri-food industry. In discussing animal products, the chapter explores growing global consumption of red and processed meats and other animal-based food products, farm animal welfare, and the development and marketing of alternatives to meat and dairy products. Media framing helps reduce climate change impacts through communication about food waste reduction, the local food movement, and alternative food networks. Communications about food insecurity issues have contributed to global efforts to curb hunger, food scarcity, and food riots. Finally, this chapter explores sustainability-related challenges in communicating about food risks, including ways that media frame the emergence of nanofoods and other food biotechnologies, genetically modified (GM) foods, food safety challenges, food risk perceptions, and social amplification of food risks.

Media framing of climate change impacts

Most of the literate global population has become aware of the climate threat over the last quarter century, mainly thanks to media warnings of peril. However, public skepticism and aversion to regulation and other mitigation strategies persist. Most people are concerned about the climate but unmotivated to push for meaningful action (Hamm, 2009).

Barriers to feeding the world include climate change, overuse of resources, population growth, and dietary norms. News coverage of food production challenges often highlights significant ethical and political questions, such as how to support adequate nutrition while upholding individual autonomy in culinary choices (Borghini et al., 2020).

News media organizations have an ethical responsibility to report on climate crisis causes and solutions. Since 1995, a strong scientific consensus of more than 2000 researchers has asserted that human activities influence the global climate. The Intergovernmental Panel on Climate Change (IPCC) consensus report in 2007 echoed this assertion, with over 200 nations and thousands of published, peer-reviewed articles concurring that human activities are more than 90% likely to be the cause of climate change and that severe impacts will be abrupt and irreversible.

Fifteen years ago, the IPCC began arguing annually that if urgent steps are taken, climate change can still be addressed at a reasonable cost. Today, the IPCC (2022) recommends 43 climate mitigation strategies. Reducing the conversion of forests and grasslands for crops and other purposes is the second-most impactful climate mitigation measure, behind solar power. Protecting undeveloped land could reduce emissions more than all transportation options combined. The fourth most impactful measure, behind wind energy, is carbon sequestration in agriculture, which includes the use of biochar and soil carbon management in croplands.

Farming relies heavily on planning for weather and seasons according to experience of past years. Changes in seasons and unusual weather patterns can lead to catastrophic loss of crops or livestock (Ngo et al., 2022). Providing more financial incentives and infrastructure to the agri-food sector for sustainable waste management, recycling and development of sustainable food technologies could further reduce emissions. Barriers to implementing climate mitigation strategies in the agri-food sector include competing demands on farmland, the economic impact of mitigation strategies on agri-food livelihoods, the complexity of farm management systems, and cultural norms (IPCC, 2022).

Efforts to improve the global food system situation include the 2015 United Nations Sustainable Development Goals, 2015 Paris Agreement climate change targets, and the 2010 Aichi Biodiversity Targets. Already threatened by current dietary patterns, these proposed targets have failed to meet short-term goals and will be further threatened through at least 2050 because of a growing population and trends toward diets with more calories, animal-source foods, and ultraprocessed foods. While dietary changes to healthier and predominantly plant-based diets are integral to meeting environmental targets, the various economic, social, and cultural barriers make systemic dietary transitions difficult (Clark et al., 2020). In addition, links between the production and consumption of livestock and anthropogenic climate change have received relatively little media coverage (Almiron & Zoppeddu, 2015).

The largest source of GHG emissions is related to electricity and heat production. Residential and commercial buildings are responsible for 60% of the global electricity demand. However, the agri-food system is also very energy- and carbon-intensive (Ricci & Banterle, 2020). Animal agriculture accounts for 15% of global emissions, about the same contributor as the transportation sector. Individual choices related to electricity, heat, transportation, and food consumption could all have strong impacts on GHG emissions. The scarcity of natural resources, exacerbated by climate change impacts, highlights the urgent need for sustainable food consumption. In addition, global meat consumption increases animal agriculture's negative impact on the environment but continues to grow (Kristiansen et al., 2021).

Ensuring an environmentally sustainable future has become a top issue on the international political agenda. However, global policy has mainly focused on reducing GHG emissions through renewable energy, emissions trading, and carbon capture and sequestration. While these policy areas are important, this emphasis has shifted attention away from other factors that heavily contribute to climate change, including meat consumption, food waste, and unsustainable food production (Brummans et al., 2016; Tjärnemo & Södahl, 2015).

Many food companies and consumers are increasingly interested in healthy and environmentally sustainable food and beverages. The food industry is an important sector of the economy because food is a necessity for human life and requires specialized handling, preparation, and logistics to be safe and nutritious for human consumption (Samoggia et al., 2020). The ecological aspects of food consumption now outweigh its economic and social considerations. Through interviews with 25 Turkish food sustainability experts, Özkaya et al. (2021) found that barriers to sustainable food consumption include consumer lack of awareness, meat production, unplanned food shopping, behavioral mistakes after consuming food, difficulties in changing lifestyles, and lack of motivation to change food consumption habits.

Agricultural impacts of climate change

Changes in global mean and extreme temperatures are triggering major changes in water, soils, pathogens, weeds, and pests in the agricultural production environment. These temperature changes are also increasing water scarcity and diminishing crop yields. Drought and lack of water are leading weather-related causes of crop loss. Flooding from excessive rain and sea-level rise, as well as saltwater advancing into aquifers and estuaries, also widely threaten agriculture and livestock. Many farmers' short-term adaptive responses to agricultural pressures and vulnerabilities have lasting negative consequences. Field crops, vegetable crops, and perennial crops have unique production requirements and diverse management requirements that farmers and ranchers need to use, to effectively adjust to evolving food production needs. The long-term planning and climate change adaptation measures that farmers can implement now could build more responsive and resilient food systems, as well as resilient communities that can cope with future climate conditions (Elias et al., 2019).

Among farmers, gain-framed, persuasive messages are more effective than loss-framed messages in shaping intentions to take climate adaptation measures, boosting self-efficacy perceptions, and in raising risk perceptions about climate change. Gain-framed messages about climate adaptation measures are most effective when they are combined with concrete-framed, loss-framed, or abstract-framed messages (Ngo et al., 2022).

Climate adaptation strategies, a major concern among food producers, reduce farm vulnerability to climate impacts. Farming adaptations include migration or abandoning coastal zones, conversion to fish farming, innovative irrigation, drought-tolerant crops, rainwater storage, and growing crops that can tolerate flooding and salt. Understanding agricultural adaptations is based on farmers' access to information and advice they find credible and feasible. For example, through interviews with Iranian farmers, Azadi et al. (2019) found that their climate risk perceptions, trust, and psychological distance often drive agricultural adaptation behaviors.

Individual attitudes about climate adaptation actions can "spill over" into their attitudes about climate mitigation. For example, negative spillover effects can occur when adoption of one pro-environmental behavior reduces the likelihood of adopting another one. Negative spillover often happens when people feel they are not personally responsible for a problem or believe the problem already has been dealt with. Weber's (1997, 2006) work on single-action bias found that farmers who adapted their crop selection or other cultivation practices—or who adopted off-farm climate adaptations such as investing in futures—were less likely to support government interventions to mitigate climate change.

The increasing CO^2 in the atmosphere endangers the earth's habitability, but organic carbon in the soil is the foundation of agricultural fertility. Emerging carbon capture technologies present promising climate adaptation strategies for farmers. Conserving and improving soil carbon through carbon sequestration can increase agricultural productivity and mitigate soil carbon loss. The perceived credibility, salience, and legitimacy of carbon mitigation in crop productivity practices often highlight the complex and contested nature of managing soil carbon. Although these ideas frequently arise in scientific community discussions, they rarely appear in media coverage. Some climate-smart agricultural methods convert waste and other biomass into biofuels or crop fertilizer, to help keep CO2 out of the atmosphere (Ingram et al., 2016).

Soil carbon enables plant infiltration and storage of water, drives nutrient cycles, and reduces the need for fertilizers. Bogs, peatlands, and other wetlands are among the largest carbon sinks on the planet, containing twice as much carbon as all the world's forests. Since global warming and severe weather sparked by climate change cause harmful loss of soil carbon, wetlands restoration can significantly reduce GHG emissions. Farm and livestock food producer organizations are powerful change agents and advocates for their industries, especially during climate and weather-related hardships. They engage with policymakers, their own members, and the public to pursue the interests of their specific communities. For example, livestock and forage organizations often focus on weather issues, while diverse national umbrella farm organizations often share common ground about climate issues involving rotational pasture grazing and other farm management topics (Tourangeau et al., 2019).

Their agenda-building efforts to influence media coverage may have been less effective, so far. Although climate-smart agriculture news coverage has explored many innovative strategies, including new technologies for crop production, animal production, fisheries, postharvest management, food safety, and climate-related marketing and management, few of these stories discuss climate change (Kutyauripo et al., 2021).

The farming press has framed many sustainable agricultural practices around economic and agronomic aspects. Even though farmers often use these publications as a source of information, they often believe these outlets are agribusiness mouthpieces. Although they are rarely motivated to try a new sustainable agricultural practice after learning about it in the farming press, these outlets do raise farmer awareness about wider agricultural topics (Rust et al., 2021).

The food − water − energy nexus

With population growth and the urgent need to significantly increase agricultural production, a changing climate is putting additional stress on soil and water resources (Delgado & Li, 2016). The term "food − water − energy nexus" (FWE), constructed by the United Nations Food and Agriculture Organization, means that water security, energy security, and food security are powerfully linked to one another—and the actions in one area often affect one or both of the other areas.

The FWE nexus has been increasingly used to inform food governance practices and to address food demand and supply challenges across various global ecosystems. However, to support climate change mitigation and adaptation, agri-food policymaking also needs to consider interdependencies among FWE sectors. To this end, the UN's 17 global Sustainable Development Goals emphasize integrated resources management that account for FWE nexus linkages (Avellán et al., 2017). Incorporating FWE strategies can help policymakers balance trade-offs and identify synergies and opportunities for promoting sustainability. Beyond situations where water and energy are needed for each other, the FWE linkages are critical considerations within many different food, ecosystem, and climate contexts (Yillia, 2016).

Media coverage of FWE nexus policies often involves framing contentious trade-offs. Authoritative interview sources for FWE stories often include managers of federal, state, local, and nonprofit organizations. These leaders often explain their approaches to managing interconnected systems in ways that are sustainable and reduce the depletion of scarce resources. Other expert sources for FWE stories often come from engineering fields, often focusing on science-based strategies for managing resources more efficiently. FWE stories frequently quote diverse experts from sectors that have never collaborated before. Food governance within the different systems responsible for agri-food resource delivery has been fragmented and decentralized. Since there is little meaningful communication across the water, energy, and food policy sectors, resource management decisions largely have been siloed. Strategies are needed to break down these silos, to improve interdisciplinary communication and collaboration (Greer et al., 2020).

The FWE trade-offs are often complex. Increasing agricultural production involves irrigation from wetland sources, but this irrigation can adversely impact fish habitats. The siting of solar energy panels can reduce agricultural acreage needed for food production. The use of pesticides that greatly increase crop yields also pollutes the environment (Gardezi et al., 2020). Sustainable food messaging can prompt nonsustainable food behaviors, such as homogenized food consumption that depletes environmental resources (Simeone & Scarpato, 2020). Public understanding of these trade-offs can drive support for food production over biofuels, water-friendly crops over meat production, and water conservation over water-intensive agriculture. For example, proecological attitudes and public confidence in the effectiveness of proenvironmental practices have driven support for access to safe drinking water and sanitation. Public preferences, values, and knowledge—often shaped through media coverage and social media conversations—also can influence FWE policy design and implementation (Gardezi et al., 2020).

Registered dietitians are on the front lines of food and nutrition recommendations. While most agree that climate change is an important issue, few feel that they should play a major role in climate change mitigation strategies. Most do not engage in activities that promote diet as a climate change mitigation strategy. Vegetarian and vegan dietitians are more likely than nonvegetarian and nonvegan dietitians to engage in climate mitigation activities (Hawkins et al., 2015).

FWE nexus discussions about how to feed the world often ignore soil resources, even though soil is necessary in 99% of the world's food production. Soil is also essential in capturing water and generating energy from biological systems. Efforts to translate scientific awareness into action strategies to promote the importance of soil resources and to improve soil management are limited. Erosion exacerbates soil degradation, which reduces food production. The soil degradation rate is expected to increase until it reaches a tipping point, when technological advances cannot overcome the impacts of a reduced topsoil depth coupled with a volatile climate (Hatfield et al., 2017).

Agri-food corporate social responsibility

The grand challenge to feed the world in ways that promote healthy people, a healthy planet, and healthy communities is often reflected in the missions and actions of food suppliers and companies across the entire supply chain. Corporate social responsibility communication, sometimes known as "conscious capitalism," is public communication about corporate engagement and activities (Veldung et al., 2022).

Consumer decisions about food often do align with actual food company messages and corporate priorities (Edge & Hermann, 2021). However, food retail companies that pursue a clear and consistent communication strategy that promotes conscious capitalism principles on social media are more likely to present a strong, active, and involved fan community, as well as interactivity and engaging content. Conscious capitalism messages about sustainable food often convey higher purpose, inclusive stakeholder orientation, conscious corporate leadership, and conscious corporate culture (Veldung et al., 2022).

Corporate social responsibility (CSR) involves international private business self-regulation that contributes to philanthropic, activist, or charitable societal goals, often by volunteering or using ethically oriented practices. CSR private regulations play a central role in food governance, and CSR reporting documents how well agri-food firms commit to their stated social responsibility objectives. However, relying solely on voluntary corporate actions and disclosures has not effectively corrected the negative impacts and social costs produced by the agri-food sector. In addition, the level of comprehensive reporting of CSR reports is typically low, with a high level found among only a few companies (Sodano & Gorgitano, 2017).

CSR disclosures on the web can provide food marketing advantages for small- and medium-sized food processing firms, regardless of their profitability or indebtedness. Although corporate web communications for larger food producers generally provide more CSR information than smaller firms, few firms of any size mention social and environmental aspects of food sustainability in their CSR disclosures. Highlighting these issues could help a firm distinguish itself from others in a market and build a more positive brand (Sommer et al., 2015).

CSR obligations have prompted many international development institutions to financially support the Sustainable Development Goals and the Paris Agreement on Climate Change. For example, in Indonesia, limited green financing practices support sustainable food crops, animal husbandry, fisheries, plantations, industrial forests, and organic food and beverages development projects (Wayan Budiasa, 2020). CSR actions to address food security for pregnant women have included increased food production, ensuring its equitable distribution, nutrition intervention, and nutrition education (Lakshmi & Indumathi, 2016).

Media coverage of food-related CSR goals has highlighted employee policies, animal welfare, fair-trade practices, consumer rights about safety, nutrition, marketing—as well as environmental issues such as climate, aquatic dead zones, and biodiversity. Corporations often follow food policies because they are required by laws, because these policies point to the right thing to do, or because the policies provide a competitive advantage. In developed nations, retailers and consumers drive food companies toward policy changes that align with company values and stakeholder views (Lähteenmäki-Uutela, 2014).

Food policy framing in the media also can affect corporate response to climate change. However, food system policies aimed at controlling GHG emissions are often weak because they rely on food corporations' voluntary, piecemeal, and sporadic CSR. Public and private climate mitigation interventions have been ineffective in addressing food sector problems exacerbated by climate change, mainly because most of these interventions have been political rather than economic and have depended upon power relationships between states and large corporations. Regulating voluntary CSR among agri-food companies, while enforcing corporate and antitrust laws, could help balance corporate rights with corporate responsibility in addressing climate-related food problems (Sodano & Hingley, 2018).

Global Reporting Initiative guidelines assess CSR commitment to sustainability among major agri-business companies operating in the seed, agrochemical, food processing, and retailing industries. Since soft regulation of agri-food organizations is not effective, Sodano and Hingley (2018) argue that government agencies and other independent third parties should monitor and evaluate CSR. One initiative that promotes impactful CSR activities is the Equitable Food Initiative (EFI), a third-party certification program that seeks to improve agricultural working conditions as well as food safety practices, environmental stewardship, and farm viability. The initiative uses innovative cross-functional labor-management teams to promote compliance with EFI standards and advocate for worker interests. For example, after farms in the United Sstates and Mexico implemented EFI programs, they contributed to collective advocacy, worker participation, empowerment, CSR, and corporate accountability (Zoller et al., 2020).

When discussions about climate and animal welfare issues become stuck in controversy or deadlocked debates, many well-intentioned CSR initiatives eventually provoke debate, conflict, and protests. The complexities of Western agriculture include legitimate moral standpoints that proliferate without the realistic prospect of a consensus. Agri-business CSR initiatives often deal with trade-offs among different sustainability issues, which may allow for enough space to navigate legal and economic responsibilities (de Olde & Valentinov, 2019). CSR practices alone do not translate into individual behavior change. For example, CSR practices among Mexican food companies do not promote proenvironmental behaviors among their workers (Astudillo et al., 2021).

Communicating about animal products

The problem with animal products

Most European social sciences and humanities studies about food sustainability initiatives have explored five themes: debating and visioning food from animals, transforming agricultural spaces, framing animals as food, eating practices and identities, and governing transitions beyond animal-based food systems (Morris et al., 2021).

There is increasing concern about the health and environmental impacts of the increasing rates of livestock product consumption, especially in high-income and middle-income countries. Tobacco, alcohol, and foods that are high in fat, salt, and sugar generate much of the global burden of noncommunicable diseases (Ireland et al., 2019). According to the World Health Organization, a diet high in vegetables may reduce the risk of coronary heart diseases, stroke, and cancers and can reduce an individual's carbon footprint. Health attitudes motivate vegetable consumption more so than environmental attitudes (Gustavsen, 2020).

Public arguments for or against livestock consumption are sometimes distorted by sweeping generalizations, vested interests, using facts out of context, or misinterpretations of evidence. For example, a fact that is seldom mentioned is that livestock support the livelihoods of many people in low-income countries. The benefits of livestock for poor livestock keepers are numerous, including supporting crop production in mixed systems, supplying nutrients and income, and fulfilling cultural roles. Livestock farming also can provide resilience against economic and climate shocks (Salmon et al., 2020).

Cultural factors play a critical role in shaping consumer attitudes about animal products. For instance, Australia's socioeconomic and cultural ties to livestock production and its widespread consumption of animal products are impeding climate mitigation (Mayes, 2016). In some countries such as Germany, there are also strong cultural links between eating meat and the celebration of special occasions, eating out, hosting guests and offering hospitality, and as an expression of high social status. Sustainable food campaigns might be more effective if they could link the consumption of plant-based dishes to already established social practices, especially since many people ignore the distinction between public and private meat consumption (Biermann & Rau, 2020).

Many English-speaking countries have a media blind spot for meat because the news media fail to address the responsibility of individual dietary choices that impact the climate. Although top 10 Italian and Spanish newspapers do cover the issue, the Mediterranean dietary background of their audiences does not affect ethical or dietary sensitivity to meat issues among readers (Almiron & Zoppeddu, 2015). The Spanish press also has underrepresented the role of animal agriculture in climate change, allocated insufficient attention to it, discussed proposed solutions that do not reflect current climatic urgency, and support the consumption of animal products more often than the defense of nonhuman animals (Moreno & Almiron, 2021).

When the news media does cover meat, it often taps into consumer fears about food safety. This coverage can have a major negative impact on meat production companies and brands. For example, after ABC World News Report aired a story series about lean finely textured beef in 2012, beef prices hit a 10-year low, and a major firm that produced it had to declare bankruptcy. Use of the "pink slime" frame in stories provokes perceived risk about ground beef but not about health risks related to consumption of red meats or processed foods (Runge et al., 2018). Similarly, news frames about a 2019 meat listeria crisis in three Spanish newspapers highlighted consequences, while a calm frame sought to soften the seriousness of the facts (Marín-Murillo et al., 2021).

Soon after scientists discovered that Bovine Spongiform Encephalopathy (BSE)—also known as "mad cow disease"—could be transmitted to humans through beef consumption, the news media amplified BSE risks. In light of their meat purchasing or meat production decisions, BSE risk perceptions among consumers and cow − calf producers were amplified or attenuated by the quantity and quality of BSE media information (Yang & Goddard, 2011). The number of published articles increased as more BSE-related incidents occurred. Most stories were based on factual reports, including neutral policy discussions, and generally emphasized agriculture, trade, commerce, and incident details. Once food safety advocacy emerged, appeals for weaker policy measures appeared most frequently in the coverage, in an effort to balance between proactive objectives and rational acceptance of health risks. Alarming stories that contributed most to the social amplification of BSE risks did not originate from any single health or industry viewpoint. The coverage demonstrated ambivalence and amplified risks less than expected. The stories also did not isolate health risks, increase public fear, or advocate for stronger government policies (Sato & Webster, 2022).

Another prion disease, Chronic Wasting Disease (CWD), recently emerged as a possible threat. Hunters and indigenous harvesters rely on their knowledge, self-efficacy, and other perceptions to weigh the risks of eating or distributing meat from deer and other wildlife that may have contracted CWD. According to the U.S. Centers for Disease Control (2022), no reported cases of CWD infection have been reported in humans. However, some animal studies suggest that CWD does pose a risk to monkeys and other primates that eat CWD-infected animal meat or come in contact with brain or body fluids from infected animals. In addition, it can take over a year before an infected animal develops symptoms, which can include drastic weight loss (wasting), stumbling, listlessness, and other neurologic symptoms.

Parlee et al. (2021) found that Canadian hunters and harvesters of moose, white-tailed deer, and mule deer expressed a high level of concern about CWD during the last decade, while also expressing a high degree of confidence in their own knowledge of CWD risk. Their self-efficacy remained high, even though the amount of information they received declined over time and the CWD cases have significantly increased.

Food retailers potentially play an important role in developing an environmentally sustainable food system because they have a powerful influence in food procurement and directly encourage consumers to buy certain foods. Many food retailers do develop environmental visions, policies, and goals and implement strategies for energy and transport efficiency and waste recycling. However, few encourage consumers to buy organic, local, and seasonal food or to minimize food waste. Virtually no food retailers help consumers reduce meat consumption because they rely on meats to attract new customers and keep loyal ones (Tjärnemo & Södahl, 2015).

Older adults see information about the health risks of red meat intake as moderately easy to understand but often struggle to understand numerical recommendations about serving size. They tend to use general arguments applicable to any context or theme as their main deliberative strategy, along with individual strategies for reducing uncertainty about red meat quantities. Health policies might help reduce consumer uncertainty about numerical scientific labeling and related food information, if they could take individual deliberative strategies into account (Gaspar et al., 2018).

Evidence of health and environmental harms of red meat is growing, but highlighting certain harms in red meat reduction messages may be more effective in motivating dietary changes. For example, Grummon et al. (2022) found that most parents are unaware that red meat contributes to prostate cancer, heart disease, early death, and other health problems—or that it contributes to water shortages, deforestation, extinction among plants and animals, climate change, and other environmental harms. Heart disease elicited the most discouragement, followed by early death, plants and animals going extinct. Parents who were younger, identified as Black, and identified as politically liberal had higher general perceptions that red meat is bad for health. Those with higher levels of usual red meat consumption were more discouraged from wanting to eat red meat, after seeing messages about health and environmental harms.

International expert groups are calling for significant reduction in the production and consumption of red and processed meat (RPM), to reduce associated health and environmental harms. The corporatization and globalization of food supply chains contribute to intensive RPM production and high consumption levels. Food system stakeholders include actors from industry, academia, civil society, and governmental organizations. Many stakeholders still value the continued production and consumption of red and processed meat, even though they also believe the status-quo situation is untenable. The news media often covers RPM as a policy issue, but public discourse about RPM is often polarizing, contentious, lacking in context and nuance, and oversimplifies harms and benefits. Policymakers have devoted limited attention (Sievert et al., 2022).

Even many of the nongovernmental organizations (NGOs) that focus on environmental, food-focused, and animal protection argue that addressing RPM consumption in a climate change context is not part of their core mission, that meat consumption has limited social and political appeal, and that their organizations should not tell people what to do (Laestadius et al., 2014).

Some nongovernmental organizations do highlight the dangers and social costs related to RPM consumption by using social countermarketing (SCM) strategies, which use commercial marketing tactics to reduce RPM consumption influences and increase prohealth and prosustainability messages. These RPM campaigns often address existing sociocultural norms, public policies, or political decision-making about the RPM industry.

Consumers are more likely to amplify SCM campaigns about red meat when messages resonate with them and when this resonance promotes social amplification and third-party endorsement of the messages. They are most likely to amplify diagnostic and prognostic messages about meat-selling practices because these frames are relevant to daily life experiences. These engaged consumers are also more likely to tweet and retweet calls for mobilization, endorse meat reduction messages, and help strengthen the position and visibility of these messages in online food-quality debates. Ultimately, consumer assumptions about responsibility or blame, regarding the impact of consuming red or processed meat, often amplify existing risk perceptions about specific meat brands (Jacobs et al., 2021).

Low public awareness about how animal food consumption affects climate change has impeded efforts to reduce meat consumption, especially in Western diets (Kristiansen et al., 2021). People often reside within the social media "bubbles" that support their own views. Food sustainability advocates must break through these digital echo chambers to reach wider audiences. One approach is to use social media advocacy strategies that engage a broad international base and raise issues about meat impacts. Other effective strategies that promote credibility and intrinsic values in meat reduction campaigns include promoting coordinated action within digital networks, featuring the environment as a complementary message frame, using high-profile experts or celebrities to promote complementary frames, and leveraging agenda melding to highlight the importance of news issues in discussions (Friedlander & Riedy, 2018).

People often resist messages about the health effects of red meat consumption by discounting the risk messages. Often, they decide the risk message does not apply to themselves, cast doubt on the credibility of the message source, or assume that others are amplifying the risk (Regan et al., 2014). When meat consumption is framed in terms of environmental protection, animal welfare, or public health, relevant policy action frames can have a major impact on public support. However, using rhetorical frames to provide direct information to individuals about plant-based diets or meat reduction has had only limited influence on policy support (Whitley et al., 2018).

Animal-based food production practices have enabled marketers to sell a variety of products that satisfy consumer demand while reducing meat and dairy production costs. However, some production practices have come under increased public scrutiny in recent years. These include the inhumane treatment of animals, use of antibiotics, feed additives and hormones, and how these are monitored, regulated, and communicated to consumers. Consumers also have raised concerns about food labeling, including information about organic farming, genetic engineering, animal welfare standards, and place of origin (Centner, 2019).

Farm animal welfare

Industrial production of animal-sourced food products involves significant exploitation of animals, labor, and the natural environment (Dickstein et al., 2020). Although the animal agriculture industry's exploitation of billions of nonhuman animals receives little media coverage (Moreno & Almiron, 2021), the media is an important potential source of information about farm animal welfare issues. The way stories represent these issues can suggest causes, solutions, or moral evaluations.

The Voiceless Animal Cruelty Index (VACI) calculates three indices for deliberate or willfully ignorant animal mistreatment, in 50 countries with high agricultural output. These indices include animal slaughter (producing cruelty), animal consumption (enabling cruelty), and laws protecting farmed animals (sanctioning cruelty). The more income inequality that exists within a country, the more weighted numbers of animals are consumed there and the higher its overall VACI rank among nations (Morris, 2021).

In 2015, the German food retail and industry sector launched an initiative known as "Initiative Tierwohl" to improve animal welfare standards. It communicated with participating companies mostly through its websites and web pages. Initiative Tierwohl campaigns mainly appealed to consumers motivated by self-transcendence and openness to change, self-enhancement, or conservation, depending on their specific information sources and needs. Most consumers looked to brochures, fliers, and interpersonal contacts as their primary sources of information about animal welfare standards (Hirsch et al., 2019).

Examining public attitudes and behavior toward the red meat industry can help inform the design of an effective media campaign exposing animal cruelty. For example, Australians who viewed a "60 Minutes" investigation piece about the live export of sheep by sea were more likely to engage in relevant discussions in the community and on social media, to perceive conditions aboard the sheep transport boats as important, and to cite social and internet media as a source of information. Despite their exposure to the expose, they still felt that the red meat industry was acceptable, still trusted farmers in the red meat industry, and did not express increased concern about red meat farming practices or about the welfare of sheep or beef cattle (Rice et al., 2020).

In addition to coverage of sheep exports, Australian mainstream press coverage of farm animal welfare coverage has largely framed governments and farm animal production industries as untrustworthy to ensure good farm animal welfare and consumers as agents to improve animal welfare through ethical consumption (Buddle & Bray, 2019). Similarly, social media activists interacted with Australian news media in 2011 to provoke public outrage and disrupt the policy agenda around the slaughter of beef cattle in Indonesian abattoirs. Extensive media coverage of this moral outrage morphed into a critique of Indonesian values and cultural practices, and then Australia banned live cattle exports to Indonesia (Small & Warn, 2020).

Normalizing controversial food often involves subtle shifts in messages over time. For example, British newspaper stories that used the word "veal" encouraged readers to engage in incrementally transformative food politics that do not involve demanding food as a right (O'Neill, 2019). Blogs covering the North American meat industry emphasize social problems with meat product consumption more often than newspaper stories do (Bateman et al., 2019).

Animal product alternatives

While consumption of animal food products continues to increase, consumer interest in alternative meat and milk products is also increasing. Alternative products include plant-based milk, plant protein products, and cultured meat and milk. Lowering GHG emissions from the livestock sector calls for marketing carbon-free dairy products to be competitive with plant-based mylk products (Lombardi et al., 2017). Efforts to promote widespread transition to meat and dairy alternatives and plant-based diets often rely on media coverage, marketing campaigns, and environmental policymaking. To stimulate a systemic transition to alternative proteins, marketing messages must appeal to a wide variety of consumer segments, diets, and identities. Media coverage can then highlight the promises and tensions of these markets, as well as consumer preferences, attitudes, behavior change policies, and the politics and ethics of protein alternatives (Lonkila & Kaljonen, 2021).

Predictably, proenvironmental consumers are consistently more positive toward meat curtailment policies than those who endorse human supremacy beliefs or are more attached to meat consumption. However, reading just one informative story about approval of a meat curtailment policy has increased consumer support for meat curtailment policies, regardless of their ideology and food consumption habits. Communicating about policymaking that reduces meat consumption also can influence consumer attitudes and increase public support for policies that promote plant-based diets (Graça et al., 2020).

More media coverage is needed about shifting to plant-based diets, to encourage engaged discussions about demand-side climate mitigation (Mayes, 2016). The UK and US elite media have devoted little attention to animal agriculture's role in climate change and the roles and responsibilities of various stakeholders in addressing the problem. However, when the issue is covered in the news, the responsibility is often attributed to consumers more often than to governments or large-scale livestock farms. Personal dietary change is far more prominent in the coverage of solutions than government policies, reforming agricultural practices, or holding major animal food companies accountable for their emissions (Kristiansen et al., 2021).

Even with an emphasis on consumer responsibility, print media sometimes acts as a "de-meatification" mechanism that is shifting the status of meat in UK society. For example, UK news reporting about a "Meat Free Mondays" campaign reflected shifting meanings of meat and debate about meat-based diets (Morris, 2018). Similarly, in 2016, Swedish climate reporting began shifting its focus from general climate change to the harmful climate impacts of meat consumption (Olausson, 2018).

Politicized framing often flows from media coverage to social media conversations. For instance, in Facebook discussions, livestock production is often legitimized through polarization frames such as livestock production versus other environmental issues, environmentally "good" versus "bad" countries, and "reliable" versus "unreliable" information (Olausson, 2018).

The nutritional choices and personal convictions of vegans, vegetarians, omnivores, and conscientious omnivores influence their responses to questions about climate change and animal welfare issues (Ploll et al., 2020). Social media framing of sustainable food can highlight the spiritual benefits of plant-based diets. For example, a humanistic nongovernmental organization, Buddhist Compassion Relief Foundation in Singapore, framed mindful food consumption and vegetarianism through discussions about affective embodiment (the connection between physical experiences and cognitive development), invocation (daily mindfulness), and transmediation, which is silent, mantra meditation (Brummans et al., 2016).

Public and scientific debates about meat consumption often include a narrative that challenges plant-based and alternative-protein narratives. Tweets about reducing meat consumption often generate more consumer engagement—including more frequent mentions and greater diversity of content—than tweets promoting sustainable or ethical meat. Commercially invested tweets posted for marketing purposes, as well as #sustainablemeat tweets, often promote a particular business or meat production, whereas #eatlessmeat tweets are more often linked to planetary issues and highlight a vegan narrative (Maye et al., 2021).

Refusing to consume animal products not only has tangible economic and social impacts, it also sensitizes individuals and communities to the sociopolitical effects of their consumer behavior. However, veganism is often seen as liberal advocacy fixated on consumerism, identity, and Western capitalist oppression and economic structures (Dickstein et al., 2020). For instance, a vegan Twitter group's posts included criticism and negative comments about veganism, while a plant-based diet Twitter group's posts were more focused on impacts of nutrition, physical activity, and consumer health (Aleixo et al., 2021). A reframing of veganism boycotts could help them align with broader commitments to social and environmental justice and with a trans-species, antiexploitation ethic, as an effective political action tactic for eroding capitalism and other systems of domination (Dickstein et al., 2020).

The absence of overt vegetarian framing can influence consumers to choose vegetarian menu items more often. Reframing the names of vegetarian food categories can influence consumer choices of dishes from restaurant menus. A proenvironmental frame such as "environmentally friendly main courses for a happy planet," a social frame such as "refreshing main courses for relaxing conversations," and a neutral frame such as vegetarian and nonvegetarian dishes within the same menu section, such as "main courses" can all increase the likelihood that consumers will select vegetarian menu choices. On the other hand, a direct vegetarian label such as "vegetarian main courses" is less likely to influence vegetarian menu choices (Krpan & Houtsma, 2020).

Alternative meat

Manufactured alternative meats that substitute for animal source food (ASF) include plant-based ingredients, animal cells grown in culture, or ingredients produced using synthetic biology. Nutritionists, geneticists, and veterinarians have successfully developed these new technologies to reduce the environmental footprint of ASF (Van Eenennaam & Werth, 2021). The alternative protein sector includes the cell-based meat industry—also known as cultured, lab-grown, clean, cultivated, in-vitro meat, or plant-based meat (Broad, 2020). Cultured meat is produced by in vitro animal cell cultures using tissue engineering techniques. Plant-based meat, also known as meat substitute, mock meat, or imitation meat, is a meat-like substance made from vegetarian ingredients such as soy or pea protein.

Cellular food agriculture, the technologies that use cell cultivation to produce agricultural products such as cultured meat, could replace conventional agriculture and reduce food insecurity while reducing environmental burdens and land use pressures. Cellular agriculture also removes the need to slaughter animals (Helliwell & Burton, 2021). Cellular agriculture has promised sustainable and ethical alternatives, as well as solutions to global food challenges. Although cultured meat is not yet available to consumers, many people form opinions after seeing media coverage about it.

Ryynänen and Toivanen (2023) identified three broad cultured meat themes in Finnish reader comments on news stories: the necessity of cultured meat, in light of environmental, animal well-being, and health considerations; anticipated product characteristics such as naturalness, potential risks, and sensory qualities; and societal effects of stakeholder decision-making and inequities.

Alt-meat is providing new opportunities for rural producers in the United States, including growing crops as ingredients for plant-based meat or feedstock for cultured meat, raising animals for genetic material for cultured meat, producing cultured meat in bioreactors at the farm level, transitioning into new sectors, new market opportunities for hybrid animal- and alt-meat products, and new value around regenerative farming or farms that emphasize animal welfare. Market threats of producing cultured meat include loss of livelihood or income for ranchers and livestock producers and for farmers growing crops for animal feed, barriers to transitioning into alt-meat production, and exclusion from alt-meat industry sectors. Universities and research organizations, government agencies, and nonprofit organizations could help maximize the benefits and minimize the risks. Alt-meat likely will capture much of the anticipated growing demand for protein, rather than displace animal meat entirely. It also may offer more markets for rural producers to sell to and more forms of production to pursue (Newton & Blaustein-Rejto, 2021).

In response to environmental, public health, and animal welfare challenges of conventional agri-food production, advocates of animal food alternatives have championed these products as key to feeding the growing global population. Some Big Food companies are beginning to produce plant-based and cell-based animal product alternatives. Some market alt-meat with a "food tech justice" approach, to appeal to environmentally conscious consumers (Broad, 2019). However, even "novel protein" production and other technology-driven responses to climate change can ultimately reinforce corporate power in the food system (Sievert et al., 2022).

Consumer acceptance of cultured meat is rooted in personal experiences and individual differences and depends on associations evoked by the technology, its perceived naturalness, and consumer trust in the food industry using the technology. Food neophobia—the irrational fear or dislike of unfamiliar foods—as well as disgust sensitivity and cultural values also informs individual differences in cultured meat acceptance (Siegrist & Hartmann, 2020).

Alternative protein discussions often ignore these consumer apprehensions and highlight the "meat is made" and "the market" metaphors. The "meat is made" metaphor aims to decouple meat from its long-standing connection to animal farming and reframe it as a set of tastes and textures that can be reconstructed through biotechnology. The metaphor of "the market" for alternative meats highlights innovation, capital investment, and insights more so than behavioral economics and marketing as the primary agents for a "post-animal bioeconomy." Both metaphors downplay public concerns about the meat sector's public health, cultural, economic, and ecological impacts (Broad, 2020).

Consumer attitudes toward alt-meat, as well as their purchase intentions, often depend on perceived effectiveness of these products in mitigating climate change and the use of figurative (vs literal) language in the message (Ye & Mattila, 2022). However, companies often convey sweeping sustainability proclamations about alt-meat to attract customers, venture capital investments, and market share. They often promise to end world hunger or reduce the amount of animal agriculture, greenhouse gases, and land and water used for food production. Rather than spread simplistic greenwashing messages, however, Van Eenennaam and Werth (2021) recommend that both alt-meat and conventional meat producers instead convey how food innovations can sustainably meet future demand. Evidence-based messages could promote greater awareness of these impacts while also counteracting misinformation, fear, and uncertainty.

If cultured meat marketing could overcome consumer uncertainty, it could address major ethical, environmental, and public health concerns associated with conventional meat production. Consumers who encounter cultured meat through a "high tech" frame tend to have more negative attitudes toward it and are less likely to consume it than people exposed to a societal benefits frame or a "same meat" frame. The high-tech frame has been dominant in media coverage of cultured meat because the technology makes it newsworthy. However, this coverage has prompted negative attitudes toward cultured meat despite its societal benefits (Bryant & Dillard, 2019).

The far-reaching effects of cellular agriculture have yet to be the topic of major public or political discussion. The effects of cultured meat production might include altering human relationships with nature and disrupting existing food systems, land use patterns, rural economies, drivers of environmental change, and biodiversity in terrestrial and aquatic ecosystems. So far, media coverage has failed to cover many underexplored and unexplored questions, contingencies, and eventualities of envisioned cellular agriculture developments. Public debate about the coverage that does exist often focuses on the consequences of cellular agriculture that could stem from the creative and destructive potential of these technologies (Helliwell & Burton, 2021).

Another "alt meat" developed through stigmatized technology is meat from transgenic (genetically modified) animals. Transgenic animals could offer cheap, healthy, and environmentally friendly sources of protein. Unlike transgenic plants, which have had DNA modified to introduce new traits, no transgenic animals have reached the American dinner table except AquaBounty's AquAdvantage Salmon. This fish is now among the most heavily regulated products in the U.S. Food and Drug Administration's history. The genetically engineered salmon has a growth hormone-regulating gene from Pacific Chinook salmon and a promoter sequence from ocean pout, to enable the GM salmon to grow year-round instead of only during spring and summer. AquAdvantage salmon has a severe negative stigma because of media framing and misinformation spread by special interest groups. To counteract this misinformation, the FDA could adopt a more proactive public outreach role in explaining what transgenic animal products are, how FDA regulates them, and why FDA feels they are safe for human consumption (Ball, 2013).

Dairy and plant mylks

The U.S. dairy industry has significantly reduced its environmental impacts since the late 1990s, primarily through improved dairy cow productivity (Capper & Cady, 2020). However, dairy products and attempts to reduce their negative environmental impacts are increasingly contested, especially when it comes to adjusting food production and consumption. Three commonly proposed trajectories of change—more milk, better milk, and less milk—represent broad but insufficient mitigation efforts (Clay & Yurco, 2020).

Dairy farm systems have intensified to meet growing demands for animal products, but public opposition to this intensification has also grown because of concerns about animal welfare and the use of GM foods and other food technologies. These concerns track with public perceptions about dairy farm practices, including cow − calf separation, the fate of excess dairy calves, pasture access, and removing horn buds from young cattle, known as disbudding (Ly et al., 2021).

Public perceptions of dairy farm practices are limited, especially about how the dairy industry is improving dairy cow productivity. Dairy system efficiency depends on farm management practices involving the metabolism and nutrient requirements of dairy cattle, herd population dynamics, resource inputs, nutrient excretion, and the effects of these factors on GHG emissions. Since 2007, the resources needed to produce 1 million metric tons of energy-corrected milk have been considerably reduced—including the number of cattle and the amount of feedstuffs, land, and water—even while overall milk production has significantly increased. In addition, the dairy industry produces less manure and manure emissions (Capper & Cady, 2020).

Small-scale dairy systems can support food security more than corporate, specialized, and family-owned dairy farms partly because they often use different practices and equipment for management of animals, land forage resources, and feed (Osuna & Barrantes, 2020). Six factors predict regional milk preferences for pasteurized or unpasteurized milk: reliance on media as information source, political philosophy, anger, food safety knowledge, institutional trust, and reliance on the influence of others (Medeiros & Le Jeune, 2016).

Plant-based milk alternatives, also known as "mylks," have surged in popularity over the past decade (Clay & Yurco, 2020). Mylk companies often use positive framing in marketing plant-based milk as wholesome and convenient, while negatively framing dairy milk as environmentally damaging and cruel. These brands present mylk as palatable but disruptive to the status quo. They encourage consumers to reach for "plant-based" milk as a way to cope with environmental catastrophe and a life out of balance. Thus, mylks can promote a flawed, neoliberal ethic that individualizes systemic problems, presents a commercial commodity as a solution, and reinforces the political economy of industrial agriculture (Clay & Yurco, 2020). Vegetarians and vegans tend to express more support for dairy alternatives, including plant-based and yeast-based milk production (Ly et al., 2021).

The Smith and Raven (2012) fit-and-conform and stretch-and-transform niche framework is based on the idea that a protective space can shield, nurture, and empower societal transition to a path-breaking innovation. Applying this concept, plant-based milk diffusion often follows a "fit" pattern in one place and a "stretch" in another. In many cases, cultural meanings of sustainability either galvanize transitions from dairy milk to plant-based milks or erode the positive dairy frames that promoted consumer dairy purchasing in the past (Mylan et al., 2019).

Food waste reduction

Waste occurs when edible food is thrown away. Food waste is a major source of local and global carbon emissions. At both the local and global levels, food waste is a major source of carbon emissions. Effective public communication about food waste is essential for promoting engagement with climate mitigation efforts (Pong, 2021). Although message framing can stimulate awareness and concern about food waste, messages rarely prompt waste reduction behavior change if they focus on environmental or societal impacts (Chen & Desalvo, 2021).

Public buy-in for food waste reduction is a major strategic communication challenge. The IPCC (2022) asserts that an adequate reduction in food waste by 2030 would be cost-effective, but only if the practice gains enough public acceptance and support. When environmental sustainability tools and messages present choices, they can more effectively help end users reduce food waste, avoid overeating, and consume nutritious, sustainable, balanced diets. A sustainable "balanced diet" consists of plant-based foods—coarse grains, legumes, fruits, vegetables, nuts, and seeds—as well as meat, dairy, and other animal-sourced foods produced in resilient, low-emission farms and other sustainable facilities (IPCC, 2022).

The Extended Theory of Planned Behavior is a useful framework for exploring consumer food waste prevention behavior. Social media usage, self-efficacy, and moral norms all shape consumer intentions to reduce food waste, except among consumers with negative attitudes about food waste. In turn, expressed intentions to reduce food waste predict actual behavior (Teoh et al., 2021).

Almost half of the total global food waste comes from households in developed countries. Household food waste reduction campaigns have been characterized by three themes: creativity, esthetics and ethics of food waste, and the interconnections among food, waste, and social media (Närvänen et al., 2018). Most household food waste is produced through food handling practices. A new citizen science method called fridge studies provides knowledge about what drives food waste, by asking consumers to document the amounts and types of food they waste. They also talk to researchers while rummaging through their refrigerator, providing thick descriptions of practices that generate food waste. They select food items, tell their story, and photograph the fridge and its contents (Heidenstrøm & Hebrok, 2021).

Companies, nonprofits, and individuals often construct the food waste issue on social media by explaining it, exhibiting different actions, and appealing for potential solutions (Sutinen & Närvänen, 2022). Food waste startups often use four key frames—salvation, thrift, innovation, and normalization—to help legitimize their operations, communicate their value, raise awareness about the problems they can solve, and create demand for innovative solutions (Närvänen et al., 2018). "Reduce waste" messages are more effective in motivating consumer behavior than "don't waste" or "stop waste" messages. However, emphasizing economic, social, or environmental consequences of food waste does not affect consumer willingness to tackle food waste (Nisa et al., 2022).

People with low mindfulness tend to have lower intention to participate in food waste reduction efforts, especially when the message is verbal (vs written) and when it is prompted before meals versus during meals. People with high mindfulness often have higher intention to participate in food waste reduction, regardless of communication modality and message presentation order. Olavarria-Key et al. (2021) found that even when people are aware of the problem of food waste, only about half act frequently to reduce food waste. Two-thirds said their family did not throw away anything from their last meal, and 82% felt uncomfortable discarding food. Those who use social media frequently are more likely to measure the exact quantity of ingredients prior to food preparation, waste less food overall, eat leftovers, and compost. However, those who used social media less often are more likely than heavy users to take action in other ways, including having fewer leftovers, checking expiry dates, being serious about food wastage, and making plans to minimize it.

Upcycled foods include ingredients otherwise wasted or previously thought inedible. A recent study found that European consumers are more likely to try upcycled foods if they are frugal and concerned about the environment, especially when the upcycled food product categories fit well with their regional food preferences. The main barrier to trying upcycled foods is food neophobia, the persistent fear of new foods (Aschemann-Witzel et al., 2022).

Injunctive, descriptive, and benefit message appeals can effectively encourage consumers to eat edible, surplus, or upcycled foods, including foods never sold to or consumed by targeted customers. Surplus foods include unattractive or off-grade fruits and vegetables, as well as processed food products that are still considered safe to eat after their expiration dates. Descriptive appeals are the most likely to persuade people to consume surplus foods if they already belong to a group or organization that promotes sustainable food consumption. Meanwhile, benefit appeals are most effective for prompting those who aspire to belong to such groups to eat surplus foods. Injunctive appeals, which convey what people believe should be done to solve a problem, are most persuasive when they use negative framing. Descriptive and benefit appeals persuade more people to eat surplus foods when the messages use positive framing (Lin, 2022).

Circular agriculture

Circular agriculture is a new food system design, in which the food chain transitions from a linear production chain to a circular system with minimal unnecessary losses (De Lauwere et al., 2022). Circular agriculture is offering new water conservation strategies for interlinked food supply sectors located in arid regions of the world. These areas, such as the Gulf Cooperation Council region, are threatened by natural scarcity and overuse of resources. Circular agriculture strategies involve restructuring food consumption and production patterns, reusing municipal wastewater, reducing food waste, and assigning monetary value to organic waste (Al-Saidi et al., 2021).

Both intrinsic and extrinsic motivations—as well as subjective knowledge, self-efficacy, and perceived risk or uncertainty—predict farmers' intention to take circular agriculture measures. Farmers who take more circular agriculture measures are motivated more by social and environmental values than economic values. Perceived barriers to adopting these measures include lack of knowledge, environmental challenges, unsuitable legislation, and lack of experience with circular agriculture (De Lauwere et al., 2022).

Circular food systems sustainably "prioritize regenerative production, favor reuse and sharing practices, reduce resource inputs and pollution, and ensure resource recovery for future uses. They close resource loops and pursue cross-sectoral synergies with water and energy systems that contribute to the resilience of a territory" (ICLEI, 2021). For example, an 800,000 egg-producing farm in Singapore is a circular food system that reduces food waste. It invested in waste-to-green energy technology that converts chicken wastes into biogas as fuel to generate electricity, converts wastewater sludge into agricultural fertilizers, and recovers waste heat from turbine exhaust for the drying of chicken feed. This green technology generates clean energy and reduces pollution and use of fossil fuels, while increasing food security by providing safer eggs for consumption (Heng, 2017).

A global transition toward circular food systems requires widespread circular food behaviors, part of a systemic circular economy view that defines consumers as "doers" or "prosumers," pursues long-term sustainability goals supported by technologies, and relies on a high level of engagement among skilled consumers. Circular food behavior research has highlighted the novelty and quick popularity of the movement, European food consumption, behaviors related to protein alternatives, food waste, the importance of communication and consumer education, and upcycled food consumption (Canto et al., 2021).

The intertwined practices, technologies, institutions, and policies of a circular economy require moral responsibilities in everyday food consumption. The transition toward a circular economy requires a change in everyday practices such as household food waste routines such as efficient use of leftover food (Lehtokunnas et al., 2020).

Culturally dominant perspectives in affluent societies often frame charitable food redistribution as a "win—win solution" to confront food poverty and food waste. Widespread public support for this approach sees charitable food redistribution as a consensus frame, in which actors work together around a short-term ethical objective. However, to be effective, food waste or food poverty policies should consider trade-offs between different framings of the problems at stake (Arcuri, 2019).

Many citizen-driven initiatives for food sustainability strive to bring about change, spread their practices, link to social movements, and promote grassroots innovation and green-consumption movements. For example, citizen-driven food waste recovery initiatives have succeeded in mobilizing material resources, displaying and reframing various rationales, and creating collaborative local networks to develop waste-prevention practices. Ideally, these food waste-prevention practices provide rationales to spark new institutional mandates to minimize food waste (Campos & Zapata, 2017).

In light of a natural resource-based view, a company's competitive advantage depends upon its relationship with the natural environment and ability to support sustainable development. A food company's ability to support sustainable development may depend in part on its food waste management capabilities. Within rapidly changing online food supply chains, food companies can have a major impact on minimizing and mitigating food waste (Rodrigues et al., 2021).

More precise numerical information can increase consumer awareness of food waste issues when it is combined with loss message framing, while less precise numerical information is more effective when combined with gain message framing (Khalil et al., 2021). Food waste mitigation messages can be more persuasive if they promote global identity such as "Identification With All Humanity" (IWAH), a concept that predicts many food waste mitigation behavioral intentions. Individuals with a strong IWAH orientation are more likely to have behavioral intentions to engage in food waste mitigation efforts, after reading a globally veresus locally framed message. Messages promoting individual food waste mitigation should communicate the severity of the issue locally and its impact on climate change globally (Pong, 2021).

Consumers often pay more attention to visual than verbal attributes of eco-design food packaging, and the health consciousness they perceive can affect their food waste choices. However, consumer food waste decisions are affected more by communication guidance for food storage than by social functions such as the pollution related to packaging or commercial functions such as food category identification (Zeng et al., 2021). Low levels of consumer knowledge about the environmental impact of food packaging lead to misperceptions and irresponsible behaviors. To counter this lack of awareness, "green marketing" messages about the benefits of sustainable food consumption should be more informative about food packaging impacts and account for existing consumer perceptions and behaviors (Tseng et al., 2021).

Food security

Food insecurity framing

By the year 2050, agricultural production is expected to feed nearly 10 billion inhabitants (Hashem et al., 2021). Food insecurity has become a challenge for both developing and high-income countries. United Nations' Sustainable Development Goal number 2 is to end hunger by 2030 (United Nations, 2015). However, many experts argue that achieving food security in the face of climate change is unachievable, especially in light of the lack of media coverage of agricultural innovation (Kutyauripo et al., 2021). The UN touted its final Millennium Development Goals report as "the most successful anti-poverty movement in history," asserting that global poverty and global hunger have been cut in half since 1990.

Even though the media has widely repeated this good-news narrative, UN's claims about poverty and hunger are misleading and inaccurate. Statistical manipulation made it appear that poverty and hunger trends have been improving, when in fact they have worsened. The Millennium Development Goals also define poverty and hunger in ways that dramatically underestimate the likely scale of these problems. Around 4 billion people remain in poverty today, and around 2 billion remain hungry—more than ever before in history and between two and four times what the UN estimated (Hickel, 2016).

In some regions, climate change has a multiplier effect on food insecurity. For example, the Pacific islands are especially at risk because of their limited land availability, population pressures, and low-lying topography that makes them vulnerable to sea-level rise. While climate change is a pressing concern for those governments and their communities, food insecurity also continues to worsen (Cauchi et al., 2019). Climate change is exacerbating food insecurity, malnutrition, and resistance to diseases among pregnant women, and in some regions, these problems have led to increased morbidity, apathy, lethargy, and reduced working efficiency in this population (Lakshmi & Indumathi, 2016). Ensuring food security will require political will combined with improved logistics of the food supply chain. The surplus of food for some and the malnutrition of others is a basic problem facing mankind. Numerous global partnerships to combat hunger have achieved only short-term, local results. Global food security has not been achieved (Prosekov & Ivanova, 2018).

Many doomsday predictions in media coverage have focused on scarcities of food and other necessities, as well as the specter of climate change. Environmental optimists oppose these predictions, arguing that emerging food scarcities can be countered by human ingenuity, technological progress, and national and international economic and political institutions. They also insist that the long-term trend toward less violence in human affairs is unlikely to be reversed by climate change (Gleditsch, 2021). Similarly, many influential international organizations assume that climate-smart agriculture will significantly reduce food insecurity. For example, the World Bank, UN Food and Agriculture Organization, and Organization for Economic Cooperation and Development envision future food systems that prioritize technological innovation to maximize food output (Lajoie-O'Malley et al., 2020).

Although doomsday predictions stir controversy, understanding worst-case climate change scenarios is essential for effective risk management. Causal pathways to societal collapse are rarely discussed in media coverage, but stories can use visualization technologies to improve public understanding of climate scenarios. For example, one team of researchers visualized climate change pathways to food insecurity and societal collapse, using a causal loop diagram defined at global scale and national granularity. This diagram shows how climate change effects could undermine agricultural systems, disrupt food supply, and lead to economic shocks, sociopolitical instability, starvation, migration, and conflict (Richards et al., 2021).

Newspaper coverage of food insecurity issues related to climate-smart agriculture has focused on many different climate-relevant agri-food practices, including crop production, animal production, fisheries, postharvest management, food safety, value addition, marketing, and administration. However, climate-related coverage of food security rarely mentions food safety, postharvest management, or fishery management (Kutyauripo et al., 2021).

Many newspaper stories about food insecurity and food waste portray organizations and government agencies working in partnerships to reduce food loss. The stories that discuss how consumers can reduce food loss and waste tend to be more detailed than diet change coverage. Stories about diet change efforts, on the other hand, often highlight substantial disagreement including legislative efforts to restrict labeling of alternative meat and dairy products. In providing new evidence of the need to shift diets for sustainability, stories often use false balance by quoting both a lead researcher and an opponent with ties to the livestock industry. Inclusion of "both sides" sourcing is similar to the extensive coverage that has presented the existence of climate change as an open debate (Fry et al., 2022).

Several international associations have shared proposed food insecurity solutions directly with the media and public. For example, in 2010, the Soil and Water Conservation Society and the World Association of Soil and Water Conservation both issued position statements emphasizing the importance of using soil and water conservation practices to mitigate and adapt to climate change. These statements assert that without implementation of these conservation practices, the survival of all species will be in jeopardy because water access and fertile soils are critical resources for meeting the increasing demands for food (Delgado & Li, 2016).

The number of pollinators—including bees, birds, bats, butterflies, moths, flies, beetles, wasps, and small mammals—continues to decrease worldwide. Climate change, pesticides, and loss of habitat have been blamed for this decline, which in turn is harming the food supply. Although no definitive link exists between pesticides and pollinator decline, some agri-food retailers prohibit their suppliers from using certain pesticides. Media coverage of this topic has influenced everyday consumer purchasing decisions. Specifically, consumers who read negative stories that link pesticides to pollinator decline say that they intend to purchase more pollinator-friendly plants in the future, especially if the news originated from researchers at universities. However, they are less influenced by information that originated from the federal government, nursery/greenhouse industry associations, or environmental activists (Campbell & Steele, 2020).

Progress in food security is often limited by vague and conflicting media narratives about justice. When political and global elements of justice are overlooked, stories reflect polarized positions about whether individuals or structures are more responsible for promoting food justice. Food security coverage frequently fails to discuss shared responsibilities and how more stakeholders could be included in decisions that could lead to just distributions of goods and bads in a food system (Moragues-Faus, 2017). Journalists could make food insecurity stories more newsworthy by highlighting science-based reports and expert analysis. Bringing complex and nuanced food issues and facts to public audiences requires consistent advocacy efforts to inform and sensitize media practitioners (Thakurta & Chaturvedi, 2012).

Goffman defines keying as "the set of conventions by which a given activity, one already meaningful in terms of some primary framework, is transformed into something patterned on this activity but seen by the participants to be quite something else" (Goffman, 1974, p. 43). Mooney and Hunt (2009) found that the food security consensus frame in media coverage is often associated with several distinct collective action frames including hunger, community development, and industrialized food system vulnerability to accidents and agriterrorism. Each food security collective action frame and its associated keys reflect distinct interests, such as demands for different applications of science and technology or contested ownership of food security issues. Prognostic framing in media coverage could reduce risks of agricultural accidents and agricultural vulnerabilities to terrorism.

Tweets discussing food insecurity and food prices have predicted actual food insecurity and food price increases (Lukyamuzi et al., 2018). In addition, tweet volume can be used to measure reactions to food shortages over time (Kahn, 2018).

Even in affluent Western countries, where vitamin-rich foods are abundantly available, many people suffer from vitamin deficiencies because of gross misinformation spread by food faddists. Other reasons are poor eating habits and dislike of certain foods. Despite agricultural science innovations and vigorous progress, mankind's food security is increasingly precarious, and modern agriculture is no longer an open-ended producer that can overcome food supply uncertainties.

Population growth and climate change have forced most nations to supplement local food supplies with massive imports of grain (Slater, 2019). When local food consumption skyrocketed during the COVID-19 pandemic, many agri-food companies turned to social media to promote local food. Local food tweets during the pandemic often highlighted perceived proximity, including organic farming, sustainability and supporting the local economy, short distances between farm and table, food production methods that guarantee quality, freshness and taste, and the pleasure, experience, and (re)discovery of their products (Chicoine et al., 2021).

Food governance involves the intersectionality of food access and allocation, the systemic challenges it poses, and the broader political economy around it. Institutions, norms, and power affect access to and allocation of food resources. These factors include media institutions and norms (Azizi, 2020). Food security encompasses food sovereignty, food justice, and food democracy. Deliberative democratic processes—including those facilitated by civic journalism within a community—can balance these goods, respect the interests of all stakeholders, and address power imbalances. To be effective, this deliberative process must account for nutrition, health, equitable distribution, support of livelihoods, environmental sustainability, and social justice (Thompson et al., 2020).

Collective action frames, including those embedded in news coverage about food insecurity, represent actions taken by a group that shares knowledge, resources, and efforts to achieve a common purpose. For example, in the United Kingdom, the government approach to food production and food security is rooted in resilience achieved through sustainable intensification, market liberalization, and risk management. Even though local food systems are mostly sidelined within these "official" framings, collective action frames are open to debate, are evolving entities within a mobile multiorganizational political field. Adopting a more holistic and transformative perspective could broaden governmental definitions of food security beyond the quantity of food available to meet the needs of communities, households, and individuals. If concepts of incompleteness and fragility could be incorporated into public deliberation about the contributions of local food systems to food security, then the "official" government interpretation of food security could become more inclusive and incorporate social justice imperatives (Kirwan & Maye, 2013).

The media often frame East European food systems as undeveloped, traditional, and uninspiring because their global identities are often based on perceptions about their food chains, food initiatives, or diets. However, food systems such as these often have a rich diversity of food practices, the potential to achieve true sustainability, and unique, embedded meanings that go beyond Western food system conventions (Jehlička et al., 2020).

Educating citizens about gathering and harvesting wild foods can improve food security and food sovereignty, promote socioecological resilience, and create communities of belonging. However, foraging is often prohibited by regulations governing public lands. Increased public interest in food forests suggests that policymakers and land managers may reconsider whether foraging should continue to be broadly prohibited. Foraging frames in news media coverage, especially during economic recessions, likely influences land managers' deliberations about it. Prevalent foraging media frames represent it as a self-provisioning practice or as a source of luxury commodities and experiences, and economic uncertainty affects how often these frames appear. Foraging frames in the news also may influence future regulations and other policymaking (Sachdeva et al., 2018).

COVID-19 impacts on agri-food

Food systems are complex and fragile. The COVID-19 pandemic disrupted livestock production and exacerbated many other food supply chain vulnerabilities—as well as existing climate, biological, economic, and policy shocks and stressors. Social distancing and labor shortages significantly impacted the sustainability of agro-livestock farming systems (Hashem et al., 2021).

Eventually, agricultural information and communication technology (ICT) solved many pandemic production problems related to precision farm management, product marketing, and access production inputs. However, the relationships among food suppliers, producers, and consumers ultimately could not maintain a viable food production chain. The agricultural communication technologies used most often during the pandemic were Facebook, WhatsApp, and Internet services. Robotic vehicles and drones were not used frequently (Hashem et al., 2021).

An analysis of international news coverage of the global chicken meat system during the earliest months of the pandemic showed that restrictions on product movements contributed to greater vulnerability than exposure to the virus. Social media played a bigger role than news coverage in amplifying, mitigating, and mediating pandemic impacts on the chicken meat system (Chapot et al., 2021).

Perceived food risks also include fears about food insecurity. Pandemics trigger panic buying of groceries. During the COVID-19 pandemic, community leaders expressed frustration and bewilderment about widespread panic buying. Pandemic-related panic buying often arises when people are told to go into self-isolation as part of containment interventions. Episodes of panic buying typically last 7 − 10 days, initiated by an anxious minority of fearful shoppers who spark fear contagion among other shoppers. Exploitative or selfish overpurchasing is often motivated by "dark" or psychopathic personality traits. Social media dissemination of images and videos of panic buying and empty shelves in stores amplifies panic buying behavior. A snowballing effect arises where fear of scarcity creates real, short-term scarcity, and "don't panic" messages from community leaders are ineffective and counterproductive (Taylor, 2021).

During the pandemic lockdown, consumers who used social media more frequently were also more likely to become neurotic, waste food, and engage in impulse buying (Lahath et al., 2021). Nevertheless, consumer fears about food shortages were founded because globalized food supply chains became increasingly susceptible to food shocks and other systemic disruptions. Natural, social, and economic food shocks in one region can potentially lead to price spikes and food supply changes at the global scale. Food system projections commonly adopt a "business as usual" approach that fails to account for shocks or unpredictable events that can have dramatic consequences, as seen during the COVID-19 pandemic.

Increasing connectivity through social media can heighten volatility and vulnerability to food shocks because it shapes and amplifies consumer attitudes and behavior. Food producers that rely on limited crops or automation technologies throughout a food system also could contribute vulnerabilities (Hamilton et al., 2020). Personal income level is strongly linked to fear of food insecurity, especially in less developed countries. Consumer interest in a particular food such as bread during a crisis can highlight characteristics of a population. For example, people who already relied on cheap, white-flour bakery products in their regular, prepandemic diet increased their consumption of these bakery products during COVID-19 (Ladaru et al., 2021).

COVID-19 has profoundly affected the food trade and dietary and lifestyle-related behaviors around the world. As people search for immune-boosting nutrients/herbs, they replace outdoor activities with sedentary indoor behaviors. An analysis of Google Trends relative search volumes before and during the pandemic, as compared with daily confirmed COVID-19 cases, revealed that the most frequent keywords during the spring lockdown period were related to food security, dietary behaviors, outdoor-related behaviors, and immune-related nutrients, herbs, and foods. Popular food-related keywords included food shortage, food bank, free food, food delivery, restaurants, take-away, food delivery, vitamin C, vitamin A, zinc, immune, vitamin E, garlic, omega-3 fatty acid, vitamin D, and turmeric (Mayasari et al., 2020).

News articles and media organizations used saliency of the COVID-19 pandemic to highlight links between animal agriculture, pandemic risks, and other widespread public health threats. For example, some messages used the saliency of the pandemic to highlight the risk of disease transmission from factory farms or to highlight the threat to worker's health created by factory farms. These pandemic-related stories promoted beliefs about negative consequences of meat consumption, as compared with more traditional narratives that highlight the environmental, personal health, or animal welfare implications of factory-farmed meat consumption. However, the pandemic-related food sustainability narratives did not motivate behavioral intentions to reduce meat consumption or to choose plant-based alternatives. Individuals face numerous perceived barriers to behavior change including cost, taste, and social factors, so messages that highlight personal health benefits of reduced meat consumption and address barriers to behavior change could be more effective at increasing public trust in the message source (Niemiec et al., 2021).

The COVID-19 pandemic provided new opportunities for positive changes in the food system. It made food waste more salient among young people because they anticipated economic recession. As a result, more young people reduced food waste, increased their awareness of food waste ethics, and increased their awareness about the environmental consequences of food waste (Burlea-Schiopoiu et al., 2021). The necessary diversification of food products and marketing channels contributed to the economic success of some small-scale agricultural producers involved in short food supply chains. For example, one in five small-scale farmers in Estonia, Hungary, Portugal, and Romania increased their sales of fruits and vegetables during the pandemic, and farm-gate sales were among their most important marketing channels. The use of digital marketing of produce and home food delivery also increased. These diversification strategies paid off, in terms of marketing channels and product categories. However, the advantages generated by diversification rapidly tapered off during the pandemic, temporarily for food sales and permanently for use of digital marketing channels (Benedek et al., 2021).

The pandemic highlighted the unique role of farmers' markets in providing a local, sustainable food source that is less susceptible to food shocks. Farmers' markets support local agriculture, local culture, relationships, and healthy diets. For example, consumer motivations to attend the Canadian Yellowknife Farmers Market during the COVID-19 pandemic included eating dinner, enjoying the atmosphere, and supporting local businesses. Most patrons attend as couples and spend over half of their time talking to others. Instead of moving online, this farmers' markets implemented a small-scale, physically distanced "shop, don't stop" market that received positive coverage in local media. Technology challenges and loss of the event-like atmosphere were perceived barriers to moving the market online. Although online farmers' markets can support local food by facilitating purchasing and knowledge sharing, they cannot replicate the open-air or social experience. Online farmers' markets may become useful as a complementary strategy to cope with pandemics and other future stressors and to support the resiliency of local food systems (Radcliffe et al., 2021).

New grocery sector policies that swiftly emerged during the COVID-19 pandemic were often communicated through newspaper stories, company websites, and social media posts. New policies addressed employee health and well-being issues, including financial and emotional support, health recommendations and protocols, and new employee guidelines. Other pandemic policies addressed community support, food supply issues, operational measures, and safety measures including sanitation, personal protection, transmission prevention, considerations for high-risk populations, physical distancing, and limiting access. Indicators of policy success included compliance and enforcement, employee teamwork, and support for employees (Riediger et al., 2022).

Food riots

An upsurge in food rioting has drawn broad attention to the role of news media and social media in mobilizing these events. Food prices and food access also affect the mobilization of food riots and other collective violence. Incidents of violent conflict tend to increase during times of high global food prices (Heslin, 2021), as well as perceived insecurities about daily staples. The effects of food and water insecurities are mutually reinforcing, in compelling citizens to take to the streets (Koren et al., 2021). Exploring media interpretations of what drives food riots can illuminate the factors that shape increasing food insecurity around the world.

The news media is more likely to highlight food system governance when covering how nongovernmental stakeholders address climate-related food insecurity during a food price crisis. When global food prices spiked in 2007−2008, Africa experienced more food riots than anywhere else in the world (Heslin, 2021). In two emerging economies, Brazil and South Africa, food governance coverage has focused on how food retailers have adapted their practices to cope with the societal consequences of climate change (Pereira et al., 2013).

During the global food price spike of 2007−2008, news coverage about food insecurity presented Western geopolitical anxieties of the "threatening rise of Asia" and "fast growing" Asian appetites as culprits of the crisis. In the context of market-driven journalism, the media ultimately contributed to a rapid and uncritical circulation of elite-based racial interpretations and neoliberal geopolitical approaches to food security (Gong & Le Billon, 2014). During the concurrent global food crisis of 2007—2012, international media overly simplified food-related protests using a "hungry man is an angry man" frame. In turn, this framing provoked more anger and violence. The frame also dominated public discourse, informed global policymaking, and promoted policy research about food riots. In general, the coverage called for global policy action and political intervention. Different kinds of media content produced different audience effects, depending on whether articles were intended to inform, analyze, or advocate. Although the international media silenced or subdued certain voices, food rioters in the developing world often were treated with more sympathy than rioters in the North might expect and with more sympathy than they received in their own national media (Hossain, 2018).

Media accounts of food riots often highlight food access as the primary concern of food rioters. However, actors can leverage reduced food access in order to mobilize protest around a wide range of grievances, including some issues unrelated to food access. Efforts to address the causes of food-related instability, including media coverage of food riots, will not promote peace if they focus solely on food access.

To effectively promote peace, media coverage must also address the context of food riots, including the motivating grievances and how food access relates to those grievances (Heslin, 2021). For example, international news coverage of 14 African countries where food riots have occurred has portrayed poverty and hunger as factors that link the incidence of food price rises with the occurrence of riots. Meanwhile, the African media framed the food riots as caused by a more complex set of factors, including citizen dissatisfaction and people's ability to mobilize protests (Sneyd et al., 2013).

Local food and alternative food networks

High-profile failures of the mainstream, industrialized, globalized food system have sparked interest in alternative food networks (AFNs), which promote "local food," urban agriculture, community-supported farms, fair trade, and gardens for homes, schools, and communities. Home gardening and other food self-provisioning (FSP) practices are often economically motivated and need-related, even though proponents often frame FSP and urban agriculture practices as ethical sustainability strategies. Even in less affluent countries, FSP is often primarily motivated by the desire to obtain fresh and healthy food less expensively and to engage in a pleasurable activity. Gardening messages that highlight these desires can also promote broader sustainability-compliant behaviors and attitudes (Jehlička et al., 2021).

Despite their diversity, emergent alternative food networks all share a common vision that is rooted in social and environmental justice. Thus, the success of AFNs depends on their ability to reembed food within a broader moral economy (Zerbe, 2010). The news media tends to frame AFNs positively, in light of their democratic participation and social resistance values, while using vertical frames to discuss the corporate food system, in which food is commodified for profit at the expense of health and the planet (Mann, 2019).

Through citizen participation, communities can resist corporate actors and democratize their own foodscapes. Participatory food justice occurs across rural veresus urban and North veresus South divides. It promotes greater indigenous knowledge, gender equity, and food justice. It seeks to mobilize food growers and eaters, to collectively contest the power structures that shape food environments and to focus on community-based climate justice solutions (Mann, 2019).

Food justice research focuses on the role of food policy in society, including social movement activism, development of alternative food practices, analyses of inequalities in conventional and alternative food systems, and historic and present-day inequalities of race, class, and gender. Agro-ecology and land tenure are areas of growing interest in the food justice field (Glennie & Alkon, 2018).

Fair trade is a recent hot topic in the food justice public arena. In recent decades, consumers in the global North have increasingly looked to fair or alternative trading systems to promote ecologically and socially sustainable agricultural production. While fair trade has historically been limited to international commodity networks, U.S. agri-food activists have recently started building a domestic fair-trade movement. Third-party certification and labeling could promote social justice for farm workers in U.S. agricultural production. Fair-trade principles reflect a growing need for small- and mid-sized farmers to garner food price premiums.

Erosion of organic food price premiums, the failure of organic food certification to advance a holistic vision of food sustainability, and the embrace of voluntary food regulations as an alternative to public regulation and collective bargaining have all promoted fair-trade frames that have undermined the movement. These frames have hurt the movement's ability to address social relations of agri-food production, to prioritize the "family-scale" farm and food localism, and to obscure farm worker roles in fair trade (Brown & Getz, 2008).

Since the 1990s, alongside industrialized agriculture and other conventional food-supply systems, AFNs have supported local development, social innovation, and sustainability objectives. One approach within this new framework in Europe is a bottom-up, spatially bound, rural development method known as Liaison Entre Actions pour le Development de l'Economie Rurale (LEADER). Focusing on short food supply chains, LEADER local action groups seek to manage alternative food systems. Official websites of LEADER action groups reflect local action group activities that support short food supply chains. Local action groups serve as hubs of social innovation in local foodscapes, mainly in less developed areas where other nonprofit or for-profit organizations are unable to manage food supply chains (Ruszkai et al., 2021).

Food sovereignty prevails when the same people who produce, distribute, and consume food have the right to define their own local food and agriculture systems rather than corporations. Food sovereignty narratives in the media often focus on the access and control small farms have over food production resources and diversification. These narratives often do not address food-related political anxieties at the national level. Food sovereignty, food self-sufficiency, and food security narratives can also support large-scale food chains—as long as they acknowledge the livelihood and food poverty challenges facing marginalized, rural populations (McCarthy & Obidzinski, 2017).

Civic journalism, which seeks to engage citizens in public deliberation and problem solving, can tap into city food governance systems that create "spaces of deliberation." For example, German nonprofit organization Foodwatch informs consumers about problematic food industry practices through food governance campaigns. It works with a wide variety of key issues and publics that address food governance topics through media and food-related expertise (Schneider et al., 2019).

Many stakeholders from civil society, private groups, and local governments come together to seek solutions to food insecurity and other food sustainability challenges. Conflicts of interest, situations where some win and others lose, and institutional deadlocks all can affect the effectiveness of local food governance and the ways it is framed in public discourse. Different partnerships can mobilize equality, participation, inclusion, and freedom; coproduce knowledge, values, and self-reflection within governance spaces; champion the knowledge, needs, and experiences of those living at the margins; and create different forms of connectivity and autonomy to transform urban foodscapes. Over time, different food partnerships and the urban foodscapes they create can evolve (Moragues-Faus, 2020).

Urban food strategies (UFSs) seek to create more sustainable food systems but do not openly refer to concepts such as food security, food justice, food democracy, and food sovereignty in shaping food policy. Implicit, fragmented, or unspecified social justice objectives in UFSs may hold back public consciousness, debate, and collective action to address food system inequalities, or these objectives can be disregarded in policy budgeting, implementation, and evaluation.

Fraser's (2005) three-dimensional theory of justice involves economic redistribution, cultural recognition, and political representation. In light of this framework, urban food stakeholders could incorporate social justice into their strategies, ambitions, motivations, practices, policy trajectories, food-related resource allocations, values, target groups, and decision rules (Smaal et al., 2020). Some urban, community-based organizations have used public health interventions in poor neighborhoods to explore the social and environmental determinants of local food insecurity. These organizations often characterize urban poverty by racial and gender disparities. The main reason people often choose to participate in public health interventions is to obtain food (Gadhoke et al., 2019).

U.S. newspapers often construct hunger in Africa by highlighting the relationship between U.S. citizens and foreign famine sufferers. The stories often present polarized frames of African hunger, such as presenting the problem as either irrelevant or pertinent to the public sphere, the victims as either worthy of political action or as removed from it, and the readers as either political agents or as impotent (Kogen, 2015). Media coverage of food insecurity can influence individual attitudes, perceived risks, and decisions. These media effects involve sources, transmitters, and receivers of food insecurity information, and media channels can amplify perceived food risks, even when their impact depends on individual predispositions, trust, and public understanding of scientific arguments (Isernia & Marcolin, 2018).

Food security debates in the news media and on social media platforms inform conversations about local food security solutions. Alternative food networks (AFNs) are local-level organizations that work with small-scale producers to increase agricultural sustainability. AFNs seek to strengthen local food security, build a "local food" movement, create more sustainable and just food systems, and develop drivers, initiatives, and policies that support alternatives to the dominant industrialized food system and its detrimental environmental and socioeconomic impacts (Cerrada-Serra et al., 2018). For example, common AFN initiatives include farmers' markets, veggie boxes, food waste reduction, excess food redistribution, local food restaurants, and organic and fair-trade goods. The "local food" movement is built on the ethics of sustainability, social justice, animal welfare, and the esthetic values of local food cultures. AFNs oppose the standardizing pressures of the corporate mainstream, with its "placeless and nameless" global food supply networks (Goodman et al., 2012).

The viability of AFNs is more important than ever because of increasing disruptions in global supply chains and shifting populations in different regions (Ashtab & Campbell, 2021). As the era of "cheap food" draws to a close, and as fears of food security and other climate change impacts intensify, AFNs may play an important role in helping communities transition to more sustainable food systems (Goodman et al., 2012). For example, many European AFNs have linked food security outcomes to specific places, created hybrid processes that can synthesize alternative and conventional food system approaches, and promoted advocacy and collective action within different levels of the community (Cerrada-Serra et al., 2018).

Perceptions and expectations about "good food" communicated through news media and social media are diverse and nuanced. These perceptions can inform conventional and alternative food supply chain distribution strategies and support diverse food producers, their communities, and urban food policy goals. Rural farmers, large-scale food producers, urban stakeholders, and local economies can support and be supported by metropolitan food initiatives. Policies also can support local food procurement and facilitate dialog between urban food policymakers and rural producers to understand and mitigate potential tensions and capitalize on opportunities (Jablonski et al., 2019). In framing food sustainability challenges, solutions, and motivations, news media sometimes quote activists who omit important information or use terms such as "sustainability" and "awareness" to mean different things. This conflicting coverage can impede food system transformation policymaking (Baldy, 2019).

Through their alliances with labor unions, AFNs use confrontational politics to demand greater food justice and economic justice within the conventional agri-food system. For instance, in struggles against Walmart, they have used the "Good Food, Good Jobs" slogan to build alliances between alternative food activists and labor activists to address root causes of food insecurity and food deserts. This campaign is committed to increasing the power and health of food chain workers and the communities they live in, by rejecting unjust trade-offs between food and jobs (Myers & Sbicca, 2015). News coverage sometimes highlights the various tactics that food movement activists use to challenge and subvert the agri-food structures they encounter. Some food movement activists operating in less-robust AFNs knowingly engage in high-risk activism, to actively resist the agricultural status quo. They see their own regional food systems as restrictive to the food movement, so they make tactical choices to maneuver around various constraints that hinder their activism. They also use various framing devices to explain the risks of these tactical workarounds (Raridon et al., 2021).

The media drive food system transformation. AFNs, and the ways they are represented publicly, often focus on the concept of food justice. Food justice addresses the root causes of food inequalities, as part of a transformative food justice agenda. The concept of food justice has shifted within different geographies, practices, and worldviews, within the contexts of its place-based and political characteristics. To date, the food justice agenda encompasses a vision, movement, narratives, frameworks, and potential outcomes (Moragues-Faus, 2018). Activist movements motivated by environmental, animal rights, health, and anticorporate agendas have long had concerns that are now central in the "food politics" of primetime television cooking shows, mobile apps, and social media. News, digital media, food advertising, and food labeling highlight the marketing practices of AFNs, conventional, and alternative food producers—and they highlight the relationships between food industries, media, and the public (Phillipov & Kirkwood, 2018).

When the media conveys moral values about food to the public, sometimes these values become embedded in local food markets to help create social value within communities. This cultural entrepreneurship can help create cross-sector partnerships, legitimize local food procurement by large, established organizations, and enable scaling of local food markets. Cultural entrepreneurship via news media and social media communications also can communicate locavore values and valorize stories and activities that promote sustainable food procurement practices (Hedberg & Lounsbury, 2021).

Social media can widely promote food sustainability issues, offer opportunities for the self-organization of food movements, generate data for new forms of food governance, and form a new force and resource in the governance of agri-food. Social media also brings food system stakeholders together for the governance of food sustainability. While social media platforms can be used to disrupt food governance, they also serve as a resource for powerful players to regain control (Stevens et al., 2016). Media and social media communications about AFNs connect local small-scale food producers with urban end consumers. Intermediary organizations can help AFNs develop models of democratic control and ownership, as well as economic arrangements where created value is shared among shareholders (Rosol & Barbosa, 2021). Urban food activists often promote their agenda on Twitter and to the news media, including discussions about scaling up urban agriculture. This Twitter network is a loose constellation of different communities that post materials about creating a shared, normative picture of urban food. This narrow emphasis in the past may have contributed to a lack of connections among potential allies and a lack of newspaper coverage about urban food issues (Reed & Keech, 2018).

Media coverage of AFNs reflects the politics of building alternatives to the conventional agri-food system. A small segment of the movement engages in confrontation with the conventional system. However, the movement often ignores race and class inequality within the agri-food system, in favor of promoting environmental sustainability and supporting small farmers. As a result, corporate agribusinesses have coopted the movement's consumer-centric and health-centric framings, to legitimize low-wage, big-box retail development in low-income urban communities. In response, the movement does not always acknowledge the unique language and economic development tactics of low-income urban communities (Myers & Sbicca, 2015).

Consumption of local food is increasing, in part because of its improved convenience, self-identity in supporting the environment, and positive perceptions of foods sold by a social media-based local food distribution system. Several factors shape consumer attitudes toward local food and their brand love for local foods. Negative food product perceptions and prices are common reasons that consumers do not select local foods (Kumar et al., 2021). Top factors influencing consumer decisions to buy local farm products are food attributes, personal support for community economic development, and an authentic buy-local experience that allows them to meet the farmers (Ashtab & Campbell, 2021).

Locality comprises not only food but also place, people, and a cultural context that can convey local identity and national and global elements. For example, Finnish customers at a pop-up food tourism event framed locality in terms of immediate surroundings, national ethos, and global discourses of food enthusiasts. Their feedback suggests that local food events can be promoted to locals, nearby residents, and tourists by redesigning the eating environment though music and visuals and by highlighting regional identity (Aaltojärvi et al., 2018). Local green initiatives also appeal to some food tourists. A four-item green equity measure, which evaluates tourist satisfaction and other perceptions of a destination's environmental initiatives, has shown that local environmental initiatives such as recycling, reducing energy consumption, and other greening efforts elevate tourist satisfaction with a destination's food and transportation offerings (Wong et al., 2021).

The effectiveness of a message to persuade tourists to consume healthier and more ecological food from a hotel buffet is influenced by message content, an identifiable message sender, and characteristics of those who received the message. Volgger, Cozzio, and Taplin (2021) found that health-related and localized messages, as well as tourists' attitudes and habits, influence tourist food choices, persuade them to eat in a considerate manner from hotel buffets, and counter hedonistic consumption. These messages are especially influential among tourists with existing food waste reduction attitudes and habits.

Culture, as communicated through news media and social media, plays a significant role in shaping emerging AFNs and other local food initiatives. For example, Slow Food's Terra Madre forum and the Otago Farmers' Market demonstrated the role of culture in the places, practices, and politics of AFNs. Media coverage of these networks has reflected the cultural values that can protect and enhance AFNs (Parkins & Craig, 2009).

Communicating food risks

Food safety framing

Food-borne infections cause numerous illnesses, which heavily affect healthcare systems (Santeramo & Lamonaca, 2021). The term "food safety" often evokes concerns among both consumers and food policy stakeholders. Understanding consumer concerns about food safety problems is crucial to formulating effective health communications and improving food security (Çelik, 2016).

Exposure to news stories can influence individual responses to food safety threats. For example, during Florida's 2018 Red Tide outbreak, exposure to news coverage of the outbreak and perceived consequences predicted both in-state and out-of-state visitor behavior (Cahyanto & Liu-Lastres, 2020). News stories can accurately and thoroughly reveal the causes of and solutions to food safety problems. However, social media use usually has a greater influence than news stories on behavioral intentions to use food safety precautions. The types of media people use to obtain news, attributions of safety responsibility to individuals or the government, risk perceptions, and behavioral intentions have all shaped their food safety concerns about food additives, agrochemical residues, and pesticides in dairy products. Food safety news consumption increases government blaming, which predicts individual risk perceptions, which in turn predicts food safety behavioral intentions (Kim et al., 2019).

Information seeking about food safety issues on Facebook is often linked to individual risk perception, emotion, and social trust (Wu, 2015). Meanwhile, Twitter and Wikipedia often provide information about health and food-related incidents that could develop into a potential food safety crisis (Meyer et al., 2015). Emotional responses to food safety incidents predict food safety risk perception and prevention action more so than awareness of food safety incidents and facts. For example, Chinese use of microblog Weibo has contributed to various cognitive and behavioral responses to food safety concerns, while their access to other online and offline news and information outlets has had no effect (Mou & Lin, 2014).

With increased popularity of organic food production, more information about the risks attached to organic food products has become available. Media coverage helps consumers make sense of this information, interpret it in terms of risks and benefits, and choose whether to buy different organic foods. Story frames also influence risk perception and sense-making about organic foods and encourage dialog about organic foods with others. In one experiment, simulated chats about organic food with an expert, peer, or anonymous partner included a gain, loss, or uncertainty frame. Chatting with partners perceived as experts was associated with lower levels of risk perception about eating organic food while chatting with partners perceived as similar was associated with higher levels of information need, intention to take notice, and search for and share information. The more positive consumers are in discussing organic food, the lower their risk perception and the higher their need for information, and intention to take notice of, search for, and share information after the chat (Hilverda et al., 2017).

When people consume fresh produce contaminated with pathogens such as fungi, viruses, or bacteria, they can get food poisoning that can lead to severe health damage or even death. The quality of crisis communication about these incidents largely depends on how well accurate and relevant information is conveyed to the public. Quick decisions and reactions from public authorities and privately owned companies are important during any food crisis. To earn consumer trust, risk managers must identify and track back contaminated products, withdraw them from the market, and inform stakeholders about potential threats and recent developments. When a food-borne outbreak is first detected, information about the scope is often unknown, and the demand for information is high. Slow or ineffective crisis communication can result in widespread health damage and a major loss of trust in the entire food system. For example, during the outbreak of Shiga Toxin-producing *Escherichia coli* in Germany and neighboring countries in 2011, nearly 4000 people became ill and more than 50 died. When consumers learned about the outbreak, their trust in the safety of fruits and vegetables decreased sharply (Meyer et al., 2015).

Social media has witnessed exponential growth in use and influence in recent years, democratizing the process of communicating food risks (Regan et al., 2016). However, effectively managing food safety crises often depends on stakeholder willingness to use social media effectively. Stakeholders holding frontline positions in managing and communicating about food risks often use social media for one-way risk communication in a food safety crisis. These risk managers disseminate crisis information, in order to protect their organization's reputation, protect consumer health, educate and inform the public about a particular food risk, and prevent confusion and alarmism (Regan et al., 2016).

To craft broad-reaching and engaging communications, food safety organizations also use data analytics to identify popular online food safety queries. These data help them optimize search engines, to ensure that their content is discoverable. Some also leverage online games, mobile apps, and viral social media posts to reach and engage public audiences (Schiro et al., 2020).

Ironically, perceived health risks from pollution exposure can promote safer eating behaviors. For example, the southeastern United States is one of the top five global regions exposed to high mercury stress because of mercury deposition and methylation. The average mono-methyl mercury concentration in fish tissues there exceeds the limits considered safe for human consumption. The region is vulnerable to increased mercury stress in climate change scenarios. Even though many residents there are aware of the health hazards associated with fish consumption, only the women of reproductive age are willing to adopt safe fish consumption habits (Ferreira-Rodríguez et al., 2021).

A food safety message frame in media coverage can influence food purchasing intentions, especially after consumers read headlines and food safety information in news stories. The intensity of the framing effect depends on prior knowledge of a particular food safety message. People with less knowledge about it have greater variation in purchase intentions after they are exposed to different message frames, and they are more likely to panic after reading media reports about a food hazard. Overall, informed consumers have less dramatic responses to food safety news than less informed people (Jin & Han, 2014).

The preparation of animal food products has led to an increase in severe food-borne infections (Mayes, 2016). Various food scandals have pushed the meat sector into the public spotlight in recent years (Cordes et al., 2015). For example, in a 2014 newspaper interview, a researcher from the Norwegian Institute of Public Health said she has never touched chicken with her bare hands. The story sparked a media storm and a dramatic drop in chicken sales. Some consumers questioned the research behind the news, while others compared the food scare's danger with other risks. Even though consumers did contemplate the facts in the food scare article, their preexisting attitudes and emotions affected their behavior more systematically than their reflections did (Veflen et al., 2017).

In the aftermath of a food contamination incident, food risk and food safety are extremely sensitive issues to communicate to public audiences, both directly and through the media. Among German consumers who read a fabricated article about a fictitious meat processing scandal, most felt the meat company should actively respond to negative media attention by apologizing or denying responsibility, to help counteract reputation damage and financial loss (Cordes et al., 2015).

In China, the government typically conveys dairy-related food safety incidents to the public early, while the news media play a complementary role in food safety governance (Zhu et al., 2019). Chinese government agencies that use social media to convey food risks, as well as traditional and face-to-face communications, can more effectively convey information about food safety regulations to public audiences. In China, more educated young urban dwellers who live alone often perceive a higher level of food safety (Han & Liu, 2018).

Food safety incidents often receive extensive media coverage that conveys risk messages designed to minimize harm and prevent unnecessary boycotts. Once a food risk is declared to be eliminated, communication efforts must rebuild trust among consumers. For example, many consumers still believe that foods from Fukushima are radiologically contaminated. This region of Japan has battled stigma since a 2011 nuclear accident. However, a detached, numerical messaging approach promotes significantly higher perceived credibility and message acceptance than a narrative message alone or a narrative message that includes numerical information. Consumers with a more positive stance toward nuclear energy more readily accept the message (Wolf et al., 2020).

Similarly, U.S. consumers changed their food purchasing and food preparation habits after reading print news coverage of the 2010 Iowa egg recall, the largest in the nation. The stories framed the recall as a failure of government oversight and the consequence of poor farm production practices. Proposed responses to the recall coverage included consumer behavior change to minimize the immediate risk of infection and public support for legislative and regulatory food safety reforms to minimize the risk of future outbreaks. The news mainly focused on the U.S. Food Safety Modernization Act and the FDA Egg Rule but devoted little attention to industrial agriculture as a causal factor or the purchasing of "alternative" eggs as a potential response. The coverage conveyed the policy relevance of the recall but failed to fully contextualize the outbreak within the history of previous outbreaks and food safety concerns—nor did it convey the relationship of the outbreak to the current system of industrial agriculture (Laestadius et al., 2012).

Food contamination outbreaks, controls, alerts, and nutritional risks are the most widely covered food safety topics in online information sources. Thematic online sources devote major attention to nutritional topics, while national news sources are more likely to cover food risks, especially during food emergencies. Although social media reader interest is usually higher for nutritional topics and animal welfare issues, traditional online sources still publish a great amount of content related to food risks and safety (Tiozzo et al., 2020). Social media use and online social capital influence social trust and risk perception of a food safety crisis (Mou & Lin, 2017).

Best practices in food risk communication about acute incidents consider the characteristics of the target population, the information content, and the information sources. Chronic risk communications also need to consider audience perceptions and relevant behaviors, to provide concrete and actional recommendations for behavioral change (Frewer et al., 2016).

Food recalls can have substantial, negative impacts on corporate reputation and marketing. Current knowledge, risk perception, perceived channel beliefs, and perceived information-gathering capacity predict whether a person perceives the need for food safety information and the intention to seek information about food recalls (Liao et al., 2018).

Consumers typically underestimate or exaggerate food safety risks, based on what they read about. Public trust in scientists, experts, doctors, and representatives from expert groups plays a significant social role in supporting food safety messages. Food safety risk communication goals include promoting full use of social media and other communication channels for disseminating information to target audiences, a stable attitude toward self-preserving behavior among people, improved interactions among stakeholders about issues related to population health and food products safety, and involving experts with the community (May et al., 2018).

Artificial intelligence/machine learning applications can predict food safety risks that quality assurance teams can use to prevent product recalls, financial loss, and damage to a brand's reputation (Stoitsis et al., 2023). One global organization that promotes food safety is the One Health Platform, created by medical and public health associations. It communicates with public audiences about the complex interconnections among the health of humans, animals, and the environment, and their impact on food safety and food security. The movement incorporates perspectives from disease ecology, conservation, and veterinary and human medicine to reduce disease and poor health, through a focus on systems and the upstream factors that contribute to poor health outcomes. It also monitors the spread of food-borne diseases and contaminants in the environment and in animals, to prevent or reduce the transfer of these diseases to humans and improve food supply sustainability (Neeley, 2017).

Food risk perceptions

Media coverage of food scares can trigger public "outrage," rooted in emotional factors that influence risk perception. Food risks that are considered involuntary, industrial, or unfair are often given more weight than factors that are seen as voluntary, natural, or fair. Although bioterrorism is usually a high-outrage threat, media coverage of bioterrorism generally does not increase concern about the safety of the food supply unless the stories specifically discuss pathogens that threaten food safety (Brinkley et al., 2007). In one high-outrage situation in Italy, media attention triggered a prolonged food scare in 2008. Raw milk was blamed as the cause of several cases of Hemolytic Uremic Syndrome (HUS) due to shiga-toxin-producing strains of *E. coli* that can be lethal in children. The news highlighted an urgent order from Italy's Minister of Health, to boil raw milk before consumption. The stories implied that raw milk is unsafe and blamed farmers for selling dangerous milk. In this inverted policy approach, which tried to reassure citizens while providing protective risk management, the government allowed raw milk sales to continue under rigid conditions. The coverage strongly influenced the policy response and limited deliberation about various policy options (Finardi & Pongolini, 2014).

When news media amplify food safety concerns, especially during a food scandal, the coverage provokes public outrage. However, news reporting about food scandals shapes food risk perceptions on more of a national than regional level. The more media exposure someone has, the more concerned they likely will be about food safety risks.

Where outbreaks of food-related health hazards have become increasingly rampant in developing countries such as China, regulatory authorities can work with the food industry and NGOs to advance food safety education and to help mitigate media framing and spillover effects of food scandals and rumors (Liu & Ma, 2016). Continuing outbreaks of avian influenza in China are linked to the consumption of poultry products. These outbreaks have caused great financial loss to agribusiness, provoked serious fear and distrust, and undermined consumer confidence in poultry products. Media reports, including misleading stories about poultry safety and quality, have greatly influenced consumer intentions to purchase poultry products. Specifically, potential health threats and uncertainty about the origin of poultry products raise perceived risk among consumers, causing them to avoid purchasing poultry products (Wen et al., 2019).

Perceived knowledge about food safety predicts protection motivation among consumers, which prompts individuals to appraise relevant threats and coping strategies. Protection motivation, trust in food safety management, and concern about food recalls all predict protection behavior intention in response to major food recalls (Liao et al., 2020).

Another psychological process, elaborative information processing, is more influential than active reflection in people's learning about food safety from the news. This information processing involves a careful and thoughtful consideration of the merits of the food safety information, which requires a high level of understanding of key arguments. Other factors that predict learning about food safety include attention to local television, gender, education, ethnicity, and perceived safety of the local food supply (Fleming et al., 2006). Cognitive dissonance and confirmatory bias impair consumer ability to adequately process food safety information. Even when they have increased risk perceptions, these biases lead consumers to pay limited attention to signals and make purchase decisions that lean toward their initial choices. Existing habits and psychological biases also make consumers less responsive to public food safety information (Cao et al., 2015).

Selective media exposure to traditional and social media, as well as high levels of news media consumption, influences food risk perceptions more than perceptions about media credibility. Even in cultures where citizens do not avoid uncertainty, people who use media often are more likely to have higher risk perceptions (Niu et al., 2022). Cognitive risk perception, which includes perceived susceptibility and severity, is shaped by exposure to news media and social media communications about food risks. The more people worry about food risks in response to what they read about, the less likely they are to consume foods they perceive as risky.

When consumers pay more attention to relevant food risk news, this has a stronger effect on their perceived risk and worry than mere exposure to news. For example, when people think about food safety risks of processed foods imported from China and seafood from Japan, their cognitive risk perception and affective risk perception (worry) lead them to avoid these foods (Shim & You, 2015). People use different information sources for food safety information, based on their age, gender, education, and race. The use of some information sources predicts greater perceived susceptibility to food-borne illnesses and greater perceived severity of food contamination (Nan et al., 2017).

The intensity and recency of newspaper coverage about food safety issues often predict how well consumers can recall food safety incidents (De Jonge et al., 2010). Many food risk stories convey alarmist messages rather than try to inform and mitigate risks. Comparing newspaper coverage of food risks with RASFF Portal food safety alerts can verify whether stories include adequate food risk information. However, most food risk stories do not mention RASFF data, suggesting gatekeeping bias (Tiozzo et al., 2019).

Individuals have a higher level of perceived food risk after they have been exposed to health claims framed with political appeals that do not match their values (incongruent appeals). For example, when referring to patriotism and traditional values, health claims have been less effective and have resulted in higher risk perceptions among liberals, while references to social justice and sense of community increased risk perceptions among conservatives (Boeuf, 2019).

Another common food risk concern is food allergies and intolerances. Professional organizations around the world are increasingly helping individuals with food allergy risk management by offering support, developing legislation, and enforcing laws. For example, after the 2015 release of new food allergen rules in the United Kingdom, the Daily Telegraph published a letter and article endorsed by 100 chefs criticizing the legislation. The chefs argued that food innovation and creativity were harmed by the requirement to state the presence of 14 allergens in the dishes they cooked. Many concerned consumers used social media to share views about the legislation. Tweets, responses to news articles, and online forum comments illustrated how different stakeholders positioned themselves and others in the online debate and how the debate was framed. The emergent frames emphasized various medical concerns around managing risks associated with food allergy/intolerance, assignment of responsibility, fairness of access, the political nature of the debate, and the financial implications involved (Hamshaw et al., 2017).

The Social Amplification of Risk Framework (SARF) has been used to illuminate how risk signals and societal responses amplify some risks to provoke public concern while attenuating other risks so they are seen as trivial. News media and social media are central in disseminating and framing risk information, sometimes hyping risks and increasing public fear. Media is a key variable in SARF research because the public often perceives risks in social life through relying on, using, and trusting the media (Niu et al., 2022). Peter Sandman's (1993) risk communication framework highlights 15 outrage factors that contribute to the social amplification of risk. For example, in government health institution press releases about a food safety controversy, the most salient outrage factors were controllability, trust, voluntariness, familiarity, and human origin. Perceived catastrophic potential often contributes to an increase in news coverage. Food provider behaviors that the public does not trust can influence journalistic attention to certain food risks (Ju et al., 2015).

No human activity is risk-free, including eating and other essential activities needed for survival. Understanding food risk perceptions is essential for designing clear, transparent, and scientifically sound strategies for communicating food safety risks. Various factors impact risk perceptions within a population, including whether people see the risk as voluntary, known, brings benefits, and whether the risk information is provided by trusted sources (Sandman, 1993). According to SARF, region, cultural aspects, gender, and age can influence food risk perceptions. In most countries, consumer perceptions of food safety risks are high when considering GM crops, pesticides, and food additives. However, public audiences lack confidence in the food industry and governmental authorities to control these risks. As a result, foods considered more natural, such as organic food and "GM-free," are often perceived as less risky and healthier (Rembischevski & Caldas, 2020).

The widening gap between food producers and consumers in the developed world has increased the need for consumer trust in the food supply. When media construct stories about invisible food risks, the stories can amplify risk and create consumer anxiety about the safety of the food system. Food regulators and public health professionals should frame messages that decrease this anxiety (Henderson et al., 2014).

The news media sometimes amplifies food safety risks in public debates. For example, in government press releases and stories about the 2011 *E. coli* outbreak in Germany, the media contributed to ripple effects caused by their escalation of risk. The use of dominant media frames shifted, while the affected stakeholders amplified the risk the most. At the start of the debate, the media framed the infection risk as a public health issue and linked it to medical-scientific progress. However, as the debate developed, the media devoted more attention to the political and economic consequences of the outbreak, and then the discussion was layered with other risk-related events (Raupp, 2014).

Politicization also amplifies perceived food risk in media coverage. For example, in summer 2017, several European Union member states were involved in a food alert about the presence of fipronil pesticide residues in chicken eggs. Italian media amplified the fipronil alert, framing it as a political scandal, as different social actors exploited the case. However, online information sources correctly communicated that consumer risks were low, reporting mainly what Italian health authorities had communicated (Ruzza et al., 2020). Similarly, when the Philippine mainstream media reported about weevil-infested rice for 21 days, this coverage persisted much longer than for most local agriculture-related topics and mainly focused on government inefficiency in handling the situation. A mix of negativity, attachment, cooccurrence with other major contentious issues, and high social relevance provoked the extended media coverage and its social amplification effects (Manalo et al., 2021).

Some food industry strategies try to minimize or conceal certain food risks rather than amplify them. For example, the French association of salt producers tried to counteract public health messages claiming that salt posed a danger to health. The association undermined and intimidated opponents and used indirect communication, concealment, denial, and diversion to manufacture ignorance (Déplaude, 2015).

Social diffusion of information also can amplify perceived food risks, through the birth, death, and distortion of message content. Dread risks involve uncontrollable, fatal, involuntary, or catastrophic outcomes such as food contamination, pandemics, terrorist attacks, and nuclear accidents. Dread risk is susceptible to media amplification and sensationalism because of psychological biases involved in avoiding these risks. The more negative statements a message or story contains, the more widely it is transmitted, especially about high dread topics (Jagiello & Hills, 2018). Perceived benefits and dread have a greater influence on consumer risk perceptions about Chinese processed foods and Japanese imported seafood than media use, trust, and the favorability of a food's country of origin (You & Ju, 2017).

People who avoid food risks often have lower perceived trust in risk messages and perceive higher catastrophic potential. In reading about various food risks—including Benzopyrene in noodle soup, Norovirus in school cafeteria food, misuse of plastic coffee bags, and Salmonella in infant food—most people had higher levels of perceived risk when they saw these risks as dreaded, potentially catastrophic, or framed by untrustworthy sources. Some people who saw the risks as familiar or harmful to children also had heightened risk perception (You et al., 2019).

The more "bad news" (negative information) a person receives, the greater their perceived risk and the more negative messages they convey. Their level of knowledge mitigates this effect. Reexposure to initial information does not reduce bias, which highlights the danger of transmitting risk information through media and other social channels (Jagiello & Hills, 2018).

Past experiences with a particular food category can affect the level of perceived risk. For example, in 2008, a case of intentional food poisoning involving Chinese imported dumplings resulted in mass panic in Japan. Within a context of sensitive foreign relations and Japanese agriculture in decline, the media were key to the enhanced risk perception among the public. In comparing the incident's coverage with a similar event in 2014 involving domestic produce, Walravens (2019) found that the media construction of both incidents led to two different levels of perceived risk, references to former experiences and symbolic connotations, and labeling of the food risks as domestic or imported.

Nanofoods and other biotech foods

Biotechnology has evolved into a precise science that promises to help solve some of the globe's most pressing challenges, including food insecurity and environmental degradation (Ball, 2013). The approaching limits of conventional agricultural production are sparking a serious world effort to develop and commercialize new synthetic foods and synthetic nutrients. These are nonliving products made by chemical reactions or biochemical processes from nonliving raw materials (Slater, 2019). In the case of nanofoods, microscopic particles are incorporated into foods to change how they taste, get absorbed by the body, or stay fresh.

The future of nanofoods largely hinges on public perceptions and consumer willingness to accept this novel technology. The potential health risks of nanofoods, as well as the difficulty in assessing low-dosage nanoparticle risks, have sparked media coverage that emphasizes debate and values-based objections. Some models that predict public attitudes and intention to purchase nanofood products account for consumer attitudes about perceived threats (Cummings et al., 2018). Consumer acceptance of nanofoods is necessary for effective risk regulation governance and achieving the full potential of the nanofood industry. This acceptance is rooted in a complex feedback structure supported by institutions, consumer trust, stakeholder empowerment, shared knowledge, and informed perceptions of human environmental health risks (Yawson & Kuzma, 2010).

European Union regulations have allowed nanofood products to enter the market freely, while consumers have had limited access to information needed to evaluate risks and make informed choices. perceived nanofood benefits will not outweigh the perceived costs unless the media highlights the power struggles and political issues at stake. In the future, direct regulation of nanofoods may include premarket authorization, mandatory labeling, and establishment of a public register of products and producers (Sodano, 2015).

For nanofoods and other food biotechnologies to become embedded within food systems, complex and unpredictable processes of mutual adjustment must occur between the technology and various social contexts. Debates about food biotechnology, which influence its development and diffusion, are often framed in light of existing concerns and interests (Berkhout, 2002). Factors that influence public support for labeling and banning of nanofoods include attitudes toward nanofoods, preference for natural products, science knowledge, and risk perceptions. Attention to food safety media coverage and attention to nanofood news on social media also predict public support for labeling but not banning nanofoods. On the other hand, benefit perceptions predict support for banning but not labeling nanofoods. In addition, underlying individual values, expressed attitudes, and behaviors involving consumer acceptance of nanofoods can serve as potential barriers to nanofood commercialization (Chuah et al., 2018).

Introducing nanofoods has induced worry and concern among many consumers, but social media is helping people find information and share insights and concerns about it.

An abundance of viewpoints about nanofoods has guided individual acceptance and decision-making. Kuttschreuter and Hilverda (2019) found that nanofood-supporting, non-expert comments (social proof) on a fake Facebook image predicted lower risk perception, higher benefit perception, and more positive attitudes toward nanofoods. Surprisingly, initial feelings of dread along with higher levels of risk perception predicted willingness to buy nanofoods, while initial feelings of optimism had no effect.

Genetically modified foods

Genetically modified (GM) foods were widely commercialized nearly three decades ago. Public debate persists about the use of agricultural biotechnology, despite global scientific consensus on their safety and usefulness. Ever since the first food biotech product became commercially available in 1995, GM crops have become the fastest growing agricultural technology, dominating large shares of the global market. The development of GM crops has given rise to public debate about human and ecological safety, culminating in local and global political battles. Monsanto has used many different communication strategies to legitimize GM crops, including concealing details about actors and actions, reflecting trends among experts in global sustainability discourse, and reshaping narratives to promote the company, its products, and biotechnology in general (Lamphere & East, 2017).

In a "big data" analysis of English-language media articles and social media posts, Evanega et al. (2022) found that the volume of traditional media coverage of GM increased significantly between 2018 and 2020. However, the volume of social media posts about GM foods dropped more than 80%, which may indicate a drop in the salience of the GM debate among the wider population. Favorability increased over time in both traditional media and social media content, especially in social media, indicating that both social and traditional media are moving toward a more favorable and less polarized conversation on ag biotech.

A longitudinal study of the GM debate in the Swedish general press and agricultural press found that the debate was most intense in the mid-1990s, after which the frequency of reporting on GMs declined overall and the debate steadily became less negative. Farmer perspectives received more attention than expected, but smallholder farming and food security in the Global South received relatively little coverage (Fischer & Hess, 2022).

Media attention shapes public attitudes and social norms and predicts individual intent to consume nanofoods (Ho et al., 2020). Media frames describing GM foods first emphasized science and business, then safety concerns. Despite the rapid adoption of GM crops, some governments have developed mandatory labeling and other restrictive procedures. Media coverage also may have contributed to greater restrictions on GM foods. For example, after a regulation shift, Russian news coverage of GM foods became more negative until 2016, when they were banned (Bickell, 2019).

Although GM crops have been a feature of food production for over 30 years, the public and many politicians remain concerned about GM crops despite extensive science-based risk assessment. Much of this concern stems from exposure to media coverage and social media discussions. Biosafety legislation and regulatory frameworks identify and regulate GM food risks, to ensure human health and environmental safety. These regulatory frameworks align to international scientific risk assessment methodologies on a case-by-case basis. Regulatory risk assessments of GM crops and foods have all reached the same conclusion: they provide no greater risk to human health or the environment than non-GM crops and foods.

The science about the safety of GM crops and food appears conclusive, and societal benefits have been globally demonstrated. However, the use of innovative GM products has contributed only minimal improvements to global food security. For more than 15 years, politically motivated regulatory barriers to genome editing have prevented possible improvements in food security. Global food insecurity continues to rise because of climate variability, along with biotic and abiotic stresses, lack of access to innovative technologies, and political interference in decision-making processes. Political interference in the regulatory approval process of GM crops is also adversely affecting the adoption of innovative, yield-enhancing crop varieties (Smyth et al., 2021).

Interview sources influence the news framing of GM foods through their power status and embedded interests in issues. U.S. newspaper coverage has framed GM issues in terms of risk delineation, but sources with embedded interests in the production and consumption of GM foods are typically not associated with public interest frames and risk delineation frames (Li & Wu, 2020). The agri-food industry uses green genetic engineering, including GM crop technologies, to create new plant species that are highly resistant to pests and that contain higher levels of nutrients. Monsanto claims that it develops plants that are more valuable and cost-effective, more environmentally friendly, and more productive. However, newspapers have reported the negative aspects of green genetic engineering more frequently than positive aspects. Stories also frequently mention the agri-food industry as a proponent of green genetic engineering (Demke & Höhler, 2019).

British and Canadian GM foods news coverage has not adequately contributed to public understanding of GM food science because it failed to construct food risks and lacked expert knowledge. The coverage also did not explain the complexities of relevant political, social, and economic issues connected to GM foods (Chénier, 2009). Hungarian media has used anti-GM threat framing in its coverage of GM crops. These stories framed GM crops as a food and health risk, focused on risks generally, blamed foreign companies and accused them of being "just out for profit," and used images of corn and deterioration rhetoric. Other media frames focused on wider cultural themes, including environmental concerns (Vicsek, 2014). In contrast to coverage in some Western countries, news reporting about GM crops/foods in high-circulation Hungarian tabloids and political papers have featured GM foods only marginally. The few stories that did appear emphasized an anti-GM threat frame more frequently than in other countries, a minority pro-GM (advancement and benefits) frame, and economic arguments against GM technologies (Vicsek, 2013).

Advertisement framing of health and environmental benefits, individual knowledge, and perceived risk about GM foods all shape public understanding of GM foods. For example, when experts explain health and environmental benefits of GM foods in advertisements, these messages promote knowledge gain, acceptance of and favorable attitudes toward GM foods, and reduced perceived risk (Raza et al., 2021). Public attitudes toward GM foods have a spillover effect on how audiences make decisions about nanofoods (Ho et al., 2020).

A 2016 U.S. National Academies of Sciences, Engineering, and Medicine report summarized the scientific consensus on GM crops and their implications for the public and policymakers. The report release corresponded with reduced negativity toward GM crops in Twitter discussions and increased ambivalence in public risk and benefit perceptions of GM crops. Both shifts in public opinion mirrored the report conclusions. The attitude changes were most common among those least trusting of scientific studies or university scientists (Howell et al., 2018).

Social capital, the relationships among people in a social environment that enable it to function effectively, predicts risk information seeking about GM foods from social media. Risk perception, affective response, and perceived information-gathering ability indirectly encourage this behavior (Wen, 2020). Educated millennials, who use social media extensively, are increasingly relevant in the GM food debate. Their beliefs about the safety, environmental impact, and ethical justifications of GM food consumption are driven by scientific knowledge and trust in information providers (Russo et al., 2020).

The gateway belief model illustrates how a shift in audience perceptions about the scientific consensus on an issue can lead to subsequent changes in attitudes, and these new attitudes in turn predict changes in support for public action. For example, the use of credible expert organizations as sources of scientific consensus on GM food safety reduces public consensus misperceptions. In turn, lower misperceptions boost GM food consumption behaviors (Sander, et al., 2015). The credibility of GM food experts may increase as a result of correcting misperceptions (Bode et al., 2021).

Domestic biotechnology and biosafety regulations often determine the adoption, cultivation, and commercialization of new GM crop varieties. While most developed countries have detailed GM regulations, many less developed countries are still formulating such regulations to enhance agricultural productivity and food security. Media shape political debate about GM regulations, media market, and consumer regulation preferences, and the structure of the media market can affect which social groups become involved with GM regulation. The type and amount of news consumed and the news routines used to produce GM stories influence public perceptions of GM risks and other attitudes about GM foods (Vigani, 2017).

Asynchronous approvals of GM foods occur when there is a time gap between approval to cultivate a biotech crop and approval to import the crop, caused by a time lag in regulatory procedures and processes. In light of political economy theory, the determinants of asynchronous approval of GM foods include regulatory quality, corruption, trade relations with tough markets, size of rural populations, and the number of internet users in a country. These factors all influence asynchronous approval of GM foods across various countries. The higher the share of internet users in a country, the less often the GM food approvals are asynchronous and the less influenced the internet users are by traditional news coverage of biotechnology. The more efficiently a government implements GM regulation, the less time is needed to obtain public approval for new GM crops, and the faster the approval among public audiences (Rosa et al., 2020).

Conclusions

The recent research in this chapter highlights an urgent need for a global shift away from consumer diets dominated by meat, dairy, and eggs to mainly plant-based diets. This shift could play a major role in mitigating climate change, preserving more forest land, reducing antimicrobial resistance and the spread of zoonotic diseases, stabilizing the global food supply chain, and reducing animal suffering and diet-related human health problems. Changing behaviors and systems are grand challenges that require strategic, effective public communication.

For the most part, news media and social media do not devote attention to the welfare of billions of animals that the meat and dairy industries exploit every year. However, public information about this problem could help consumers understand these issues in the context of everyday moral and dietary choices. These audiences also need to understand the benefits of substituting animal product alternatives in their diet, including alternative meats, cultured meats, and plant mylks.

Critical reductions in global hunger rely, in part, on journalists because they can raise awareness of the effects of climate change on food insecurity and the agricultural innovations needed to feed the world. However, this food security news coverage needs to clarify shared responsibilities among all of the stakeholders in shared decision-making. Food governance coverage could illuminate how community leaders share knowledge, resources, and efforts in their efforts to create more just and resilient food systems.

In the absence of informed, transparent food governance, a food system can collapse, leading to widespread hunger and food riots. Food riots are more likely to occur when food prices spike or when consumers perceive instability in the food chain. In response to food-chain instability risks, alternative food networks (AFNs) have emerged to promote "local food," urban agriculture, community-supported farms, fair trade, and food self-provisioning practices such as home gardening. The news media often highlight the democratic participation and social resistance values of AFNs, in light of corporate food systems that commodify food at the expense of human health and the planet. AFN community activities often include farmers' markets, veggie boxes, food waste reduction, excess food redistribution, local food restaurants, and organic and fair-trade goods. Both AFNs and the "local food" movement emphasize the ethics of animal welfare.

When edible food is thrown away, it generates significant methane and CO2 emissions. News media could play a key role in gaining the public buy-in needed to reduce enough food waste to ultimately lower carbon emissions. Social media posts about food waste often use four frames—salvation, thrift, innovation, and normalization—to explain it, show individual actions, or highlight potential solutions.

The widespread consumption of animal food products has led to an increase in severe food-borne infections. Food biotechnology could greatly address global food insecurity and the environmental degradation caused by conventional agricultural production. One emerging solution is commercializing new synthetic foods and nutrients, which can be created through chemical reactions or biochemical processes from nonliving raw materials. For example, "nanofoods" use microscopic particles to change how foods taste, get absorbed by the body, or stay fresh.

The future of nanofoods, genetically modified foods, cultured meat, and other new food technologies largely hinges on health risk perceptions shaped by the media. Media attention to ag biotech shapes public attitudes and social norms and predicts individual intent to consume these products. In recent years, both social and traditional media have reflected a more favorable and less polarized conversation on ag biotech.

News coverage often frames it in light of existing concerns and interests. As a result, limited public understanding of food science and safety risks often supports existing preferences for natural products and banning unconventional foods. For decades, regulatory risk assessments of GM crops and foods have concluded that they provide no greater risk to human health or the environment than non-GM crops and foods. However, even media coverage that emphasizes science-based risk assessments and economic interests has sparked public polarization and greater regulations and other restrictions on the production of these foods.

The global COVID-19 pandemic exposed the impacts of food shocks and highlighted how complex and fragile existing food systems are. Pandemic restrictions contributed to more disruptions in the food supply chain than exposure to the virus. Lockdowns and social media hysteria triggered widespread consumer panic buying, in which fear of scarcity of groceries created real, short-term shortages that caused price spikes and food supply challenges on a global scale.

COVID-19 coverage revealed the risk of food-borne disease transmission from factory meat and dairy farms, as well as the threat to worker health created by factory farms. It also highlighted the unique role of farmers' markets in providing a local, sustainable food source that is less susceptible to food shocks. Ultimately, the pandemic profoundly affected the food trade and dietary and lifestyle-related behaviors around the world. It shone a spotlight on the urgent need for a global shift away from a conventional corporate food production to local food governance—and for consumers to adopt plant-based diets, reduce food waste, and support agri-food technologies.

News coverage of food security frames can promote global climate solutions and highlight innovations in climate-smart agricultural production, food safety risks, hunger, food waste, the impact of diet changes, disappearing pollinators, and the consequences of pesticide use and other industrial practices. Communications about positive agri-food corporate activities can help address the grand challenge of feeding the world. Even though corporate actions cannot fully counteract the social costs and negative impacts they have on the climate, information campaigns that are clear, authentic, and consistent are more likely to promote a significant behavioral shift among consumers. The most effective messages convey charitable goals or another higher purpose and reflect a corporate culture of conscious self-regulation.

Public communication about the evolving global nexus of water security, energy security, and food security can inform innovative solutions and policymaking for addressing supply and demand challenges of food production. Ideally, media coverage and social media framing of the food-water-energy nexus should highlight the complex and contentious trade-offs across food policy silos, to help mobilize citizens to take action. Social media conversations also can explore connections among the role of farmers' markets, irrigation, water quality, food miles, energy consumption, and renewable energy sources. In turn, this knowledge can equip individuals to identify their roles and opportunities within their local food ecosystem.

References

Aaltojärvi, I., Kontukoski, M., & Hopia, A. (2018). Framing the local food experience: A case study of a Finnish pop-up restaurant. *British Food Journal, 120*(1), 133–145.

Aleixo, M. G. B., Sass, C. A. B., Leal, R. M., Dantas, T. M., Pagani, M. M., Pimentel, T. C., Freitas, M. Q., Cruz, A. G., Azeredo, D. R. P., & Esmerino, E. A. (2021). Using Twitter as source of information for dietary market research: A study on veganism and plant-based diets. *International Journal of Food Science and Technology, 56*(1), 61–68. Available from https://doi.org/10.1111/ijfs.14743.

Almiron, N., & Zoppeddu, M. (2015). Eating meat and climate change: The media blind spot—A study of Spanish and Italian press coverage. *Environmental Communication, 9*(3), 307–325. Available from https://doi.org/10.1080/17524032.2014.953968.

Al-Saidi, M., Das, P., & Saadaoui, I. (2021). Circular economy in basic supply: Framing the approach for the water and food sectors of the gulf cooperation council countries. *Sustainable Production and Consumption, 27,* 1273–1285.

Arcuri, S. (2019). Food poverty, food waste and the consensus frame on charitable food redistribution in Italy. *Agriculture and Human Values, 36*(2), 263–275. Available from https://doi.org/10.1007/s10460-019-09918-1.

Aschemann-Witzel, J., Asioli, D., Banovic, M., Perito, M. A., & Peschel, A. O. (2022). Communicating upcycled foods: Frugality framing supports acceptance of sustainable product innovations. *Food Quality and Preference, 100.* Available from https://doi.org/10.1016/j.foodqual.2022.104596.

Ashtab, S., & Campbell, R. (2021). Explanatory analysis of factors influencing the support for sustainable food production and distribution systems: Results from a rural Canadian community. *Sustainability, 13*(9). Available from https://doi.org/10.3390/su13095324.

Astudillo, R. I. M., Astudillo, Y. P. M., Zavala, J. A. M., Jiménez, C. L. M., & Miller, M. X. A. (2021). Corporate social responsibility and pro-environmental behaviour in employees: Evidence in Acapulco, Mexico. *Sustainability, 13*(9), 4597.

Avellán, T., Roidt, M., Emmer, A., Schneider, P., & Raber, W. (2017). Making the water-soil-waste nexus work: Framing the boundaries of resource flows. *Sustainability, 9*(10), 1881.

Azadi, Y., Yazdanpanah, M., & Mahmoudi, H. (2019). Understanding smallholder farmers' adaptation behaviors through climate change beliefs, risk perception, trust, and psychological distance: Evidence from wheat growers in Iran. *Journal of Environmental Management, 250,* 109456.

Azizi, D. (2020). Access and allocation in food governance: A decadal view 2008–2018. *International Environmental Agreements: Politics, Law, and Economics, 20*(2), 323–338. Available from https://doi.org/10.1007/s10784-020-09481-9.

Baldy, J. (2019). Framing a sustainable local food system: How smaller cities in southern Germany are facing a new policy issue. *Sustainability, 11*(6). Available from https://doi.org/10.3390/su11061712.

Ball, K. M. (2013). Meet your meat: The argument for increasing education and public outreach regarding the regulation and safety of animal biotechnology. *Food and Drug Law Journal, 68*(3), 281–307.

Bateman, T., Baumann, S., & Johnston, J. (2019). Meat as benign, meat as risk: Mapping news discourse of an ambiguous issue. *Poetics, 76,* 101356.

Benedek, Z., Fertö, I., Marreiros, C. G., De Aguiar, P. M., Pocol, C. B., Čechura, L., Poder, A., Pääso, P., & Bakucs, Z. (2021). Farm diversification as a potential success factor for small-scale farmers constrained by COVID-related lockdown. Contributions from a survey conducted in four European countries during the first wave of COVID-19. *PLoS One, 16*(5). Available from https://doi.org/10.1371/journal.pone.0251715.

Berkhout, F. (2002). Novel foods, traditional debates: Framing biotechnology sustainability. *New Genetics and Society, 21*(2), 131–148. Available from https://doi.org/10.1080/1463677022000006989.

Bickell, E. (2019). The framing effect of the media in the regulation of GMOs: A case study of Russia Galata. *Russian Journal of Communication, 11*(3), 240–252.

Biermann, G., & Rau, H. (2020). The meaning of meat: (Un)sustainable eating practices at home and out of home. *Appetite, 153.* Available from https://doi.org/10.1016/j.appet.2020.104730.

Bode, L., Vraga, E. K., & Tully, M. (2021). Correcting misperceptions about genetically modified food on social media: Examining the impact of experts, social media heuristics, and the gateway belief model. *Science Communication, 43*(2), 225–251.

Boeuf, B. (2019). Political ideology and health risk perceptions of food. *Social Science and Medicine, 236.* Available from https://doi.org/10.1016/j.socscimed.2019.112405.

Borghini, A., Piras, N., & Serini, B. (2020). Ontological frameworks for food utopias. *Rivista di Estetica*, *75*, 120−142.

Brinkley, M., Lester, J., & Ghiselli, R. (2007). The utility of media frames as a research tool in food service operations. *Journal of Culinary Science and Technology*, *5*(1), 19−31.

Broad, G. M. (2019). Plant-based and cell-based animal product alternatives: An assessment and agenda for food tech justice. *Geoforum*, *107*, 223−226. Available from https://doi.org/10.1016/j.geoforum.2019.06.014.

Broad, G. M. (2020). Making meat, better: The metaphors of plant-based and cell-based meat innovation. *Environmental Communication*, *14*(7), 919−932. Available from https://doi.org/10.1080/17524032.2020.1725085.

Brown, S., & Getz, C. (2008). Towards domestic fair trade? Farm labor, food localism, and the 'family scale' farm. *GeoJournal*, *73*(1), 11−22. Available from https://doi.org/10.1007/s10708-008-9192-2.

Brummans, B. H. J. M., Cheong, P. H., & Hwang, J. M. (2016). Faith-based nongovernmental environmental organizing in action: Veroes' campaigning for vegetarianism and mindful food consumption. *International Journal of Communication*, *10*, 4807−4829.

Bryant, C., & Dillard, C. (2019). The impact of framing on acceptance of cultured meat. *Frontiers in Nutrition*, *6*. Available from https://doi.org/10.3389/fnut.2019.00103.

Buddle, E. A., & Bray, H. J. (2019). How farm animal welfare issues are framed in the Australian media. *Journal of Agricultural and Environmental Ethics*, *32*(3), 357−376. Available from https://doi.org/10.1007/s10806-019-09778-z.

Burlea-Schiopoiu, A., Ogarca, R. F., Barbu, C. M., Craciun, L., Baloi, I. C., & Mihai, L. S. (2021). The impact of COVID-19 pandemic on food waste behaviour of young people. *Journal of Cleaner Production*, *294*. Available from https://doi.org/10.1016/j.jclepro.2021.126333.

Cahyanto, I., & Liu-Lastres, B. (2020). Risk perception, media exposure, and visitor's behavior responses to Florida Red Tide. *Journal of Travel & Tourism Marketing*, *37*(4), 447−459. Available from https://doi.org/10.1080/10548408.2020.1783426.

Campbell, B. L., & Steele, W. (2020). Impact of information type and source on pollinator-friendly plant purchasing. *HortTechnology*, *30*(1), 122−128. Available from https://doi.org/10.21273/HORTTECH04473-19.

Campos, M. J. Z., & Zapata, P. (2017). Infiltrating citizen-driven initiatives for sustainability. *Environmental Politics*, *26*(6), 1055−1078.

Canto, N. R. D., Grunert, K. G., & De Barcellos, M. D. (2021). Circular food behaviors: A literature review. *Sustainability 13 (4)*, *1872* (1−27).

Cao, Y. J., Just, D. R., Turvey, C., & Wansink, B. (2015). Existing food habits and recent choices lead to disregard of food-safety announcements. *Canadian Journal of Agricultural Economics*, *63*(4), 491−511. Available from https://doi.org/10.1111/cjag.12089.

Capper, J. L., & Cady, R. A. (2020). The effects of improved performance in the U.S. dairy cattle industry on environmental impacts between 2007 and 2017. *Journal of Animal Science*, *98*(1), skz291.

Cauchi, J. P., Correa-Velez, I., & Bambrick, H. (2019). Climate change, food security and health in Kiribati: A narrative review of the literature. *Global Health Action*, *12*(1), 1603683.

Çelik, E. D. (2016). FoodWiki: A mobile app examines side effects of food additives via semantic Web. *Journal of Medical Systems*, *40*, 41.

Centner, T. J. (2019). Consumers, meat and animal products: Policies, regulations and marketing. *Consumers, meat and animal products: Policies, regulations and marketing* (pp. 1−232). Taylor and Francis. Available from https://doi.org/10.4324/9780429430572.

Cerrada-Serra, P., Moragues-Faus, A., Zwart, T. A., Ortiz-Miranda, D., & Avermaete, T. (2018). Exploring the contribution of alternative food networks to food security. A comparative analysis. *Food Security*, *10*(6), 1371−1388.

Chapot, L., Whatford, L., Compston, P., Tak, M., Cuevas, S., Garza, M., et al. (2021). A global media analysis of the impact of the COVID-19 pandemic on chicken meat food systems: Key vulnerabilities and opportunities for building resilience. *Sustainability*, *13*(16). Available from https://doi.org/10.3390/su13169435.

Chen, H. S., & Desalvo, D. (2021). The effect of message framing and focus on reducing food waste. *Journal of Quality Assurance in Hospitality and Tourism*, 1−19.

Chénier, L. A. (2009). Food Democracy and the Construction of Risk in the Canadian and UK Media. Master's thesis, University of Toronto.

Chicoine, M., Rodier, F., Durif, F., Schillo, S., & Dubé, L. (2021). Exploring social media data to understand how stakeholders value local food: A Canadian study using Twitter. *Sustainability*, *13*(24). Available from https://doi.org/10.3390/su132413920.

Chuah, A. S. F., Leong, A. D., Cummings, C. L., & Ho, S. S. (2018). Label it or ban it? Public perceptions of nano-food labels and propositions for banning nano-food applications. *Journal of Nanoparticle Research, 20*(2). Available from https://doi.org/10.1007/s11051-018-4126-5.

Clark, M., Macdiarmid, J., Jones, A. D., Ranganathan, J., Herrero, M., & Fanzo, J. (2020). The role of healthy diets in environmentally sustainable food systems. *Food and Nutrition Bulletin, 41*(2_suppl), 31S—58S. Available from https://doi.org/10.1177/0379572120953734.

Clay, N., & Yurco, K. (2020). Political ecology of milk: Contested futures of a lively food. *Geography Compass, 14*(8), e12497.

Cordes, H., Ermann, M., Rühmann, H., & Spiller, A. (2015). Public-oriented communications in the case of a food scandal. *Journal of the Austrian Society of Agricultural Economics, 25,* 97—106.

Cummings, C. L., Chuah, A. S. F., & Ho, S. S. (2018). Protection motivation and communication through nanofood labels: Improving predictive capabilities of attitudes and purchase intentions toward nanofoods. *Science Technology and Human Values, 43*(5), 888—916.

De Jonge, J., Van Trijp, H., Renes, R. J., & Frewer, L. J. (2010). Consumer confidence in the safety of food and newspaper coverage of food safety issues: A longitudinal perspective. *Risk Analysis, 30*(1), 125—142. Available from https://doi.org/10.1111/j.1539-6924.2009.01320.x.

De Lauwere, C., Slegers, M., & Meeusen, M. (2022). The influence of behavioural factors and external conditions on Dutch farmers' decision-making in the transition towards circular agriculture. *Land Use Policy, 120,* 106253.

de Olde, E. M., & Valentinov, V. (2019). The moral complexity of agriculture: A challenge for corporate social responsibility. *Journal of Agricultural and Environmental Ethics, 32*(3), 413—430. Available from https://doi.org/10.1007/s10806-019-09782-3.

Delgado, J. A., & Li, R. (2016). The Nanchang Communication about the potential for implementation of conservation practices for climate change mitigation and adaptation to achieve food security in the 21st century. *International Soil and Water Conservation Research, 4*(2), 148—150. Available from https://doi.org/10.1016/j.iswcr.2016.05.002.

Demke, A., & Höhler, J. (2019). Agenda-setting in the agricultural and food industries of the case of green genetic engineering. *Journal of the Austrian Society of Agricultural Economics, 29*(25), 215—223. Available from https://oega.boku.ac.at/fileadmin/user_upload/Tagung/2019/AJARS29/28_Kasparek_DOI29_26.pdf.

Déplaude, M. O. (2015). Minimising dietary risk: The French association of salt producers and the manufacturing of ignorance. *Health, Risk and Society, 17*(2), 168—183. Available from https://doi.org/10.1080/13698575.2015.1033151.

Dickstein, J., Dutkiewicz, J., Guha-Majumdar, J., & Winter, D. R. (2020). Veganism as left praxis. *Capitalism Nature, Socialism, 33,* 56—75.

Edge, M. S., & Hermann, M. (2021). Bridging the gap between corporate sustainability commitments and consumer action: Nutrition communications help consumers contribute to a more sustainable food value chain. *Nutrition Today, 56*(5), 246—252.

Elias, E. H., Flynn, R., Idowu, O. J., Steele, C., & Sutherland, C. (2019). Crop vulnerability to weather and climate risk: Analysis of interacting systems and adaptation efficacy for sustainable crop production. *Sustainability, 11*(23), 6619.

Evanega, S., Conrow, J., Adams, J., & Lynas, M. (2022). The state of the 'GMO' debate-toward an increasingly favorable and less polarized media conversation on ag-biotech? GM Crops &. *Food, 13*(1), 38—49.

Ferreira-Rodríguez, N., Castro, A. J., Tweedy, B. N., Quintas-Soriano, C., & Vaughn, C. C. (2021). Mercury consumption and human health: Linking pollution and social risk perception in the southeastern United States. *Journal of Environmental Management, 282.* Available from https://doi.org/10.1016/j.jenvman.2020.111528.

Finardi, C., & Pongolini, S. (2014). "Boil before consumption": Lessons to learn from the risk (mis)management case of raw milk in Italy. *Progress in Nutrition, 16*(3), 159—167. Available from http://www.mattioli1885journals.com/index.php/progressinnutrition/index.

Fischer, Klara, & Hess, Sebastian (2022). The Swedish media debate on GMO between 1994 and 2018: What attention was given to farmers' perspectives? *Environmental Communication, 16*(1), 43—62. Available from https://doi.org/10.1080/17524032.2021.1960406.

Fleming, K., Thorson, E., & Zhang, Y. (2006). Going beyond exposure to local news media: An information-processing examination of public perceptions of food safety. *Journal of Health Communication, 11*(8), 789—806.

Fraser, N. (2005). Reframing justice in a globalizing world. *New Left Review, 36,* 69—88.

Frewer, L. J., Fischer, A. R. H., Brennan, M., Bánáti, D., Lion, R., Meertens, R. M., Rowe, G., Siegrist, M., Verbeke, W., & Vereijken, C. M. J. L. (2016). Risk/benefit communication about food: A systematic review of the literature. *Critical Reviews in Food Science and Nutrition*, *56*(10), 1728−1745. Available from https://doi.org/10.1080/10408398.2013.801337.

Friedlander, Judith, & Riedy, Chris (2018). Celebrities, credibility, and complementary frames: Raising the agenda of sustainable and other 'inconvenient' food issues in social media campaigning. *Communication Research and Practice*, *4*(3), 229−245. Available from https://doi.org/10.1080/22041451.2018.1448210.

Fry, J. P., Stodden, B., Brace, A. M., & Laestadius, L. I. (2022). A tale of two urgent food system challenges: Comparative analysis of approaches to reduce high-meat diets and wasted food as covered in U.S. Newspapers. *Sustainability*, *14*(19), 12083.

Gadhoke, P., Pemberton, S., Foudeh, A., & Brenton, B. P. (2019). Informing the design of a food security and public health nutrition pilot intervention for clients of a community-based organization contending with urban poverty. *Journal of Hunger and Environmental Nutrition*, *14*(5), 629−642. Available from https://doi.org/10.1080/19320248.2018.1484314.

Gardezi, N. U. Z., Steel, B. S., & Lavado, A. (2020). The impact of efficacy, values, and knowledge on public preferences concerning food−water−energy policy tradeoffs. *International Journal of Environmental Research and Public Health*, *17*(22), 1−20, 8345.

Gaspar, R., Domingos, S., & Demétrio, P. (2018). Serving science to the public: Deliberations by a sample of older adults upon exposure to a serving size recommendation for meat. *Food Quality and Preference*, *66*, 85−94.

Gleditsch, N. P. (2021). This time is different! Or is it? NeoMalthusians and environmental optimists in the age of climate change. *Journal of Peace Research*, *58*(1), 177−185. Available from https://doi.org/10.1177/0022343320969785.

Glennie, C., & Alkon, A. H. (2018). Food justice: Cultivating the field. *Environmental Research Letters*, *13*(7). Available from https://doi.org/10.1088/1748-9326/aac4b2.

Goffman, E. (1974). *Frame analysis: An essay on the organization of experience*. Harvard University Press.

Gong, Q., & Le Billon, P. (2014). Feeding (on) geopolitical anxieties: Asian appetites, news media framing, and the 2007−2008 food crisis. *Geopolitics*, *19*(2), 291−321. Available from https://doi.org/10.1080/14650045.2014.896789.

Goodman, D., DuPuis, E. M., & Goodman, M. K. (2012). *Alternative food networks: Knowledge, practice, and politics*. Routledge.

Graça, J., Cardoso, S. G., Augusto, F. R., & Nunes, N. C. (2020). Green light for climate-friendly food transitions? Communicating legal innovation increases consumer support for meat curtailment policies. *Environmental Communication*, *14*(8), 1047−1060. Available from https://doi.org/10.1080/17524032.2020.1764996.

Greer, R. A., Hannibal, B., & Portney, K. (2020). The role of communication in managing complex water-energy-food governance systems. *Water*, *12*(4), 1183.

Grummon, A. H., Goodman, D., Jaacks, L. M., Taillie, L. S., Chauvenet, C. A., Salvia, M. G., & Rimm, E. B. (2022). Awareness of and reactions to health and environmental harms of red meat among parents. *Public Health Nutrition*, *25*(4), 893−903.

Gustavsen, G. W. (2020). Motivations for sustainable consumption: The case of vegetables. *International Journal on Food System Dynamics*, *11*(4), 329−339. Available from https://doi.org/10.18461/ijfsd.v11i4.58.

Hamilton, H., Henry, R., Rounsevell, M., Moran, D., Cossar, F., Allen, K., Boden, L., & Alexander, P. (2020). Exploring global food system shocks, scenarios, and outcomes. *Futures*, *123*. Available from https://doi.org/10.1016/j.futures.2020.102601.

Hamm, M. W. (2009). Principles for framing a healthy food system. *Journal of Hunger and Environmental Nutrition*, *4*(3-4), 241−250.

Hamshaw, R. J. T., Barnett, J., & Lucas, J. S. (2017). Framing the debate and taking positions on food allergen legislation: The 100 chefs incident on social media. *Health, Risk and Society*, *19*(3-4), 145−167.

Han, G., & Liu, Y. (2018). Does information pattern affect risk perception of food safety? A national survey in China. *International Journal of Environmental Research and Public Health*, *15*(9), 1935.

Hashem, N. M., Hassanein, E. M., Hocquette, J. F., Gonzalez-Bulnes, A., Ahmed, F. A., Attia, Y. A., & Asiry, K. A. (2021). Agro-livestock farming system sustainability during the covid-19 era: A cross-sectional study on the role of information and communication technologies. *Sustainability (Switzerland)*, *13*(12). Available from https://doi.org/10.3390/su13126521.

Hatfield, J. L., Sauer, T. J., & Cruse, R. M. (2017). Soil: The forgotten piece of the water, food, energy nexus. *Advances in Agronomy*, *143*, 1−46.

Hawkins, I. W., Balsam, A. L., & Goldman, R. (2015). A survey of registered dietitians' concerns and actions regarding climate change in the United States. *Frontiers in Nutrition, 2*, 21.

Hedberg, L. M., & Lounsbury, M. (2021). Not just small potatoes: Cultural entrepreneurship in the moralizing of markets. *Organization Science, 32*(2), 433–454. Available from https://doi.org/10.1287/orsc.2020.1395.

Heidenstrøm, N., & Hebrok, M. (2021). Fridge studies: Rummage through the fridge to understand food waste. *Appetite, 165*.

Helliwell, R., & Burton, R. J. F. (2021). The promised land? Exploring the future visions and narrative silences of cellular agriculture in news and industry media. *Journal of Rural Studies, 84*, 180–191. Available from https://doi.org/10.1016/j.jrurstud.2021.04.002.

Henderson, J., Wilson, A., Meyer, S. B., Coveney, J., Calnan, M., McCullum, D., Lloyd, S., & Ward, P. R. (2014). The role of the media in construction and presentation of food risks. *Health, Risk and Society, 16*, 615–630. Available from https://doi.org/10.1080/13698575.2014.966806.

Heng, D. L. K. (2017). *Bio gas plant green energy from poultry wastes in Singapore, . Energy Procedia* (Vol. 143, pp. 436–441). Elsevier Ltd. Available from https://doi.org/10.1016/j.egypro.2017.12.708.

Heslin, A. (2021). Riots and resources: How food access affects collective violence. *Journal of Peace Research, 58*(2), 199–214. Available from https://doi.org/10.1177/0022343319898227.

Hickel, J. (2016). The true extent of global poverty and hunger: Questioning the good news narrative of the Millennium Development Goals. *Third World Quarterly, 37*(5), 749–767. Available from https://doi.org/10.1080/01436597.2015.1109439.

Hilverda, F., Kuttschreuter, M., & Giebels, E. (2017). Social media mediated interaction with peers, experts and anonymous authors: Conversation partner and message framing effects on risk perception and sense-making of organic food. *Food Quality and Preference, 56*, 107–118.

Hirsch, D., Meyer, C. H., Massen, C., & Terlau, W. (2019). How different consumer groups with distinct basic human values gather, Seek and process information on meat topics: The case of the German animal welfare initiative. *International Journal on Food System Dynamics, 10*(1), 100–113. Available from https://doi.org/10.18461/ijfsd.v10i1.06.

Ho, S. S., Goh, T. J., Chuah, A. S. F., Leung, Y. W., Bekalu, M. A., & Viswanath, K. (2020). Past debates, fresh impact on nano-enabled food: A multigroup comparison of presumed media influence model based on spill-over effects of attitude toward genetically modified food. *Journal of Communication, 70*(4), 598–621. Available from https://doi.org/10.1093/joc/jqaa019.

Hossain, N. (2018). How the international media framed 'food riots' during the global food crises of 2007–12. *Food Security, 10*(3), 677–688. Available from https://doi.org/10.1007/s12571-018-0802-7.

Howell, E. L., Wirz, C. D., Brossard, D., Jamieson, K. H., Scheufele, D. A., Winneg, K. M., & Xenos, M. A. (2018). National academies of sciences, engineering, and medicine report on genetically engineered crops influences public discourse. *Politics and the Life Sciences, 37*(2), 250–261. Available from https://doi.org/10.1017/pls.2018.12.

ICLEI. (2021). *City practitioners handbook: Circular food system*. Bonn, Germany: ICLEI - Local Governments for Sustainability. Available from https://circulars.iclei.org/wp-content/uploads/2021/03/ICLEI-Circulars-City-Practitioners-Handbook-Food.pdf.

IPCC. (2022). *Climate Change 2022: Mitigation of climate change. Working Group III contribution to the Sixth Assessment Report*. Geneva, Switzerland: Intergovernmental Panel on Climate Change Retrieved from. Available from https://www.ipcc.ch/report/ar6/wg3/.

Ingram, J., Mills, J., Dibari, C., Ferrise, R., Ghaley, B. B., Hansen, J. G., & Molnar, A. (2016). Communicating soil carbon science to farmers: Incorporating credibility, salience, and legitimacy. *Journal of Rural Studies, 48*, 115–128.

Ireland, R., Muc, M., Bunn, C., & Boyland, E. (2019). Marketing of unhealthy brands during the 2018 Fédération Internationale de Football Association (FIFA) World Cup UK broadcasts: A frequency analysis. *Journal of Strategic Marketing*, 1–16.

Isernia, P., & Marcolin, A. (2018). *The role of the media in increasing awareness of food security and sustainability. In Encyclopedia of food security and sustainability* (pp. 165–171). Elsevier. Available from https://doi.org/10.1016/B978-0-08-100596-5.22411-6.

Jablonski, B., Carolan, M., Hale, J., Thilmany, M. D., Love, E., et al. (2019). Connecting urban food plans to the countryside: Leveraging Denver's food vision to explore meaningful rural–urban linkages. *Sustainability, 11*(7). Available from https://doi.org/10.3390/su11072022.

Jacobs, S. H. J., Wonneberger, A., & Hellsten, I. (2021). Evaluating social counter-marketing success: Resonance of framing strategies in online food quality debates. Corporate. *Communications*, *26*(1), 221–238.

Jagiello, R. D., & Hills, T. T. (2018). Bad news has wings: Dread risk mediates social amplification in risk communication. *Risk Analysis*, *38*(10), 2193–2207. Available from https://doi.org/10.1111/risa.13117.

Jehlička, P., Ančić, B., Daněk, P., & Domazet, M. (2021). Beyond hardship and joy: Framing home gardening on insights from the European semi-periphery. *Geoforum*, *126*, 150–158. Available from https://doi.org/10.1016/j.geoforum.2021.05.018.

Jehlička, P., Grīviņš, M., Visser, O., & Balázs, B. (2020). Thinking food like an East European: A critical reflection on the framing of food systems. *Journal of Rural Studies*, *76*, 286–295. Available from https://doi.org/10.1016/j.jrurstud.2020.04.015.

Jin, H. J., & Han, D. H. (2014). Interaction between message framing and consumers' prior subjective knowledge regarding food safety issues. *Food Policy*, *44*, 95–102. Available from https://doi.org/10.1016/j.foodpol.2013.10.007.

Ju, Y., Lim, J., Shim, M., & You, M. (2015). Outrage factors in government press releases of food risk and their influence on news media coverage. *Journal of Health Communication*, *20*(8), 879–887. Available from https://doi.org/10.1080/10810730.2015.1018602.

Kahn, L. H. (2018). *Chévere! Text-based twitter patterns from venezuelan food shortages. Proceedings - 13th International Workshop on Semantic and Social Media Adaptation and Personalization, SMAP 2018* (pp. 25–30). Institute of Electrical and Electronics Engineers Inc. Available from https://doi.org/10.1109/SMAP.2018.8501870.

Khalil, M., Septianto, F., Lang, B., & Northey, G. (2021). The interactive effect of numerical precision and message framing in increasing consumer awareness of food waste issues. *Journal of Retailing and Consumer Services*, *60*. Available from https://doi.org/10.1016/j.jretconser.2021.102470.

Kim, H., Jang, S. M., & Noh, G. Y. (2019). Is it good to blame the government for food safety concerns? Attributions of responsibility, new media uses, risk perceptions, and behavioral intentions in South Korea. *Journal of Food Safety*, *39*(1), e12570.

Kirwan, J., & Maye, D. (2013). Food security framings within the UK and the integration of local food systems. *Journal of Rural Studies*, *29*, 91–100. Available from https://doi.org/10.1016/j.jrurstud.2012.03.002.

Kogen, L. (2015). Not up for debate: U.S. news coverage of hunger in Africa. *International Communication Gazette*, *77*(1), 3–23.

Koren, O., Bagozzi, B. E., & Benson, T. S. (2021). Food and water insecurity as causes of social unrest: Evidence from geolocated Twitter data. *Journal of Peace Research*, *58*(1), 67–82.

Kristiansen, S., Painter, J., & Shea, M. (2021). Animal agriculture and climate change in the US and UK elite media: Volume, responsibilities, causes and solutions. *Environmental Communication*, *15*(2), 153–172.

Krpan, D., & Houtsma, N. (2020). To veg or not to veg? The impact of framing on vegetarian food choice. *Journal of Environmental Psychology*, *67*. Available from https://doi.org/10.1016/j.jenvp.2020.101391.

Kumar, S., Talwar, S., Murphy, M., Kaur, P., & Dhir, A. (2021). A behavioural reasoning perspective on the consumption of local food. A study on REKO, a social media-based local food distribution system. *Food Quality and Preference*, *93*. Available from https://doi.org/10.1016/j.foodqual.2021.104264.

Kuttschreuter, M., & Hilverda, F. (2019). Risk and benefit perceptions of human enhancement technologies: The effects of Facebook comments on the acceptance of nanodesigned food. *Human Behavior and Emerging Technologies*, *1*(4), 341–360. Available from https://doi.org/10.1002/hbe2.177.

Kutyauripo, I., Mavodza, N. P., & Gadzirayi, C. T. (2021). Media coverage on food security and climate-smart agriculture: A case study of newspapers in Zimbabwe. *Cogent Food and Agriculture*, *7*(1), 1927561.

Ladaru, G. R., Siminica, M., Diaconeasa, M. C., Ilie, D. M., Dobrota, C. E., & Motofeanu, M. (2021). Influencing factors and social media reflections of bakery products consumption in Romania. *Sustainability (Switzerland)*, *13*(6). Available from https://doi.org/10.3390/su13063411.

Laestadius, L. I., Lagasse, L. P., Smith, K. C., & Neff, R. A. (2012). Print news coverage of the 2010 Iowa egg recall: Addressing bad eggs and poor oversight. *Food Policy*, *37*(6), 751–759. Available from https://doi.org/10.1016/j.foodpol.2012.08.005.

Laestadius, L. I., Neff, R. A., Barry, C. L., & Frattaroli, S. (2014). We don't tell people what to do": An examination of the factors influencing NGO decisions to campaign for reduced meat consumption in light of climate change. *Global Environmental Change*, *29*, 32–40. Available from https://doi.org/10.1016/j.gloenvcha.2014.08.001.

Lahath, A., Omar, N. A., Ali, M. H., Tseng, M. L., & Yazid, Z. (2021). Exploring food waste during the COVID-19 pandemic among Malaysian consumers: The effect of social media, neuroticism, and impulse buying on food waste. *Sustainable Production and Consumption*, 28, 519–531. Available from https://doi.org/10.1016/j.spc.2021.06.008.

Lähteenmäki-Uutela, A. (2014). Legal, ethical, or responsible food. *Journal of International Food and Agribusiness Marketing*, 26(4), 235–257.

Lajoie-O'Malley, A., Bronson, K., van der Burg, S., & Klerkx, L. (2020). The future(s) of digital agriculture and sustainable food systems: An analysis of high-level policy documents. *Ecosystem Services, 45*, 101183.

Lakshmi, B. N., & Indumathi, T. S. (2016). Traditional health foods security as corporate social responsibility for pregnant women health monitoring. *In Environmental science and engineering* (pp. 549–560). Springer Science and Business Media Deutschland GmbH Vol. 0. Available from https://doi.org/10.1007/978-3-319-31014-5_33.

Lamphere, J. A., & East, E. A. (2017). Monsanto's biotechnology politics: Discourses of legitimation. *Environmental Communication, 11*(1), 75–89. Available from https://doi.org/10.1080/17524032.2016.1198823.

Lehtokunnas, T., Mattila, M., Närvänen, E., & Mesiranta, N. (2020). Towards a circular economy in food consumption: Food waste reduction practices as ethical work. *Journal of Consumer Culture*.

Li, X., & Wu, Z. L. X. (2020). Source interests, news frames, and risk delineation: A content analysis of U.S. newspapers' coverage of genetically modified food (1994-2015). *International Journal of Communication, 14*, 3633–3654. Available from https://ijoc.org/index.php/ijoc.

Liao, C., Yu, H., & Zhu, W. (2020). Perceived knowledge, coping efficacy and consumer consumption changes in response to food recall. *Sustainability, 12*(7), 2696.

Liao, C., Zhou, X., & Zhao, D. (2018). An augmented risk information seeking model: Perceived food safety risk related to food recalls. *International Journal of Environmental Research and Public Health, 15*(9), 1800.

Lin, Y. T. (2022). Moderating reference group and message framing influences on sustainable surplus food consumption advertising appeals. *Journal of Marketing Management, 38*(11-12), 1218–1244.

Liu, P., & Ma, L. (2016). Food scandals, media exposure, and citizens' safety concerns: A multilevel analysis across Chinese cities. *Food Policy, 63*, 102–111. Available from https://doi.org/10.1016/j.foodpol.2016.07.005.

Lombardi, G. V., Berni, R., & Rocchi, B. (2017). Environmental friendly food. Choice experiment to assess consumer's attitude toward "climate neutral" milk: The role of communication. *Journal of Cleaner Production, 142*, 257–262.

Lonkila, A., & Kaljonen, M. (2021). Promises of meat and milk alternatives: an integrative literature review on emergent research themes. *Agriculture and Human Values, 38*(3), 625–639. Available from https://doi.org/10.1007/s10460-020-10184-9.

Lukyamuzi, A., Ngubiri, J., & Okori, W. (2018). Tracking food insecurity from tweets using data mining techniques. *Proceedings of the International Conference on Software Engineering, 27*–34.

Ly, L. H., Ryan, E. B., & Weary, D. M. (2021). Public attitudes toward dairy farm practices and technology related to milk production. *PLoS One, 16*(4 April), e0250850.

Mahfuz, S., Mun, H. S., Dilawar, M. A., & Yang, C. (2022). Applications of smart technology as a sustainable strategy in modern swine farming. *Sustainability, 14*(5). Available from https://doi.org/10.3390/su14052607.

Manalo, J. A., Nidoy, M. G. M., & Corpuz, D. C. P. (2021). Knee deep in the Hoopla: Analyzing reportage of the weevil-infested rice issue in the Philippines. *Media Asia, 48*(2), 108–122.

Mann, A. (2019). *Voice and participation in global food politics. Voice and participation in global food politics* (pp. 1–170). Taylor and Francis. Available from https://doi.org/10.4324/9781351068888.

Marín-Murillo, M. F., Armentia-Vizuete, J. I., Marauri-Castillo, I., & del Mar Rodríguez-González, M. (2021). The listeria crisis in meat: Evolution of the dominant frames in elpais.com, lavanguardia.com and abc.es. *Estudios Sobre el Mensaje Periodistico, 27*(1), 333–347. Available from https://doi.org/10.5209/esmp.67128.

May, I. V., Lebedeva-Nesevrya, N. A., & Barg, A. O. (2018). Strategy and tactics for building up efficient risk-communications in the sphere of food products safety. *Health Risk Analysis* (4), 105–113.

Mayasari, N. R., Ho, D. K. N., Lundy, D. J., Skalny, A. V., Tinkov, A. A., Teng, I. C., Wu, M. C., Faradina, A., Mohammed, A. Z. M., Park, J. M., Ngu, Y. J., Aliné, S., Shofia, N. M., & Chang, J. S. (2020). Impacts of the COVID-19 pandemic on food security and diet-related lifestyle behaviors: An analytical study of google trends-based query volumes. *Nutrients, 12*(10), 1–12. Available from https://doi.org/10.3390/nu12103103.

Maye, D., Fellenor, J., Potter, C., Urquhart, J., & Barnett, J. (2021). What's the beef?: Debating meat, matters of concern and the emergence of online issue publics. *Journal of Rural Studies, 84*, 134–146. Available from https://doi.org/10.1016/j.jrurstud.2021.03.008.

Mayes, X. (2016). Livestock and climate change: An analysis of media coverage in the Sydney Morning Herald. *Natural resources management: Concepts, methodologies, tools, and applications* (Vols. 3–2, pp. 1216–1246). IGI Global. Available from https://doi.org/10.4018/978-1-5225-0803-8.ch059.

McCarthy, J. F., & Obidzinski, K. (2017). Framing the food poverty question: Policy choices and livelihood consequences in Indonesia. *Journal of Rural Studies, 54*, 344–354. Available from https://doi.org/10.1016/j.jrurstud.2017.06.004.

Medeiros, L. C., & Le Jeune, J. (2016). Why do consumers drink unpasteurized milk? A preliminary mental model. *Food Protection Trends, 36*(6), 428–442. Available from http://www.foodprotection.org/publications/food-protection-trends/archive/.

Meyer, C. H., Hamer, M., Terlau, W., Raithel, J., & Pongratz, P. (2015). Web data mining and social media analysis for better communication in food safety crises. *International Journal on Food System Dynamics, 6*(3), 129–138. Available from https://doi.org/10.18461/ijfsd.v6i3.631.

Monnier, L., Colette, C., El Azrak, A., Bauduceau, B., Bordier, L., Essekat, N., & Schlienger, J. L. (2020). The two faces of nutritional studies: From "fake" to "real" news. *Medecine Des Maladies Metaboliques, 14*(2), 101–113. Available from https://doi.org/10.1016/j.mmm.2019.11.001.

Mooney, P. H., & Hunt, S. A. (2009). Food security: The elaboration of contested claims to a consensus frame. *Rural Sociology, 74*(4), 469–497. Available from https://doi.org/10.1526/003601109789864053.

Moragues-Faus, A. (2017). Problematising justice definitions in public food security debates: Towards global and participative food justices. *Geoforum, 84*, 95–106.

Moragues-Faus, A. (2018). A critical perspective on the transformative capacity of food justice. *Local Environment, 23*(11), 1094–1097.

Moragues-Faus, A. (2020). Towards a critical governance framework: Unveiling the political and justice dimensions of urban food partnerships. *Geographical Journal, 186*(1), 73–86.

Moreno, J. A., & Almiron, N. (2021). Representation in the Spanish press of the role of animal agriculture in the climate crisis: Between lack of attention and carnism. Estudios Sobre. *el Mensaje Periodistico, 27*(1), 349–364. Available from https://doi.org/10.5209/esmp.73745.

Morris, C. (2018). Taking the politics out of broccoli': Debating (de)meatification in UK National and regional newspaper coverage of the Meat Free Mondays campaign. *Sociologia Ruralis, 58*(2), 433–452.

Morris, C., Kaljonen, M., Aavik, K., Vinnari, M., & White, R. (2021). Priorities for social science and humanities research on the challenges of moving beyond animal-based food systems. *Humanities and Social Sciences Communications, 8*(1), 38.

Morris, M. C. (2021). The Voiceless Animal Cruelty Index and its relationship to per capita purchasing power parity and inequality. *Kotuitui, 6*(2), 384.

Mou, Y., & Lin, C. A. (2014). Communicating food safety via social media: The role of knowledge and emotions on risk perception and prevention. *Science Communication, 36*(5), 593–616. Available from https://doi.org/10.1177/1075547014549480.

Mou, Y., & Lin, C. A. (2017). The impact of online social capital on social trust and risk perception. *Asian Journal of Communication, 27*(6), 563–581. Available from https://doi.org/10.1080/01292986.2017.1371198.

Myers, J. S., & Sbicca, J. (2015). Bridging good food and good jobs: From secession to confrontation within alternative food movement politics. *Geoforum, 61*, 17–26. Available from https://doi.org/10.1016/j.geoforum.2015.02.003.

Mylan, J., Morris, C., Beech, E., & Geels, F. W. (2019). Rage against the regime: Niche-regime interactions in the societal embedding of plant-based milk. *Environmental Innovation and Societal Transitions, 31*, 233–247. Available from https://doi.org/10.1016/j.eist.2018.11.001.

Nan, X., Verrill, L., & Kim, J. (2017). Mapping sources of food safety information for U.S. consumers: Findings from a national survey. *Health Communication, 32*(3), 356–365.

Närvänen, E., Mesiranta, N., Sutinen, U. M., & Mattila, M. (2018). Creativity, aesthetics and ethics of food waste in social media campaigns. *Journal of Cleaner Production, 195*, 102–110. Available from https://doi.org/10.1016/j.jclepro.2018.05.202.

Neeley, S. (2017). Framing food within a health policy system: One health. *The intersection of food and public health: Current policy challenges and solutions* (pp. 215–229). Taylor and Francis. Available from https://doi.org/10.1201/9781315153094.

Newton, P., & Blaustein-Rejto, D. (2021). Social and economic opportunities and challenges of plant-based and cultured meat for rural producers in the U.S. *Frontiers in Sustainable Food Systems, 5*. Available from https://doi.org/10.3389/fsufs.2021.624270.

Ngo, C. C., Poortvliet, P. M., & Klerkx, L. (2022). The persuasiveness of gain vs. loss-framed messages on farmers' perceptions and decisions to climate change: A case study in coastal communities of Vietnam. *Climate Risk Management, 35*, 100409.

Niemiec, R., Jones, M. S., Mertens, A., & Dillard, C. (2021). The effectiveness of COVID-related message framing on public beliefs and behaviors related to plant-based diets. *Appetite, 165*. Available from https://doi.org/10.1016/j.appet.2021.105293.

Nisa, C. F., Bélanger, J. J., & Scumpe, B. M. (2022). Assessing the effectiveness of food waste messaging. *Environmental Science & Policy, 132*, 224–236.

Niu, C., Jiang, Z., Liu, H., Song, X., & Li, Z. (2022). The influence of media consumption on public risk perception: A meta-analysis. *Journal of Risk Research, 25*, 21–47.

O'Neill, K. (2019). From inhumane to enticing: Reimagining scandalous meat. *British Food Journal, 121*(12), 3135–3150. Available from https://doi.org/10.1108/bfj-10-2018-0708.

Olausson, U. (2018). Stop blaming the cows!": How livestock production is legitimized in everyday discourse on Facebook. *Environmental Communication, 12*(1), 28–43. Available from https://doi.org/10.1080/17524032.2017.1406385.

Olavarria-Key, N., Ding, A., Legendre, T. S., & Min, J. (2021). Communication of food waste messages: The effects of communication modality, presentation order, and mindfulness on food waste reduction intention. *International Journal of Hospitality Management, 96*, 102962.

Osuna, A. N., & Barrantes, D. L. M. (2020). Small-scale milk production systems in Colombia: A regional analysis of a potential strategy for providing food security. *Future of Food: Journal on Food, Agriculture and Society, 8*(4), 1–10. Available from https://doi.org/10.17170/kobra-202010131948.

Özkaya, F. T., Durak, M. G., Doğan, O., Bulut, Z. A., & Haas, R. (2021). Sustainable consumption of food: Framing the concept through Turkish expert opinions. *Sustainability, 13*(7), 3946.

Parkins, W., & Craig, G. (2009). Culture and the politics of alternative food networks. *Food, Culture & Society, 12*(1), 77–103.

Parlee, B., Ahkimnachie, K., Cunningham, H., Jordan, M., & Goddard, E. (2021). "It's important to know about this": Risk communication and the impacts of chronic wasting disease on indigenous food systems in Western Canada. *Environmental Science and Policy, 123*, 190–201.

Pereira, L., Da Fontoura, Y. S. D. R., & Da Fontoura, C. F. V. T. (2013). Strategic CSR shifts towards adaptive food governance under environmental change: A comparison between South African and Brazilian retailers. *Revista de Gestao Social e Ambiental, 7*(1), 100–112.

Phillipov, M., & Kirkwood, K. (2018). *Alternative food politics: From the margins to the mainstream* (pp. 1–261). Routledge.

Ploll, U., Petritz, H., & Stern, T. (2020). A social innovation perspective on dietary transitions: Diffusion of vegetarianism and veganism in Austria. *Environmental Innovation and Societal Transitions, 36*, 164–176.

Pong, V. (2021). Global versus local framing of the issue of food waste: The role of Identification With All Humanity and the implications for climate change communication. *Asian Journal of Social Psychology, 24*(2), 221–231. Available from https://doi.org/10.1111/ajsp.12453.

Prosekov, A. Y., & Ivanova, S. A. (2018). Food security: The challenge of the present. *Geoforum, 91*, 73–77. Available from https://doi.org/10.1016/j.geoforum.2018.02.030.

Radcliffe, J., Skinner, K., Spring, A., Picard, L., Benoit, F., & Dodd, W. (2021). Virtual barriers: Unpacking the sustainability implications of online food spaces and the Yellowknife Farmers Market's response to COVID-19. *Nutrition Journal, 20*(1). Available from https://doi.org/10.1186/s12937-021-00664-x.

Raridon, A., Mix, T. L., & Einwohner, R. L. (2021). "Workarounds and roadblocks": Risk and resistance among food movement activists. *Social Currents, 8*(2), 182–198.

Raupp, J. (2014). Social agents and news media as risk amplifiers: A case study on the public debate about the *E. coli* outbreak in Germany 2011. *Health, Risk and Society, 16*(6), 565–579. Available from https://doi.org/10.1080/13698575.2014.950203.

Raza, S. H., Zaman, U., Iftikhar, M., & Shafique, O. (2021). Experimental evidence on eco-friendly advertisement appeals and intention to use bio-nanomaterial plastics: Institutional collectivism and performance orientation as moderators. *International Journal of Environmental Research and Public Health*, *18*(2), 1−16. Available from https://doi.org/10.3390/ijerph18020791.

Reed, M., & Keech, D. (2018). The 'Hungry Gap': Twitter, local press reporting and urban agriculture activism. *Renewable Agriculture and Food Systems*, *33*(6), 558−568. Available from https://doi.org/10.1017/s1742170517000448.

Regan, A., Raats, M., Shan, L. C., Wall, P. G., & McConnon, A. (2016). Risk communication and social media during food safety crises: A study of stakeholders opinions in Ireland. *Journal of Risk Research*, *19*(1), 119−133.

Regan, Á., Shan, L., McConnon, Á., Wall, P., & Barnett, J. (2014). Strategies for dismissing dietary risks: Insights from user-generated comments online. *Health, Risk and Society*, *16*(4), 308−322.

Rembischevski, P., & Caldas, E. D. (2020). Risk perception related to food. *Food Science and Technology*, *40*(4), 779−785. Available from https://doi.org/10.1590/fst.28219.

Ricci, E. C., & Banterle, A. (2020). Do major climate change-related public events have an impact on consumer choices? *Renewable and Sustainable Energy Reviews*, *126*. Available from https://doi.org/10.1016/j.rser.2020.109793.

Rice, M., Hemsworth, L. M., Hemsworth, P. H., & Coleman, G. J. (2020). The impact of a negative media event on public attitudes towards animal welfare in the red meat industry. *Animals*, *10*(4). Available from https://doi.org/10.3390/ani10040619.

Richards, C. E., Lupton, R. C., & Allwood, J. M. (2021). Re-framing the threat of global warming: An empirical causal loop diagram of climate change, food insecurity and societal collapse. *Climatic Change*, *164*(3-4), 49.

Riediger, N. D., Slater, J. J., Mann, K., Pilli, B., Derksen, H., Perchotte, C., & Penner, A. L. (2022). Policy responses to the COVID-19 pandemic in the Manitoba grocery sector: A qualitative analysis of media, organizational communications, and key informant interviews. *BMC Public Health*, *22*(1). Available from https://doi.org/10.1186/s12889-022-13654-3.

Rodrigues, V. S., Demir, E., Wang, X., & Sarkis, J. (2021). Measurement, mitigation, and prevention of food waste in supply chains: An online shopping perspective. *Industrial Marketing Management*, *93*, 545−562. Available from https://doi.org/10.1016/j.indmarman.2020.09.020.

Rosa, M. B., De Faria, R. N., & De Castro, E. R. (2020). Political and economic determinants of asynchronous approval of new GM events. *World Trade Review*, *19*(1), 75−90.

Rosol, M., & Barbosa, R. (2021). Moving beyond direct marketing with new mediated models: Evolution of or departure from alternative food networks? *Agriculture and Human Values*, *38*(4), 1021−1039. Available from https://doi.org/10.1007/s10460-021-10210-4.

Runge, K. K., Chung, J. H., Su, L. Y. F., Brossard, D., & Scheufele, D. A. (2018). Pink slimed: Media framing of novel food technologies and risk related to ground beef and processed foods in the U.S. *Meat Science*, *143*, 242−251. Available from https://doi.org/10.1016/j.meatsci.2018.04.013.

Russo, C., Simeone, M., & Perito, M. A. (2020). Educated millennials and credence attributes of food products with genetically modified organisms: Knowledge, trust and social media. *Sustainability*, *12*(20), 8534, 1-20.

Rust, N. A., Jarvis, R. M., Reed, M. S., & Cooper, J. (2021). Framing of sustainable agricultural practices by the farming press and its effect on adoption. *Agriculture and Human Values*, *38*, 753−765.

Ruszkai, C., Tari, I. P., & Patkós, C. (2021). Possible actors in local foodscapes? Leader action groups as short supply chain agents—a European perspective. *Sustainability*, *13*(4), 1−21, 2080.

Ruzza, M., Tiozzo, B., Rizzoli, V., Giaretta, M., D'Este, L., & Ravarotto, L. (2020). Food risks on the web: Analysis of the 2017 Fipronil Alert in the Italian Online Information Sources. *Risk Analysis*, *40*(10), 2071−2092. Available from https://doi.org/10.1111/risa.13533.

Ryynänen, T., & Toivanen, A. (2023). Hocus-pocus tricks and moral progressions: The emerging meanings of cultured meat in online news comments. *Food, Culture & Society*, *26*(3), 591−620.

Sachdeva, S., Emery, M. R., & Hurley, P. T. (2018). Depiction of wild food foraging practices in the media: Impact of the Great Recession. *Society and Natural Resources*, *31*(8), 977−993.

Salmon, G. R., MacLeod, M., Claxton, J. R., Ciamarra, U. P., Robinson, T., Duncan, A., & Peters, A. R. (2020). Exploring the landscape of livestock 'Facts. *Global Food Security*, *25*. Available from https://doi.org/10.1016/j.gfs.2019.100329.

Samoggia, A., Riedel, B., & Ruggeri, A. (2020). Social media exploration for understanding food product attributes perception: The case of coffee and health with Twitter data. *British Food Journal*, *122*(12), 3815−3835.

Sander, V. D. L., Leiserowitz, A., Feinberg, G., & Maibach, E. (2015). The scientific consensus on human-caused climate change as a gateway belief: Experimental evidence. *PLOS One, 10*(2), e0118489.

Sandman, P. (1993). Responding to community outrage: Strategies for effective risk communication. *American Industrial Hygiene Association.*

Santeramo, F. G., & Lamonaca, E. (2021). Objective risk and subjective risk: The role of information in food supply chains. *Food Research International, 139.* Available from https://doi.org/10.1016/j.foodres.2020.109962.

Sato, H., & Webster, A. (2022). Mixed effects of mass media reports on the social amplification of risk: Frequencies and frames of the BSE reports in newspaper media in the UK. *Journal of Risk Research, 25*, 48−66.

Schiro, J. L., Shan, L. C., Tatlow-Golden, M., Li, C., & Wall, P. (2020). Healthy: Smart digital food safety and nutrition communication strategies—A critical commentary. *NPJ Science of Food, 4*(1). Available from https://doi.org/10.1038/s41538-020-00074-z.

Schneider, T., Eli, K., McLennan, A., Dolan, C., Lezaun, J., & Ulijaszek, S. (2019). Governance by campaign: The co-constitution of food issues, publics and expertise through new information and communication technologies. *Information Communication and Society, 22*(2), 172−192. Available from https://doi.org/10.1080/1369118X.2017.1363264.

Shim, M., & You, M. (2015). Cognitive and affective risk perceptions toward food safety outbreaks: Mediating the relation between news use and food consumption intention. *Asian Journal of Communication, 25*(1), 48−64. Available from https://doi.org/10.1080/01292986.2014.989242.

Siegrist, M., & Hartmann, C. (2020). Consumer acceptance of novel food technologies. *Nature Food, 1*(6), 343−350. Available from https://doi.org/10.1038/s43016-020-0094-x.

Sievert, K., Lawrence, M., Parker, C., & Baker, P. (2022). What's really at 'steak'? Understanding the global politics of red and processed meat reduction: A framing analysis of stakeholder interviews. *Environmental Science & Policy, 137*, 12−21. Available from https://doi.org/10.1016/j.envsci.2022.08.007.

Simeone, M., & Scarpato, D. (2020). Sustainable consumption: How does social media affect food choices? *Journal of Cleaner Production, 277*, 124036.

Slater, L. E. (2019). Synthetic foods: Eliminating the climate factor. *Climate's impact on food supplies: Strategies and technologies for climate-defensive food production* (pp. 207−243). Taylor and Francis. Available from https://doi.org/10.4324/9780429048630-13.

Smaal, S. A. L., Dessein, J., Wind, B. J., & Rogge, E. (2020). Social justice-oriented narratives in European urban food strategies: Bringing forward redistribution, recognition and representation. *Agriculture and human values.* Springer.

Small, V., & Warn, J. (2020). Impacts on food policy from traditional and social media framing of moral outrage and cultural stereotypes. *Agriculture and Human Values, 37*(2), 295−309. Available from https://doi.org/10.1007/s10460-019-09983-6.

Smith, A., & Raven, R. (2012). What is protective space? Reconsidering niches in transitions to sustainability. *Research Policy, 41*(6), 1025−1036.

Smyth, S. J., McHughen, A., Entine, J., Kershen, D., Ramage, C., & Parrott, W. (2021). Removing politics from innovations that improve food security. *Transgenic Research, 30*(5), 601−612. Available from https://doi.org/10.1007/s11248-021-00261-y.

Sneyd, L. Q., Legwegoh, A., & Fraser, E. D. (2013). Food riots: Media perspectives on the causes of food protest in Africa. *Food Security, 5*(4), 485−497.

Sodano, V., & Gorgitano, M. T. (2017). Assessing reporting practices in the food sector. *Quality-Access to Success, 18*, 419.

Sodano, V., & Hingley, M. (2018). Corporate social responsibility reporting: The case of the agri-food sector. *Economia Agro-Alimentare, 20*(1), 93−119.

Sodano, V. (2015). Regulating food nanotechnologies: Ethical and political challenges. *Know your food* (pp. 36−41). Wageningen Academic Publishers. Available from https://doi.org/10.3920/978-90-8686-813-1_4.

Sommer, F., Klink, J., Senkl, D., & Hartmann, M. (2015). Determinants of web-based CSR disclosure in the food industry. *Journal on Food System Dynamics, 6*(1).

Stoitsis, G., Papakonstantinou, M., Karvounis, M., & Manouselis, N. (2023). The role of big data and artificial intelligence in food risk assessment and prediction. In M. E. Knowles, et al. (Eds.), *Present Knowledge in Food Safety: A Risk-Based Approach through the Food Chain* (pp. 1032−1044). Cambridge, MA: Elsevier Academic Press.

Stevens, T. M., Aarts, N., Termeer, C. J. A. M., & Dewulf, A. (2016). Social media as a new playing field for the governance of agro-food sustainability. *Current Opinion in Environmental Sustainability, 18*, 99–106. Available from https://doi.org/10.1016/j.cosust.2015.11.010.

Sutinen, U. M., & Närvänen, E. (2022). Constructing the food waste issue on social media: A discursive social marketing approach. *Journal of Marketing Management, 38*(3-4), 219–247.

Taylor, S. (2021). Understanding and managing pandemic-related panic buying. *Journal of Anxiety Disorders, 78*, 102364.

Teoh, C. W., Koay, K. Y., & Chai, P. S. (2021). The role of social media in food waste prevention behaviour. *British Food Journal, 124*(5), 1680–1696.

Thakurta, P. G., & Chaturvedi, S. (2012). Food and nutrition justice: How to make it more newsworthy? *IDS Bulletin, 43*(1), 58–64. Available from https://doi.org/10.1111/j.1759-5436.2012.00347.x.

Thompson, M. S., Cochrane, A., & Hopma, J. (2020). Democratising food: The case for a deliberative approach. *Review of International Studies, 46*(4), 435–455.

Tiozzo, B., Pinto, A., Neresini, F., Ruzza, M., & Ravarotto, L. (2019). Food risk communication: Analysis of the media coverage of food risk on Italian online daily newspapers. *Quality and Quantity, 53*(6), 2843–2866.

Tiozzo, B., Ruzza, M., Rizzoli, V., Giaretta, M., & Ravarotto, L. (2020). Biological, chemical, and nutritional food risks and food safety issues from Italian online information sources: Web monitoring, content analysis, and data visualization. *Journal of Medical Internet Research, 22*(12), e23438.

Tjärnemo, H., & Södahl, L. (2015). Swedish food retailers promoting climate smarter food choices—Trapped between visions and reality? *Journal of Retailing and Consumer Services, 24*(C), 130–139. Available from https://doi.org/10.1016/j.jretconser.2014.12.007.

Tourangeau, W., Sherren, K., Kent, C., & Macdonald, B. H. (2019). Of climate and weather: Examining Canadian farm and livestock organization discourses from 2010 to 2015. *Weather, Climate, and Society, 11*(1), 95–111.

Tseng, M. L., Lin, C. W. R., Sujanto, R. Y., Lim, M. K., & Bui, T. D. (2021). Assessing sustainable consumption in packaged food in indonesia: Corporate communication drives consumer perception and behavior. *Sustainability (Switzerland), 13*(14). Available from https://doi.org/10.3390/su13148021.

United Nations. (2015). Resolution adopted by the General Assembly, Transforming Our World: The 2030 Agenda for Sustainable Development. UN Department of Economic and Social Affairs Division for Sustainable Development Goals. Available at: https://www.un.org/ga/search/view_doc.asp?symbol = A/RES/70/1&Lang = E.

U.S. Centers for Disease Control. (2022). *Chronic Wasting Disease*. Atlanta, GA: CDC. Available from https://www.cdc.gov/prions/cwd/index.html.

Van Eenennaam, A. L., & Werth, S. J. (2021). Animal board invited review: Animal agriculture and alternative meats − learning from past science communication failures. *Animal, 15*(10). Available from https://doi.org/10.1016/j.animal.2021.100360.

Veflen, N., Storstad, O., Samuelsen, B., Langsrud, S., Hagtvedt, T., Ueland, Ø., Gregersen, F., & Scholderer, J. (2017). Food scares: Reflections and reactions. *International Journal on Food System Dynamics, 8*(2), 155–164. Available from https://doi.org/10.18461/ijfsd.v8i2.826.

Veldung, S., Kowalczyk, P., & Otto, K. (2022). Holistic dialogical corporate communications in the food retailing industry: The importance of conscious communication in social networks. *Journal of Media Ethics, 37*(1), 53–71.

Vicsek, L. (2013). Gene-fouled or gene-improved? Media framing of GM crops and food in Hungary. *New Genetics and Society, 32*(1), 54–77.

Vicsek, L. (2014). GM crops in hungary: Comparing mass media framing and public understanding of technoscientific controversy. *Science as Culture, 23*(3), 344–368.

Vigani, M. (2017). The role of mass media and lobbies in the formulation of GMO regulations. *In Genetically modified organisms in developing countries: Risk analysis and governance* (pp. 200–212). Cambridge University Press. Available from https://doi.org/10.1017/9781316585269.018.

Volgger, M., Cozzio, C., & Taplin, R. (2021). What drives persuasion to choose healthy and ecological food at hotel buffets: Message, receiver or sender? *Asia Pacific Journal of Marketing and Logistics, 34*(5), 865–886.

Walravens, T. (2019). Recalibrating risk through media: Two cases of intentional food poisoning in Japan. *Food and Foodways, 27*(1–2), 74–97. Available from https://doi.org/10.1080/07409710.2019.1568852.

Wayan Budiasa, I. (2020). Green financing for supporting sustainable agriculture in Indonesia. *IOP Conference Series: Earth and Environmental Science, 518*(1), 012042.

Weber, E. (1997). Perception and expectation of climate change: Precondition for economic and technological adaptation. In M. H. Bazerman, D. M. Messick, A. Tensbrunsel, & K. Wade- Benzoni (Eds.), *Psychological Perspectives to Environmental and Ethical Issues in Management* (pp. 314–341). San Francisco: Jossey-Bass.

Weber, E. (2006). Experience-based and description-based perceptions of long-term risk: Why global warming does not scare us (yet). *Climatic Change, 77*(1-2), 103–120.

Wen, N. (2020). Understanding the Chinese public's risk perception and information-seeking behavior regarding genetically modified foods: The role of social media social capital. *Journal of Risk Research, 23*(10), 1370–1386.

Wen, X., Sun, S., Li, L., He, Q., & Tsai, F. S. (2019). Avian influenza: Factors affecting consumers' purchase intentions toward poultry products. *International Journal of Environmental Research and Public Health, 16*(21), 4139.

Whitley, C. T., Gunderson, R., & Charters, M. (2018). Public receptiveness to policies promoting plant-based diets: Framing effects and social psychological and structural influences. *Journal of Environmental Policy and Planning, 20*(1), 45–63.

Wolf, H. V., Perko, T., & Thijssen, P. (2020). How to communicate food safety after radiological contamination: The effectiveness of numerical and narrative news messages. *International Journal of Environmental Research and Public Health, 17*(12), 4189, 1-19.

Wong, I. A., Lu, M. V., Ou, J., Hu, R., & Wang, H. (2021). Destination green equity and its moderating role of travel satisfaction. *Journal of Vacation Marketing.*

Wu, C. W. (2015). Facebook users' intentions in risk communication and food-safety issues. *Journal of Business Research, 68*(11), 2242–2247.

Yang, J., & Goddard, E. (2011). The evolution of risk perceptions related to bovine spongiform encephalopathy: Canadian consumer and producer behavior. *Journal of Toxicology and Environmental Health, 74*(2-4), 191–225.

Yawson, R. M., & Kuzma, J. (2010). Systems mapping of consumer acceptance of agrifood nanotechnology. *Journal of Consumer Policy, 33*(4), 299–322. Available from https://doi.org/10.1007/s10603-010-9134-5.

Ye, T., & Mattila, A. S. (2022). The impact of environmental messages on consumer responses to plant-based meat: Does language style matter? *International Journal of Hospitality Management, 107.* Available from https://doi.org/10.1016/j.ijhm.2022.103298.

Yillia, P. T. (2016). Water-Energy-Food nexus: Framing the opportunities, challenges and synergies for implementing the SDGs. *Osterreichische Wasser- und Abfallwirtschaft, 68*(3-4), 86–98.

You, M., & Ju, Y. (2017). A comprehensive examination of the determinants for food risk perception: Focusing on psychometric factors, perceivers' characteristics, and media use. *Health Communication, 32*(1), 82–91.

You, M., Lim, J., Shim, M., & Ju, Y. (2019). Outrage effects on food risk perception as moderated by risk attitude. *Journal of Risk Research, 22*(12), 1522–1531.

Zeng, T., Durif, F., & Robinot, E. (2021). Can eco-design packaging reduce consumer food waste? An experimental study. *Technological Forecasting and Social Change, 162,* 120342.

Zerbe, N. (2010). Moving from bread and water to milk and honey: Framing the emergent alternative food systems. *Humboldt Journal of Social Relations, 33*(1), 4–29.

Zhu, X. Y., Huang, I., & Manning, L. (2019). The role of media reporting in food safety governance in China: A dairy case study. *Food Control, 96,* 165–179. Available from https://doi.org/10.1016/j.foodcont.2018.08.027.

Zoller, H., Strochlic, R., & Getz, C. (2020). Agricultural workers' participation in certification as a mechanism for improving working conditions: The Equitable Food Initiative. *Journal of Applied Communication Research, 48*(6), 654–674.

Further reading

Pappa, I. C., Iliopoulos, C., & Massouras, T. (2018). What determines the acceptance and use of electronic traceability systems in agri-food supply chains? *Journal of Rural Studies, 58,* 123–135.

Pollack, C. C., Gilbert-Diamond, D., Emond, J. A., Boyland, E. J., & Masterson, T. D. (2021). Twitch user perceptions, attitudes and behaviours in relation to food and beverage marketing on Twitch compared with YouTube. *Journal of Nutritional Science,* e32.

Sodano, V., & Hingley, M. (2013). The food system, climate change, and CSR: From business to government case. *British Food Journal, 115*(1), 75–91.

PART 2

CHAPTER

6

The Chefs' Manifesto

Paul Newnham

Chief Executive Officer, SDG2 Advocacy Hub, London, United Kingdom

Communicating technical and scientific messages to everyday people: working with Chefs as leaders in conveying messages and shifting the way we eat and see food.

With one in three people malnourished globally today, and food production systems that are damaging the planet, it is crucial that chefs, farmers, policymakers, and world leaders work together to deliver a sustainable, nutritious, and delicious food future.

However, many of the conversations around food, nutrition, and agriculture are technical and use language that is often misinterpreted by people without special expertise in these fields. Despite approximately one in three photos on platforms such as Instagram featuring food, or the broader public watching a variety of cooking shows, our enthusiasm for food doesn't translate into engagement on food systems challenges due to a language barrier. The world is obsessed with food, yet our advocacy fails to connect the dots. This is why I decided to bring in chefs as advocates and connectors—they bridge the gap between farm and fork and are a stakeholder group we can all be inspired by, through food.

Today, the Chefs' Manifesto is a chef-led network bringing together more than 1500 chefs from 93 countries to explore how they can help deliver sustainable food systems. It seeks to profile chefs' work, shared learnings, and best practice as well as empower them to be advocates for sustainable food systems everywhere. The Chefs' Manifesto Action Plan, consisting of concrete, simple, and practical areas and actions chefs can work on in their kitchen, creates a clear framework for chefs to deliver on the United Nations Sustainable Development Goals (SDGs).

This chapter aims to demonstrate what, in my experience, is critical to create action and the key elements to building a community of influential chef advocates. We will examine their motivations and our decisions around how we supported and empowered them with clear, digestible, and accessible language to drive action locally, in a globally relevant way.

Food Sustainability and the Media
DOI: https://doi.org/10.1016/B978-0-323-91227-3.00004-4

6. The Chefs' Manifesto

How the Internet has changed food messaging

Food features in our lives every day. It is not only integral to our survival, it is also one of the most common social pastimes: to sit together around food can be found in almost every culture globally. In food systems, sitting in between the growing of raw ingredients and how we ultimately eat these ingredients is cooking. Both culture and tradition fuse with location in influencing the ingredients and techniques used to cook, as well as people's taste. In the past, food was heavily influenced by one's location, the seasons, and culture. People gained knowledge from senior family members and, in the more recent past, from books. "Cookery the Australian Way" has been passed down from generation to generation in my wife's family, and many of the recipes are still used today.

Over time, television started to slowly expand this knowledge and awareness of food, alongside the increase of travel for pleasure. Yet a global food culture was still limited and cross-cultural food knowledge uncommon. Concurrently, television and expanding print media also introduced an awareness that food security—access to safe, nutritious, and affordable food that is available with consistent, sufficient supply—was not equitable around the world. The Ethiopian Famine in the 1980s, one of the worst humanitarian crises of the 20th century[1], garnered worldwide attention due to streamed images on our TVs, and everyday people were questioning why and how this could occur. The COVID-19 pandemic also opened much of the world's eyes to the impacts on food supply, as transport ground to a halt, labor migration stopped and trade slowed, and access to certain ingredients became harder and harder. This growing awareness of aspects of food security coincided with an expanding global food culture.

Migrant communities who have carried their food cultures across the world, including through running local restaurants where we were able to see, taste, and experience other cultures' delicacies, but this was a treat reserved for special occasions. Growing up, we would go to the local Chinese restaurant and eat honey chicken, special fried rice, and peppered beef. How different my understanding is now of the food scene in China! Fast food has brought certain cuisines to the masses such as Chinese, Italian, and Mexican, spreading broadly through big chains. Migration has also ensured different cuisines reach all corners of the globe through food fusions becoming more and more commonplace. Our local British pub served pie and chips, bangers and mash, and a Sunday roast, as well as a wicked green curry and Nasi Goreng. In Timor-Leste, one of my favorite restaurants combines Lebanese with Thai food. For Christmas last year, I found myself in a beachside town in Thailand, enjoying a Swiss/Thai/English fusion meal. As the world has shifted and global connectivity has increased, so have our eating patterns and habits. The Internet and particularly social media have emerged as new platforms to interact with and be inspired by food.

"In 1991, the world welcomed the World Wide Web... Since the release of the iphone just over a decade ago, in 2007 the social impacts of technology have become even more wide spread. Technologies unknown just a decade or two ago are now treated as essential appendages."[2]

---

[1] World Vision. (2018).1980s Ethiopian Famine: Facts, FAQs, and how to help. Available at: https://www.worldvision.org/disaster-relief-news-stories/1980s-ethiopia-famine-facts.

[2] Eve Turow- Paul - hungry - 2020 - p12.

Smart phones have given us access to the world in a way that was previously unthinkable. In our homes, our kitchens, and when we are on the move, we can access information anytime, anywhere. You're not sure what to cook for dinner? Google recipes for a particular cuisine and you will be inundated with suggestions. Or perhaps think back to the last time you scrolled through your social media feed and saw images of dishes that looked amazing, possibly eaten in stunning locations around the world. Food envy has become a real thing! Seeing real time photos of people eating delicious food, cooked in their homes or by a chef, has become commonplace on social media, and there are thousands of sites dedicated to this topic. In this social media space, certain voices gain credibility and trust through consistency and authenticity in combination with knowledge. Chefs, for example, can now not only curate and share physical food experiences in their restaurants, but also online, with a broader audience and no geographical limits.

"The Internet has enabled recipe swapping on a scale and at a speed that is dizzying. Where our grandparents (in the anglo saxon world at any rate) sat down dutifully to plates of under seasoned meat and two veg, we have developed unexpected new global palates: for spicy Turkish eggs sprinkled with sumac or vibrant salads of green mango and lime."[3]

A short history of food messaging

Food messaging was, in the past, largely dependent on the location in which you lived. Whether you had access to education, newspapers, television, even the type of media controls placed on reporting, determined the messaging you received. This impacted education about food security and global issues affecting food systems and sustainability, as well as what and how to cook.

The influence of food messaging—both in relation to food safety, accessibility, affordability, and consistency and what to cook at home nutritionally—is no longer just tied to the domain of the restaurant, the realm of health professionals, reporters, publishing houses, or high school food tech classes. Instead, we see influencers from all facets of life, reaching into our homes through social media. Chefs, for example, Jamie Oliver, Gordon Ramsey, Rachel Ray, Lorna Maseko, and Curtis Stone, have curated food entertainment, culminating in cooking as a major presence and genre on all television globally and more recently online. This is reinforced through social media platforms, as people share and engage from home, recreating dishes from around the world. This accelerates the spread of techniques, ingredient trends, and food-related messaging around the world.

What we previously knew about food we had learnt in our homes, from our families, in our schools, from journalists and new programs, and sometimes even through the medical profession. The reach was limited and our knowledge reflected this. Big brands seeded their messaging through advertising in print, through radio jingles, and on television. My wife can still sing every word of the Happy Little Vegemite song! While some of the bigger brands would go on to work with chefs who were well known and trusted at the national level, chefs' growing prominence at the global level is a relatively recent phenomenon.

[3] Bee Wilson (2019). The Way we EAT now. p 20.

The impact of a new platform

The Internet and particularly social media have emerged as new platforms, making food and chefs' work more visible and thereby more influential than ever. What chefs choose to profile today can directly drive future demand, as seen with ingredients ranging from quinoa to rocket. Not only can it serve to make food more enticing, more visible, and more desirable from everyday social media posts, the Internet makes chefs' work far more visible than ever before. Thus, elevating their influence more than ever. Food companies are very aware of this and as a result have formed partnerships with key chefs to promote and bring their products to life and pair recipes and experiences with products and ingredients. This means that advertising messages reach into people's consciousness in new ways. It is often hard to distinguish advertising from other content. Algorithms further reinforce messages so people feel like everyone is thinking and talking like them. This in turn increases the perceived credibility of these messages. Credibility and science can be questioned quickly and swayed by the crowd leading to misinformation feeling more credible.

The Internet has also given readers access to often competing and conflicting messages from fitness professionals, scientists, nutritionists, and health professionals. Competing messaging exists from the political arena around climate adaptation, food security, and sustainability, as well as specifically on nutrition. For the consumer, diets, foods to avoid, sustainable cooking, reducing carbon footprints through what you eat, eat seasonally, locally sourced, certified organic, fair-trade, fasting, vegan, vegetarian, lactose-free, gluten-free, grain-free, raw, pescatairan, free range, grass-fed, GMO, plant-forward, plant-based... the list of messages around food in terms of what and how to eat is seemingly endless and can be overwhelming.

With personal trainers now weighing in on nutrition too, frequently offering fitness and eating programs together, their voice has been elevated in a space where not all of these professionals are studied nutrition experts. Trainers' lack of nutritional knowledge is a well-known and documented global problem of the fitness industry for more than two decades. One study from Switzerland showed that 60% of fitness instructors self-assessed their nutrition knowledge as insufficient while another study from Australia concluded that there is a considerable gap between levels of nutrition education and the level of knowledge needed.[4] It is both a challenge and an opportunity of the Internet that advice can spread quickly. This makes accessibility of information far easier for people, but in turn this can also create challenges in understanding what is accurate versus not. Focusing on the positive, when expert data and evidence are paired with the right communicators, it can quickly become a powerful tool to enact change and shift behaviors.

Why do we listen to people who aren't scientists or credible health practitioners?

The three key drivers of our food choices are cost, convenience, and taste. Similar to some personal trainers, social media influencers and companies who are selling a particular style of eating

---

[4] Available at: https://www.ncbi.nlm.nih.gov/pmc/articles/PMC7146185/.

are not necessarily health or science professionals, hence don't necessarily have enough knowledge to assess and judge the true impact of food on our body. They may mean well, but the information they are disseminating is often misleading and wrong. However, what they do understand is the impact of cost, convenience, and taste on consumers' decisions, and through utilizing these drivers, they are undeniably good at convincing consumers of their products' worth.

The flip side is that many people who are knowledgeable on the science of food, nutrition, and agriculture do not understand the key drivers that impact peoples' food choices. Their messages often get lost when they hone in on healthy eating and nutritional content and focus exclusively on their expertise, oftentimes missing the human connection to food. The advocacy around food security and sustainability often focuses on the same areas—health and the human impact. People don't like to be told no, nor do they want to hear that they need to be healthier or cut foods out, yet they are desperate for trusted information that they can engage with easily.

The world is obsessed with food, yet our advocacy fails to connect the dots. Despite the described broad spread and incredible scope of food messaging on our platforms today, our enthusiasm for food doesn't translate into engagement on food systems' challenges. The reason for this is a "language" barrier. This is why we decided to bring in chefs as advocates and connectors—they bridge the gap between farm and fork and are stakeholders we can all be inspired by, through food. Using public messaging with consumers who have the privilege of choice that incorporates terminology such as food security, food systems, and sustainability and also tells them to adapt their diets to contribute to ensuring accessibility of good food for all people, is a sure-fire way to encourage consumers to tune out and turn off.

Chefs bridging the gap

Chef's know and understand what drives people to want to eat their food. They understand that taste, accessibility, convenience, cost, and engaging the senses are all aspects that are crucial to a chef's successful business model. Instead of ignoring taste, Chefs have a deep understanding of what people consume, they daily get very quick reactive feedback to what tastes good or what does not, this is quite literally a chefs' bread and butter. This means that chefs are well placed to help drive connections around food. Chef José Andrés recently wrote a while back in The New York Times,

A plate of food represents much more than a recipe. In a crisis, it says that someone cares. In more normal times, it brings family and friends together. Its components can make us healthy, or make us sick. The offerings on that plate are shaped not just by a cook, but also by politics and business.[5]

Chefs, at their core, are in the business of hospitality. Hospitality is defined as "the friendly and generous reception and entertainment of guests, visitors, or strangers."[6] As a result, it is of no surprise they are able to relate to people and connect in a way that

[5] Andres, J. (2020). The New York TImes, Opinion, We Need a Secretary of Food, Available at: https://www.nytimes.com/2020/12/08/opinion/covid-pandemic-food-crisis.html?referringSource = articleShare.

[6] Available at: https://www.dictionary.com/browse/hospitality?s = t.

enables trust. This is a key trait of the industry. It is true some will focus more on the science of cooking and preparing food, but at the heart of restaurants and chefs is hospitality.

Restaurant is a word that also comes from a deeper meaning. "The word derives from the French verb 'restaurer' ('to restore,' 'to revive') and, being the present participle of the verb, it literally means "that which restores."[7] This link to restoration when paired with hospitality helps to understand some of the history and potential explanations as to why chefs are such great connectors and communicators in general. They understand how things come together and how to interact with people. These traits and learnings are all essential for linking awareness, knowledge, and action.

Chefs are also placed at the very heart of food systems. They sit between consumers and suppliers. They can drive demand for certain ingredients, grown in particular ways, and in turn, produce dishes that can educate and influence consumers on bigger issues.

While the influence on our food choices continues to increase, thanks largely to the many voices on our social media channels, all forms of malnutrition globally are also continuing to rise. So despite an obsession with food, we have a major challenge in the way our food systems operate and function. The misuse of food is a major contributor to poor health for both people and the planet. We are in the very midst of a crisis created by inequitable and incorrect food consumption—and as such food can be seen as both the vehicle to heal, yet also the problem.

The global food crisis

As many as 783 million people around the world face some level of food insecurity while more than two billion people around the world are unable to access food that is safe, affordable, nutritious, and sufficient year-round. Achieving food security for all is a key tennant of the SDGs—but I would argue that it's not enough to just feed people. We need to nourish them too with food that is delicious, culturally sensitive, and good for the planet.

While more people are hungry now than at the start of the SDGs in 2015, 14% of the world's food was lost between the harvest and retail stage of the food chain. A significant amount more was wasted within retail and in people's homes. This "ethical outrage" as put by António Guterres, United Nations Secretary-General, is likely to have worsened due to COVID-19. Restrictions resulted in food being left to rot in fields, milk being poured down the drain and produce rotting idle in fields and factories, meanwhile an additional 122 million more people are expected to experience hunger due to the pandemic. Surprisingly, COVID-19 has brought to the forefront major inefficiencies in our food systems that many experts have been fighting to spotlight for years. In the USA alone, the food supply chain was severely broken as "farmers began destroying their crops and their animals. Hundreds of thousands of unhatched chicken eggs, millions of gallons of milk, millions of pounds of potatoes—all wasted"[8]. The issue has become topical, as everyday people question why their grocery stores ran out of certain staples, which led to

[7] Available at: https://en.wikipedia.org/wiki/Restaurant.

[8] Andres, J. (2020). The New York Times Opinion, We Need a Secretary of Food. Available at: https://www.nytimes.com/2020/12/08/opinion/covid-pandemic-food-crisis.html?referringSource = articleShare.

discussions about whether crop yields will sustain supplies in the coming year, or if manufacturing and production plants can catch up with demand. Through social media, people have also shared innovative food loss and waste hacks and with the ability to produce quick recipe video clips.

The global food loss and waste situation accounts for 8% of global greenhouse gas emission (GHG). If food loss and waste was a country, it would be the third largest GHG emitter after China and the USA. In total, global food systems account for approximately 30% of all GHG emissions and utilize 70% of the world's fresh water reserves. Depleting environmental resources is only one side of the coin. Current food systems are also fueling a health pandemic of rising malnutrition as well as diet-related diseases. Lack of accessible and nutritious food, the cost of nutritious food, and the availability of nutritious food all lead to forms of malnutrition.

Diets are increasingly consisting of highly processed foods; of the 30,000 edible plants, we eat just 150 of them and 60% of the calories we eat come from just four crops—wheat, maize, rice, and potato. This is fueling an unhealthy population with now more than one in three people classified as overweight or obese. The lack of diversity in our diets is also the cause of the biodiversity crisis that threatens the existence of our whole food systems.

Meanwhile, the double burden of malnutrition is growing. Stunting and wasting figures are still alarmingly high in under-5 year olds while the global number of children living with overweight is rising to 37 million. This double burden risks irreversible damage to the growth of a child's brain and body and impacts their potential in school and employment possibilities in later life.

To achieve systematic change throughout food systems, to deliver healthy diets that are accessible and affordable for all, provide equitable livelihoods, while within the planetary boundaries, a change in policies, funding, and incentives is required: not only in the agricultural sector but in every sector. However, policies and institutions are hard to change, fundamentally they do what the customer/public want. That is why the role of the media, to inform people to make empowered decisions, is so important.

Introduction to the Chefs' Manifesto

There is space on our media platforms to address issues of food security and sustainability in an informed way based on agreed science, introducing a network of voices that are aware, knowledgeable, active, and relatable. A network of voices that speak to the consumer or eater in a way they understand, yet can also champion the immediate need to change what and how we eat food through direct action. The Chefs' Manifesto was born out of this desire to disrupt, interrupt, and drive good change in relation to food, for all people. The Chefs' Manifesto is now a chef-led network bringing together more than 1500 chefs from 93 countries to explore how they can help deliver sustainable food systems. It seeks to profile chefs' work, shared learnings, and best practice as well as empower them to be advocates for sustainable food systems everywhere. The Chefs' Manifesto Action plan, consisting of eight concrete, simple, and practical action areas chefs can work on in their kitchens, restaurants, and communities, provides a clear framework for chefs to deliver on the SDGs.

Building the network and guiding principles

To build the network, we capitalized on the concept of decentralization and shifting narratives from previously recognized corporate and sectoral voices, to alternative influencers with powerful media platforms and stories.

"Decentralisation has been lying dormant for thousands of years. But the advent of the Internet has unleashed this force, knocking down traditional businesses, altering entire industries, affecting how we relate to each other, and influencing world politics. The absence of structure, leadership and formal organization, once considered a weakness, has become a major asset."[9]

This relational shift and potential to influence from previously ignored avenues motivated a closer analysis of social movement approaches during the conceptualisation of the Chefs' Manifesto. The concept was based on social movement theory as well as personal experience.

The first guiding principle we defined was to **work from a low resource base**. This means that we wanted to build something that did not require a big budget or staff to succeed. The question we asked ourselves was how to make this replicable and not reliant on a central organization and resources to drive action and advocacy. We initially planned to have a toolkit and narrative with examples, case studies, and a common framework to leverage.

The second key principle we built the Manifesto around was **ownership**. The inception of the Chefs' Manifesto was very carefully crafted to be an ongoing evolutionary process so that all participants could see their part to play in the network, no matter when they joined. From community development best practice, we know that for organizing political and social change, uniting people together in collective action and giving voice to communities and individuals are essential to being empowered to have ownership and choice[10]. As such, we needed to develop a way for this to be deeply ingrained in the ethos. The tag lines — 'By Chefs, For Chefs' and 'Chef-Led' — became key pillars we used in all practices to drive developmental actions as well as in all wider communication. This means we always have chefs in key roles at events and in dialogs often alongside facilitators and experts. The key was that chefs were given a platform otherwise inaccessible to them, to share their messages.

A third principle was **to translate the technical jargon** used at UN level, to a language accessible to and usable by chefs. This was a translation issue easily replicated through education and ownership. The technical and scientific regularly creates a barrier to acceptance and understanding of broader, global issues. With chefs we sought to translate the technical jargon, yet still connect essential SDG principles into culinary-friendly terms and create relatable points of connection. This was critical for ownership as it enabled a common vocabulary and understanding. It also built a network right from the beginning that

---

[9] Ori Brafman and Rod A Beckstrom - The Starfish and the Spider - The unstoppable power of leaderless organisations- 2006 - P6.

[10] Kenny, S. (2011). Developing Communities for the Future.

focused on linking an awareness of global food-related issues, to knowledge-based principles that we could direct to practical actions.

Translating the technical jargon to accessible and usable language for chefs was a process that chefs led right from the beginning. Gathering chefs from around the world at London's stunning Omved Gardens, the first ever Action Hub was held in June 2016. Through a series of large and small group sessions, chefs and industry leaders debated language, meaning, inference, and importance. As understanding and awareness grew, the participants narrowed down the key components of SDG2 that they wanted to incorporate into their own document, the Chefs' Manifesto. In its final version, the Manifesto is centered on eight thematic areas:

1. Ingredients grown with respect for the earth and its oceans.
2. Protection of biodiversity and improved animal welfare.
3. Investment in livelihoods.
4. Value natural resources and reduce waste.
5. Celebration of local and seasonal food.
6. A focus on plant-based ingredients.
7. Education on food safety and healthy diets.
8. Nutritious food that is accessible and affordable for all.

These thematic areas are the pillars that chefs can use to take action in their respective spaces. They can choose which areas speak to them, and how they would like to align their work with the Manifesto, to—ultimately—achieve the SDGs by 2030.

Chef engagement

Initially, chefs engage with us because they are curious. As creative beings, they are often motivated to find new ways to curate, to stand out, and to find a point of difference in their work. A network that is global and directly tied to the UN SDGs, using language that is accessible for all, is an immediately appealing pathway, if for no other reason than curiosity. Our strategy lies in building and nurturing relationships, growing an awareness of all the actors in the global arena, working toward achieving the SDGs by 2030, educating them on how they can play an integral part in enacting real change, and then providing platforms and pathways for ongoing engagement that motivates, recognizes, and inspires.

Our events span continents, bringing relevant people together that are using different techniques, ingredients, and approaches. This is increasingly important in our globally connected world. We provide opportunities to showcase and highlight chefs' outstanding innovations and creativity, profiling them globally as examples of leading change-agents and advocates for the SDGs, as well as elevating their work to help them gain recognition through their point of difference in a highly competitive industry. They are able to join global events and dialogs as chefs and are recognized for their valid contributions. What follows is global recognition for their efforts—instead of only at the domestic or community level. Generating pathways for chefs to be present and have access to such diverse platforms also enables them to learn from cutting-edge research and relevant experts.

The Chefs' Manifesto gives the chefs access, to bridge the previous communication divide and help both sides see the impact the other can have on their respective efforts.

We have also found that chefs are generally keen to contribute to a broader effort. Sustainability efforts can often be lonely: chefs work in the kitchen and often don't have a wider platform to explain and showcase some of the lengths they go to make an impact. Through connecting to our network, they gain and give credibility. Connecting them to a globally recognized framework in the form of an Action Pan and eight thematic areas enables chefs to see how running a zero-waste kitchen, for example, directly ties to SDG 12.3—responsible consumption—or how showcasing forgotten ingredients on a menu ties directly to SDG 2.5, which involves securing biodiverse crops.

How chefs spread the message

Different chefs engage in different ways. The beauty of the Manifesto and its strategic design is that it can be interpreted in multiple ways by any number of chefs according to their interests, lens, cultural identity, and capacity. Practically, there are four ways Chefs further spread the message through:

- Direct collaborations—Working together to align key work they are involved in to the narrative and framework. This predominantly involves menu creation, local projects, and writing.
- Attending events—We partner with a range of global platforms from the UN, civil society, and the private sector. With these partners, we look to bring in the voices of chefs to engage and present their viewpoints and messages.
- Social media platforms—We encourage and support chefs in using their own platforms to amplify content, key messages and also to create and integrate messages into their content that is rooted in the UN SDGs.
- Action Hubs—This is a way we gather chefs to discuss and collaborate in the kitchen and around the table on more national-based issues. Action hubs can focus further on local food system issues, such as sustainability and food insecurity, as well as offer actionable opportunities.

The following case studies highlight the ease with which the thematic areas have been incorporated and showcased by chefs from around the world.

Megha Kohli is one of India's youngest, successful chefs, names as Times Chef of the Year 2020. Previously the head chef of Lavaash By Saby, the Armenian restaurant in Delhi's artsy Mehrauli area, Megha is now the Executive Chef of Mademoiselle Cafe, Cafe Maz, The Wine Company and Chef Partner at Mezze Mambo. Deeply influenced by her grandmother's cooking, Megha started pottering around in the kitchen when she was barely 5 years old. Megha believes firmly in using only local, regional, and sustainable produce. Her restaurant was one of the first in India with a menu that uses 100% local ingredients. Megha's mantra is that you can serve the most perfectly prepared dish, but if you haven't connected with the food and farmers when making it, it's an effort gone to waste. Focusing on local and seasonal Future 50 Foods—a combined report released by Knorr and the WWF-UK highlighting 50 plant-based, nutritious ingredients, with minimal impact on the environment[11]—Megha linked this directly to Chefs' Manifesto thematic areas 2 and 5, the protection of biodiversity and improved animal welfare, and a celebration of local and seasonal food. Megha created an online and in-restaurant program that took a Future 50 ingredient, ran restaurant specials and recipe workshops all season. She shared these in her social platforms, particularly Instagram, which amplified the content to a much broader network in a relevant and accessible way.

Conor Spacey is a passionate Irish chef who works with the seasons. Always seeking out great Irish produce, foraging, and meeting Irish producers, he is the Author of 'Wasted' and runs 'Wasted' pop-ups throughout the UK and Europe, showcasing ingredients that would otherwise be thrown away. Conor also helped establish FoodSpace, a company that brings together restaurants and cafes around Ireland, which prioritize fresh and seasonal produce—in line with Chefs' Manifesto thematic area 5. FoodSpace works to build relationships with local producers and growers. One of the organization's key principles is "loyal to local," encouraging local sourcing of fresh produce within a 50-mi radius in rural and a 100-mi radius in urban areas. Conor has also chosen to incorporate a zero-waste principle in his kitchens, focusing his advocacy efforts predominantly on thematic area 4—valuing natural resources and reducing waste. Conor has also worked to optimize his advocacy efforts to influence the press in Ireland, to discuss sustainability further. He has amplified and shared content on all his online channels, then featured this in the press to spread the message even further.

Manal Alalem is an Egyptian professional and TV chef. Throughout her career, Manal has encouraged women's involvement in the culinary industry by leading specialized culinary courses for women as well as setting up the Manal Club—a network of 6000 women from across the Arab world who take part in culinary festivals, shows, and competitions. Manal is a champion of Middle Eastern and Arab cuisine and regularly posts cooking videos of traditional dishes on her Youtube channel. Her Youtube channel has had over 300million views and combined with other platforms, her reach is similar to many broadcast networks. Additionally, Manal has published several cookbooks such as Manal Alalem Kitchen and features in a number of newspapers and magazines. Manal also works as an ambassador for the UN World Food Program. Manal is a strong advocate across all eight thematic areas, with a particular focus on thematic area 8—nutritious food that is affordable and accessible to all.

Pierre Thiam was born and raised in Dakar, Senegal, and pays homage to traditional West African cuisine through his thoughtful approach to food. He is the Cofounder of Yolélé Foods, a purpose-driven African food company launched in 2017 that specializes in African superfoods. By working directly with African smallholder farmers, Pierre has created a value chain for the ancient, highly nutritious, climate-smart, and delicious grain Fonio. Pierre is also the Executive Chef at the celebrated restaurant Nok by Alara in Lagos, Nigeria. His Pan-African fine-dining menu brings an elevated experience of the African diaspora onto the table. Based in New York City, Pierre Thiam Catering introduces contemporary interpretations of ethnic flavors to a diverse, savvy clientele, including UN General Secretary Ban Ki Moon, among others. He has authored multiple

---

[11] Knorr and WWF-UK introduce future 50 foods. (2019). Available at: https://www.unilever.com/news/news-and-features/Feature-article/2019/knorr-and-wwf-uk-introduce-50-future-foods.html.

groundbreaking cookbooks and has been featured on numerous TV programs and radio shows. Since joining the Chefs' Manifesto, Pierre focused his advocacy efforts on thematic area 2—protection of biodiversity, and area 3—investment in livelihoods.

Alejandra Schrader is a Venezuelan American chef involved in multiple charity projects, offering cooking demonstrations, working special events, and teaching complimentary cooking classes to children, youth, and women of all ages. Honored to be a "Sisters on the Planet" Ambassador for Oxfam America, Alejandra believes she has a real responsibility to help find an effective plan to end world hunger and to develop more sustainable food systems. A hard-working entrepreneur, Alejandra has launched her own line of seasoning blends and personally blends her products in small batches. She continues to offer private chef services to families and corporations and is the author of 'The Low-Carbon Cookbook' whilst working on multiple exciting new media and television projects. Alejandra is passionate about plant-based cooking, openly sharing her own personal story to connect directly to thematic area 6—a focus on plant-based ingredients—and thematic area 8—nutritious food that is accessible and affordable for all.

Arthur Potts Dawson is a British chef who has been cooking for over 30 years. Arthur has worked at Kensington Palace, as head chef at the River Cafe, and as executive head chef for Jamie Oliver's Fifteen Restaurant and Piccadilly Diner. His two restaurants Acorn House and Water House have won numerous awards for their excellent food and sustainable practices, proving the profitability of an eco-friendly approach. Additionally, Arthur founded The People's Supermarket, a supermarket that connects the urban community with the local farming community by stocking high-quality and environmentally friendly produce from trusted, local suppliers. As an advocate for sustainable food production, Arthur often speaks on issues relating to global hunger and food waste in his role supporting the UN World Food Program. Arthur has been instrumental in the creation of the Chefs' Manifesto and cofacilitates the Chefs' Manifesto London Action Hub, located at OmVed Gardens in Highgate. Arthur has begun a new venture in the past few years, with Feed Me Seymour, offering a full-circle healthspan solution, based on science, nutrition, and flavor. With sustainability pillars at its core. Chef Arthur focuses on area 1—ingredients grown with respect for the earth and its oceans, area 4—value natural resources and reduce waste, and area 5—celebration of local and seasonal food.

Overall, our engagement is relational and values-based. We gather together using a shared language, allowing space and flexibility to bring in personal interests, build awareness and knowledge, and then create opportunities to find connection to common agendas and shared actions. Chefs are busy and to engage them outside their day-to-day businesses you therefore have to have flexibility. Business comes first for sustainability. The way the Manifesto is designed and how we work with the chefs allows their level of engagement to fluctuate throughout the year depending on the time they have on hand and the stage their businesses are at.

Drastic action is required now, to enable a future that has a food system that will thrive for people and the planet. The consistent supply of food is decreasing. The affordability of food and access to safe and nutritious food is facing a critical decade. Without action now, our food system will continue to drive potentially irreversible impacts to people and planet. Malnutrition in all its forms will continue to increase, as the misuse of food at both ends of the spectrum—stunting and starvation, as well as obesity—escalates.

To generate change immediately, and communicate technical and scientific messages to everyday people, alternative influencers must be empowered to drive action and alter the existing narratives. Empowerment requires translation and bridging investments. The media has a powerful role to play in providing pathways that enable the prolific spread of accurate messaging through such alternative influencers.

After careful analysis of the many existing voices involved in discussions around food security, nutrition, and sustainability, a deliberate decision was made by the SDG2 Advocacy Hub to establish a network of chefs with whom to raise and increase awareness, build knowledge, and encourage advocacy actions. Through leveraging their media-based platforms, greater audiences can now be reached to educate on the UN SDG issues such as food security, nutrition, and sustainability.

Through the decentralization of power, ownership, the translation of technical jargon, and empowerment through education, consciousness-raising, guided increased awareness, expert and lived knowledge as well as practical actions, the Chefs' Manifesto provides a pathway for like-minded chefs to participate in driving forward good change. To truly impact a broken system, it is necessary to disrupt the current trajectories and insert new voices, new faces, and new ways of engaging with the wider public. The chefs who are active participants in the Chefs' Manifesto are part of a community who share a common nontechnical language base that enables them to enact real change through their platforms. They are empowered to use their voices to bridge the gap, to connect the human side of food—the cultural, the stories, the heart, and soul—to the wider issues facing our world. Their intricate understanding of food systems, sitting at the center between farm and fork—coupled with their ability to leverage media platforms to far-reaching audiences, is the very reason chefs are perfectly placed to drive forward action throughout the coming decade.

chapter 7

Enabling sustainable, healthful eating in the cafeteria setting through education and social engagement: the SU-EATABLE LIFE project

Laura Bouwman[1], Leah Rosen[1], Marta Antonelli[2], Simona Castaldi[3] and Katarzyna Dembska[2]

[1]Chair Group Health and Society, Social Sciences Group, Wageningen University, Wageningen, The Netherlands [2]Fondazione Barilla, Parma, Italy [3]Dipartimento di Scienze e Tecnologie Ambientali Biologiche e Farmaceutiche, Università degli studi della Campania Luigi Vanvitelli, Caserta, Italy

Introducing the SU-EATABLE LIFE project

The SU-EATABE LIFE project (2018–2022) aims to demonstrate that we can achieve a substantial reduction in greenhouse gas (GHG) emissions and water usage through a series of activities, which engage EU citizens to adopt a sustainable, healthy diet at university and company cafeteria.

A third of global GHG emissions comes from the food system. The EU food system is particularly carbon-intensive, accounting for 30% of EU greenhouse gas emissions and ranking among the top six emitting economies. About half of the GHG emissions is CO_2, mainly linked to land use change and energy while one-third is methane (CH_4) due to livestock and rice production. Simultaneously, average dietary habits are unhealthy, inequitable, and unsustainable for the environment in wealthy and developed countries (Fanzo & Davis, 2019; Willett et al., 2019). In the EU, over half of the adult population is overweight, contributing to a high prevalence of diet-related noncommunicable diseases and related healthcare costs, and European diets are not in line with national dietary recommendations. As such, it is critical to reverse the rise of overweight and obesity rates across the EU by 2030. Moving to a more

Food Sustainability and the Media
DOI: https://doi.org/10.1016/B978-0-323-91227-3.00005-6

plant-based diet with less red and processed meat and more fruits and vegetables will reduce both the risks of life-threatening diseases and mitigate the environmental impact of the food system. Tackling this issue could significantly contribute to the European Commission's target of 40% cuts in GHG emissions by 2030 (compared with 1990 levels) as well as the wider European Green Deal objective of making Europe the first climate-neutral continent by 2050. Furthermore, the Farm-to-Fork Strategy, which is at the heart of the Green Deal, comprehensively addresses the challenges of sustainable food systems and recognizes the inextricable links between healthy people, healthy societies, and a healthy planet.

Diets fundamentally determine environmental outcomes, with plant-based diets holding large potential in terms of climate change mitigation, land use, and water use (Springmann et al., 2018; Tilman & Clark, 2014; Willett et al., 2019). There is a need for a significant change in dietary habits. These changes need to be both beneficial to human health and contribute to keeping most of the critical environmental 2050 targets within the so-called "planet boundaries," that is, "the global biophysical limits that humanity should operate within to ensure a stable and resilient Earth system—that is, conditions that are necessary to foster prosperity for future generations"(Willett et al., 2019). Different sustainable dietary models can deliver up to a 55% reduction in GHG emissions and avert over 10 million deaths per year (Springmann et al., 2018; Tilman & Clark, 2014; Willett et al., 2019). More recently, governments and citizens are becoming increasingly involved in global environmental issues, driven by the dramatic evidence surrounding environmental crises such as global warming and climate change, the unprecedented extent of plastic pollution in the oceans, water scarcity, soil loss and degradation, and loss of land productivity. In this context, a global call for sustainable diets is timely and answers several of the 17 Sustainable Development Goals (SDGs), which the United Nations have set as key development strategies to be reached by 2030 to reach a sustainable and equitable world (2015). Sustainable diets could also contribute to meeting the strict requirements of the Paris Agreement to keep global warming well below 2°C and achieve carbon emissions neutrality by 2050. As such, they are an integral component of the Farm-to-Fork strategy launched by the European Commission in May 2020, which will translate into a legislative framework for implementing food sustainability from farm to fork by 2023 (2020).

SU-EATABLE LIFE project approach

In response to this call for action, the interdisciplinary SU-EATABLE Life project team designed a multilevel, multistrategy approach in collaboration with universities and companies in Italy and the United Kingdom (UK). This approach included activities at the food service and consumer level. At the food service level, staff and management were invited to revise their methods of food procurement and preparation and align their food offering with the SU-EATABLE criteria for sustainable, healthful meals. These eight, science-based criteria were identified and translated into actions, which any citizen or stakeholder within the food sector (retailers, caterers, restaurants, cafeterias, municipalities) could apply to reduce CO_2 emissions and water use, based on changes in dietary choices. At the consumption level, cafeteria customers were invited to join on-site and online learning activities about the why, how, and what of sustainable food practices. They were invited to purchase a labeled sustainable meal in the cafeteria and exposed to banners and posters providing information on sustainable food consumption. Online, the GreenApes mobile application provided a medium where they could participate in different types of learning challenges and earn various rewards in return. The monthly challenges comprised a set of tasks to be performed and validated on the GreenApes app. Contrary to traditional approaches, which tend to only emphasize cognitive capacities (i.e., educational/informative approach), these challenges also aimed to develop other capacities with simple tasks that stimulate reflection, trigger search for information, encourage social participation, and simplify the selection and consumption of sustainable meals.

Early 2020, these tools were launched at seven university and company cafeterias in the UK and Italy. The UK sites were selected based on an inventory of willingness to participate among university- and business-cafeteria members of the Sustainable Restaurant Association. Italian sites were selected from eligible contacts of the Barilla Center for Food and Nutrition. Due to COVID-19 restrictions, the on-site activities came to an early halt in February (Italy) and March (UK). With the gradual closure of locations and drop in customer attendance, the intervention was put on pause after running 1 month in the UK and 2 months in Italy instead of the original 4 and 6 months. In this chapter, we present the project design process and our learnings from the first stage of project activities. The chapter closes with reflections on the implications of our findings for the next stage of the SU-EATABLE project.

Project design process

The first, pre-COVID version of the SU-EATABLE project was designed based on insights from literature reviews and surveys combined with theoretical perspectives on how to initiate engagement in change processes toward more healthful, sustainable dietary practices (Section 2.1). To support this, science-based guidelines were developed to establish the basis for healthy and sustainable dietary practices (Section 2.2).

Science-based design rationale

Making the environment a priority starts at the individual level. Along with government policies, international agreements, corporate measures, and technological innovations, the choices that people make every day play a pivotal role in enhancing sustainability. Hence, consumer engagement is often put forward as the preferred outcome of efforts in the area of sustainability (Bryngelsson et al., 2016; Willett et al., 2019). In response, a wide array of behavior change techniques have been applied to reach more engagement. A pivotal role has been delegated to the food sector to facilitate such a behavioral shift, notably by creating a variety of palatable and attractive plant-based meals (n.d.-b).

Literature reviews and desk studies uncovered the following key characteristics to be included in project activities aiming to shift to more sustainable practices: (1) use clear and consistent terms to avoid confusion; (2) provide credible, relevant, and applicable information and tools; (3) address cultural and habitual food practices that are (un)sustainable; (4) connect the values of health and sustainability; (5) address individual investments needed to make sustainable choices (e.g., money, time); (6) trigger intrinsic motivation by calling upon people to commit and act while rewarding them for doing so; and (7) create an enabling environment that considers individual, social, and environmental factors and where customers can learn about sustainable diets and take control of their learning process (Bacon & Krpan, 2018, 2016a; Böhme et al., 2018; Filimonau et al., 2017; Maher & Burkhart, 2017; Oostindjer et al., 2017).

In addition to these insights, the design integrated the Salutogenic Model of Health (SMH), the everyday-life perspective and gamification principles (Sardi et al., 2017; Van Woerkum & Bouwman, 2014). Based on these perspectives, the project design incorporated:

- *Learning experiences that increase an individual's coping capacities.* The SMH highlights the active role that people themselves play in coping with challenging situations to maintain a healthy orientation. This ability to cope is termed the "Sense of Coherence" and comprises three capacities—motivation to cope with the challenges (i.e., motivational), ability to understand the challenges (i.e., cognitive), and ability to identify and use resources to cope with the challenges (i.e., actionable). Learning experiences aim to help strengthen these capacities.
- *Activities that create an enabling environment, which facilitates and supports individuals in coping with the challenges of engaging in sustainable diets.* Based on the everyday-life perspective and the SMH, engagement entails multiple eating practices that are embedded in various practical and social activities. Project activities aimed to create a cafeteria environment that facilitated the identification and use of a broad range of resources, which could be used by food providers and customers in their learning process toward more sustainable practices.
- *A positive, practical, and enjoyable approach where the focus is less on traditional risk-oriented communication and more on a positive salutogenic and gamified approach.* This was done through educational content (e.g., videos) incorporating practical actions that individuals could take to make a positive impact. These were combined with gaming elements and social incentives within the project app to engage participants to complete challenges, win points, and claim rewards.

Lastly, a survey disseminated in Spring 2019 in Italy and the UK provided insights into perceptions regarding sustainable food and areas to target with the intervention. Findings indicated that while respondents were aware of environmental issues such as climate change, they were not knowledgeable about the impact of different foods on the environment. As such, they did not fully understand how to mitigate their environmental impact by changing their daily food consumption habits. Clarifying these concepts was therefore at the core of the intervention activities.

Science-based SU-EATABLE principles for sustainable diets

The SU-EATABLE project adopted the definition of sustainable diet as defined by the Food and Agriculture Organization (2012), in line with the most recent criteria of food system sustainability by the EAT-Lancet Commission (Willett et al., 2019). "Sustainable diets are those diets with low environmental impacts which contribute to food and nutrition security and to healthy life for present and future generations. Sustainable diets are protective and respectful of biodiversity and ecosystems, culturally acceptable, accessible, economically fair and affordable; nutritionally adequate, safe and healthy, while optimizing natural and human resources" (2012). It would be complex to define a single set of recommendations for sustainable initiatives, because sustainability has different dimensions. The project's focus was on the impact of food choices on climate and water resources, translated as the carbon and water footprint (CF and WF) of food in the intervention analysis. The project elaborated eight key principles based on scientific literature evidence, the basis for a healthy and sustainable diet guiding the actions of citizens and food sector stakeholders (retailers, caterers, restaurants, cafeterias, municipalities). To quantify the carbon and water footprints of food commodities, the project created a specific multilevel carbon and water footprint database (Petersson et al., 2021).

The eight principles in brief:

1. *Champion plant-based food in your diet: Make vegetables, legumes, nuts, and fruit the basis of our daily diet.*

   Emphasizing nutrient dense, plant-based foods is associated with significantly better health outcomes (n.d.-a; Satija et al., 2017; Satija et al., 2016; Toumpanakis et al., 2018; Tuso et al., 2013). A healthy diet includes: fruit, vegetables, legumes, nuts, and whole grains, with at least 400 g (i.e., five portions) of fruit and vegetables per day, excluding starchy roots such as potatoes (2018). Diets rich in plant-based food also have a lower environmental impact. Plant-based food grown in open fields has a lower average impact of one to two order of magnitude compared with animal-based food (Petersson et al., 2021), with the consumption of red meat marking the biggest difference. Substitution of red meat with legumes can lead to savings as much as 2.5 kg CO_2 equivalents and 1000 liters of WF per meal. To date, individuals in high-income countries tend to consume too much sugar, oils and fats, red meat as well as milk and cheese combined, while not enough fruits and vegetables (Vanham & Bidoglio, 2013; Willett et al., 2019).

2. *Enjoy meat in moderate amounts: Enjoy meat in moderate amounts, especially red meat and processed meats.*

   Meat production significantly affects climate change, as the "impacts of the lowest-impact animal products typically exceed those of vegetable substitutes." At the highest end of CO_2-eq emissions is ruminant meat, i.e., beef and lamb, which can be almost 10 times higher than poultry, eggs, rabbit, and most fish and four to five times higher than pork (Petersson et al., 2021; Poore & Nemecek, 2018). In terms of pressure on freshwater resources, meals that include beef have three times the WF as those that include other meat protein sources (pork, chicken) or legumes and twice the WF of eggs (Mekonnen & Hoekstra, 2010; Petersson et al., 2021). Daily intake of processed meats increases, and high intake of red meat (i.e., beef, veal, pork, and lamb) increases risk of cancer (2015b), cardiovascular disease (Abete et al., 2014), stroke, and type 2 diabetes (Chen et al., 2013; Feskens et al., 2013). Processed and unprocessed red meat was linearly associated with total mortality, suggesting that optimal intake should be low (Willett et al., 2019). Such risks can be significantly diminished by substituting other protein sources for red meat (Pan et al., 2012), in particular plant sources of protein and consuming no more than 100 g of red meat per week. Consumption of poultry has been associated with better health outcomes and can be in the order of 200 g per week, up to 400 g per week (Willett et al., 2019).

3. *Enjoy dairy products in moderation: Consume dairy products, including milk, in moderation.*
   Dairy products, in particular cheese, may have a CF and WF higher than eggs, poultry, and fish (Petersson et al., 2021). Cheese has a neutral impact on health if the intake is moderate, that is, no more than three servings per week (Guo et al., 2017). Plain yogurt and milk have a lower impact on the environment compared with cheese. Butter has a CF about four times higher compared with nontropical vegetable oils and conveys a higher risk of cardiovascular disease. Hence, nontropical vegetable oils should be preferred whenever possible for cooking and food preparation (Willett et al., 2019).

4. *Avoid too much food: Avoid having too much food on your table and in your daily diet to avoid excess calories and food waste.*
   Every kilogram of food carries the embedded environmental impact generated during its production, processing, transportation, and packaging (Clune et al., 2017; Ecrin et al., 2011; Vanham et al., 2018; Willett et al., 2019). To reduce the pressure on the environment, we need to not only balance the composition of our diet but also to reduce the portions of food we consume. A dietary shift combined with reducing our food waste might help us to reach important sustainability targets in terms of climate change and water use (Willett et al., 2019). Energy intake should be in balance with energy expenditure (2018). Excess energy intake (i.e., energy-dense unhealthy dietary patterns) is the most significant dietary factor related to weight gain and the development of obesity (Roberts et al., 2002) and contributes to increasing the environmental impact of the food sector.

5. *Celebrate variety: Vary your diet with the seasons and enjoy regional products and local varieties.*
   Agrobiodiversity is an important strategy to face the challenges of climate change, as it offers more flexible climate adaptation strategies, thus supporting food security and resource saving (Fischlin et al., 2007). Coupling variety while highlighting seasonal and local products could help boost local economies and save resources. Seasonality has significant implications for C footprint impact, with the CF of vegetables and fruits grown in heated greenhouses being up to fourfold (Clune et al., 2017; Petersson et al., 2021) than that of open-field vegetables and fruits. Generally, only when local markets are affected by seasonal constraints, which strongly limit food variety (cold climates), long-distance supply chains may be more advantageous compared with greenhouse food production or with long-term food storage throughout the seasons (Wakeland et al., 2012). Dietary diversity has been recognized as a key element of high-quality diets and is also a proxy for nutrient adequacy of the diet of individuals (Ruel, 2003). Agrobiodiversity is an underexplored avenue for giving both food producers and consumers access to greater dietary diversity (Johns & Eyzaguirre, 2006).

6. *Fresh is best (for you and for the environment): Favor fresh and naturally prepared food in your diet*
   Food processing can add further pressure on the environment. The more complex the processing, the higher amount of energy and materials required, which leads to higher CO_2-eq emissions and resource consumption per kilogram of product (Notarnicola et al., 2017). The increasing proportion of ultraprocessed foods in diets has been identified as a driver of excess energy intake (Hall et al., 2019). Also, the evidence

thus far shows that displacement of minimally processed foods and freshly prepared dishes and meals by ultraprocessed products is associated with unhealthy dietary nutrient profiles (Monteiro et al., 2018).

7. *Drink tap water whenever possible and safe: Drink plenty of water, choosing tap water over bottled water whenever it is possible and safe.*

When safe tap water is available, water in plastic bottles places an unnecessary burden on the planet. Bottled water creates additional CO2 emissions (Botto et al., 2011) and water consumption (Niccolucci et al., 2011). Bottled water in plastic material poses a further environmental burden, as plastic production relies heavily on oil (Foolmaun & Ramjeawon, 2008), and the amount of plastic recycled in many EU countries is still less than 50% of the production (PlasticsEurope, 2017). Good hydration is vital for good health and well-being. Water safety and quality are fundamental to human development and well-being, and 71% of the global population uses a safely managed drinking-water service (2019). One of the reasons for increased bottled water consumption seems to be consumer dissatisfaction with tap water characteristics that affect taste, odor, and sight, even in countries where tap water quality is considered excellent (Doria, 2006).

8. *Reduce single-use: Reduce, reuse, and recycle food packaging by bringing your own bags, cups, and cutlery whenever possible. Choose food with minimum packaging and recycle and reuse materials whenever possible.*

Food packaging represents an important fraction (25%) of the plastic materials that leak into the environment. The estimated annual input of mismanaged plastic waste to the ocean is about 13 million metric tons (Jambeck et al., 2015). This plastic remains in the ocean for centuries, causing harm to natural systems. Shopping for food in supermarkets has shown to have three times the CF per food basket than shopping in local food markets, due to the higher amount of packaging in supermarket food. Retailers could save a great deal of CO_2 emissions by helping to reduce food packaging, while users could significantly reduce the environmental footprint of their food basket by recycling or bringing their own shopping bags (Sanyé et al., 2012).

SU-EATABLE project design and evaluation—stage I

The design rationale and SU-EATABLE sustainable diet principles led to the first stage of activities at food service and customer level in seven cafeterias in the UK and Italy with the intention of running 4–6 months. The activities aimed to create an enabling environment for sustainable food service and consumption by initiating positive and practical learning opportunities while implementing the dietary principles in the canteens.

Engagement activities

Food service level

Several engagement activities were conducted at the food service level (Fig. 7.1). Firstly, staff and management were invited to revise their methods of food procurement and preparation and align their food offering with the SU-EATABLE criteria for sustainable, healthful meals. A series of meetings were held with cafeteria managers and chefs to understand the current food offering and menu structure after which menus were analyzed according to the eight principles for sustainable diets. In this way, cafeteria managers and chefs were involved in the sustainable learning trajectory. In particular, aspects such as the offer of red meat dishes, seasonality of ingredients, and offer of low environmental-impact healthy dishes were assessed. One-to-one proposals for changes in the food offering were discussed for the intervention phase. The majority of changes involved protein foods, such as the substitution of red meat with poultry, the reduction of cheese as a recipe ingredient, the inclusion of more legume-based dishes, as well as the inclusion of eggs and fish with lower environmental impact. In addition, the carbon and water footprints of each dish were calculated to attribute the sustainable label to dishes sold in the cafeteria and validate the chef recipes for the communication strategy with cafeteria users. Workshops were held with cafeteria chefs, managers, and staff in order to provide the scientific background for the eight principles, cocreate the experimental phase, and discuss sustainable menus and recipes. Following the establishment of new sustainable recipes with cafeteria chefs, all of them were compiled in a webpage to create a digital cookbook, inviting cafeteria users to choose the dish in the cafeteria when it was offered, and prepare it at home with family and friends. Fig. 7.1

Customer level

Customer-level activities were launched early 2020 in all intervention cafeterias (Fig. 7.2). A baseline customer survey was administered before the start of activities.

In the cafeteria setting, customers were invited to purchase the sustainably certified meal and exposed to banners and posters providing information on sustainable food consumption to promote an enabling environment toward the consumption of the new dishes.

Online, the GreenApes app provided a learning medium where users could participate in different types of challenges and earn various rewards in return. This approach embedded the learning in a positive, practical, and enjoyable way. Monthly challenges comprised a set of tasks to be performed and validated on the GreenApes app. Contrary to traditional approaches, which tend to only emphasize cognitive capacities (i.e., educational/informative approach), these challenges also aimed to develop other capacities with simple tasks that stimulate reflection, trigger search for information, encourage social participation, and simplify the selection and consumption of sustainable meals. Each monthly challenge introduced customers to different aspects of sustainable diets, and six out of the eight SU-EATABLE principles were selected to be covered over a period of 6 months of

FIGURE 7.1   **Stage 1 engagement activities at the food service level.** This figure overviews the engagement activities deployed at the level of food service.

experiments. An exception was made in two universities, where only the first 4 months of experiments would be conducted, matching the duration of the school semester.

Each month, challenge activities targeted the three elements of an individual's sense of coherence: (1) comprehensibility—understanding the characteristics of sustainable food choice, using knowledge and tools to make informed choices (cognitive capacity); (2) meaningfulness—being aware of and able to construct and assign meaning and value to sustainable food choice, being motivated to invest in sustainable food choice and intend to make this choice (motivational capacity); and (3) manageability—perceived self-efficacy and control over required social and practical skills needed to eat sustainably (actionable capacity). These three capacities are closely interlinked, where an increase in one capacity is likely to drive the other(s). For instance, increasing one's ability to understand what comprises a sustainable diet (cognitive capacity) could also result in an increased ability to select more sustainable options among different cafeteria dishes (actionable capacity). To complete a challenge, participants were invited to (1) watch a short educational video that addresses the topic in a positive light and shares a few practical tips, (2) consume an increasing number of sustainable meals each month by validating their purchase in the app, and (3) share a user story in response to a question or activity that requires reflection or action. Completing a challenge was rewarded with a physical, social, or collective reward to test the effect of different rewarding systems (Fig. 7.2).

Intermediate evaluation stage I

Due to COVID-19 restrictions, the on-site activities came to an early halt in February (Italy) and March (UK) after running 1 month in the UK and 2 months in Italy instead of the original 4 and 6 months. An intermediate evaluation was conducted hereinafter including the analysis of: (1) the baseline customer survey, (2) the baseline extended customer survey (survey the users filled in after downloading the app), (3) sustainable meal purchases during the intervention period, and (4) engagement in challenges during the intervention period (Table 7.1). In addition, cafeteria customers and staff were interviewed to gain insights into participation and engagement during the short initial launch (Table 7.1).

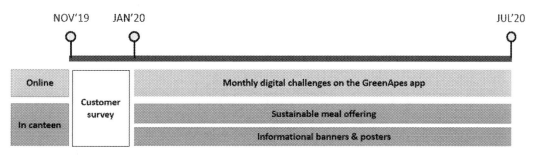

FIGURE 7.2 **Stage I intervention timeline for cafeteria users.** This figure overviews the timeline of intervention activities for cafeteria users.

TABLE 7.1   Baseline and intermediate monitoring and evaluation methods.

	Month 0 Baseline Evaluation	Month 2–3 Intermediate Evaluation UK: Jan-Mar 2020 (2 months) IT: Feb-Mar 2020 (1 month)
Sales & procurement data	*Baseline sales and procurement data*: collected in Nov–Dec 2019 before online baseline surveys were conducted	*Monthly sales and procurement data*: collected at the end of each month of experiments
Baseline survey	*Baseline survey (n = 595)*: food choice motivations, current sustainable food practices and capacities underlying a change towards sustainable diets	Analysis of *baseline survey*
GreenApes app	*Baseline extended survey (n = 46)*: food choice motivations, sustainable eating and lifestyle, self-identity and perceptions on sustainability	Analysis of: Baseline extended survey Sustainable meal purchases: monthly tracking of sustainable meals validated with GreenApes app *Engagement in challenges*: completion of challenges
Interviews and consultations	N.A.	Informal interviews and consultations with cafeteria staff Interviews with cafeteria customers in the UK (n = 18) and IT (n = 3) to evaluate participation and engagement

Food service level

Based on interviews and consultations with cafeteria staff, there was generally high interest and support from management and caterers. However, several challenges emerged during the initial 2 months of experiments (Table 7.2). The intermediate evaluation emphasized the importance of fostering a shared leadership in the implementation of project activities, reducing complexity and providing easy-to-follow informational resources. In addition, it underlined the importance of integrating project activities into the job routines of workers involved in the intervention deployment. Applying the learnings outlined in Table 7.2 is important to make the intervention more sustainable and the personnel more resilient to the challenges of implementing a change in their food offering (Table 7.2).

Customer level

Customer-level data included insights from the baseline survey, baseline extended survey, and GreenApes analytics. The following main insights were uncovered from the analysis of the two baseline surveys:

- *Country differences in food motivations*: consumers in Italy found it most important that food is healthy and good for the mood, whereas consumers in the UK considered it should be affordable, satisfying, and enjoyable. Promoting sustainable food should therefore address specific top drivers of food choice in each country.
- *Affordability of food in universities*: affordability is very important to cafeteria-goers in universities, where economic incentives could be most effective to promote sustainable foods.

TABLE 7.2 Challenges and learnings on engagement of cafeteria staff and management.

Challenge	Learning
1. *Complexity of project elements* While the main site contact person is a close collaborator and understands the project elements, effort is required as there can be a lack of understanding concerning the importance of certain project elements (e.g., menu)	Simplification of project elements and steps to be taken for participation Clear materials explaining the project that can be handed out to staff
2. Internal coordination issues Changes in key coordinating person over the course of the project could result in less internal drive to realize project activities, as well as a loss of information	Clarity in expectations and information should be established in the beginning, with supporting materials
3. *Knowledge of sustainable diets* Cafeteria staff and chefs demonstrated very limited and inaccurate knowledge about sustainable diets. This contributed to a lack of understanding about the "OPP" and "MP4TF" sustainability meal labels and sometimes led to menu changes that the labels were not suited to	Single workshops are insufficient to establish understanding and autonomy. Provide caterers and chefs with simple tools that improve their understanding and enables them to be flexible with meal offerings and to champion sustainable diets
4. *Role of the chefs* Chefs determine what food ends up being served and how tasty it is, but time, motivation, skills and understanding of sustainable diets seem to be lacking. This needs to be addressed if the key motivations of cafeteria customers when picking meals (i.e.: perception meals are "satisfying & enjoyable," "healthy" and "good for mood") are to be fulfilled	Providing incentives for chefs to expand their repertoire and serve more tasty sustainable meals. Source opportunities for inspiring chefs on the possibilities without creating additional load into their daily jobs
5. *Food culture & consumer demand* The catering business is strongly driven by consumer demand, and when meat-based dishes are perceived as comfort food (e.g., lasagna in Italy), it is difficult to replace them on the menu	Explore and add local plant-rich dishes to the menu (many of which are already popular amongst consumers) and expand the choice range rather than limiting meat.
6. *Time* The catering service is fast paced. Project activities required additional time investments (e.g., separate counter to scan QR code, weighing and calculating food) and are not sustainable over time for replication	Technological capabilities for automation should be a requirement for participation. Additional time investments should be avoided or established upfront so that additional manpower can be assigned
7. *Internal investment of resources* Management staff welcomed the project due to the alignment with health and environmental goals of their organization, but did not channel budget or resources	Establish a win-win relationship from the start where a commitment of specific manpower and financial resources should be negotiated
8. *Lunch vs dinner* Particularly in Italian cafeterias where dinner is also served, there is less of a time pressure compared to lunch. However, project activities were geared towards the lunch crowd	Consider implementing project activities also for dinner time in cafeterias where dinner is served

- *High interest but low actionability*: the high level of awareness and interest in sustainable diets were accompanied by a relatively lower perceived scores in social desirability of sustainable eating and ability to select sustainable foods. This suggested that more attention could be given to creating an environment and tools that support people in integrating sustainability into their everyday life.

- *Learning with low effort*: learning modes that involved more effort (e.g., taking part in challenges) and social interaction (e.g., peers, discussion events) were less preferred. Hence, for experiential and social learning where active participation is required, fun elements, meaningful rewards, and food enjoyment could help to trigger interest and intrinsic motivation.

The data from the GreenApes app from the first months of the intervention showed that only a small fraction of cafeteria users had downloaded the app (n = 244 across the seven cafeterias). Approximately half of those who registered scanned their first meal (n = 111), but only a third of them (n = 38) proceeded to complete the welcome challenge ("Challenge 0"). Of these users, only five proceeded to complete Challenge 1 of which four claimed the reward after completing the challenge. Increasing the value of rewards for completing the second challenge did not result in significant improvement in participation levels. This trend was consistent across all cafeterias and shows low level of engagement and participation.

These combined findings signaled that the app content needed to be adapted and the reach increased for stage II of the intervention. They also indicated the areas to work on in order to make the challenge content more relevant for the users. In addition, important new insights were gained from the interviews with cafeteria customers, see Table 7.3.

The way forward: stage II of the SU-EATABLE LIFE project

The interruption of experiments due to the COVID-19 pandemic provided an opportunity for the project team to revisit the approach and adapt the engagement activities to the context, perceptions, and challenges of the cafeteria stakeholders and customers. The current events highlighted the importance of prioritizing the replicability of experiments in different settings, fostering internal ownership, and providing the necessary tools and information for stakeholders to lead project activities internally with minimal support from the SU-EATABLE LIFE project team.

In line with other research and initiatives to promote sustainable behavior in citizens (James, 2010), lessons from the evaluation also showed the importance of positive framing, knowing the audience, and simple and concrete messaging. Next to that, fitting project activities within the routines of customers and staff, and providing them with the tools for flexibility and creativity without restricting their freedom of choice could be an important step toward empowerment and engagement. Moreover, food choice in the cafeteria was shown to be largely driven by taste, health, and affordability. Other studies have shown that successfully reducing the offer of animal-based products in food service requires chefs to reinvent meals and give consumers a sense of how tasty sustainable cuisine can be (Lopez et al. 2019). Table 7.4 summarizes the focus areas for the adaptation of engagement activities for the relaunch.

Based on the learnings, the project team is working toward several adaptations to the project activities to ensure replicability, increase participation and engagement, and adjust to the uncertainties of COVID-19 times:

TABLE 7.3 Evaluation of project activities and learnings for adaptation.

Dimension to be addressed	Learnings to be applied
1. *I come to eat not to read* People do not especially take the time to read, process and understand multiple guidelines and steps they need to take to participate	Very simple messaging is needed Communication materials should be located at decision-making points or places where people have time to read
2. *A place for relaxation and enjoyment* People do not want to be disturbed and do not want to make much additional effort during their lunch time	Participation should be integrated as much as possible into the regular lunch routine Decrease material to be filled
3. *Seeking tasty, healthy, familiar foods* Food choices are driven by food preferences and habit. Although the labeling of sustainable meals do trigger a few people to choose something different, sustainability is not usually considered	Offer a broad range of popular plant-rich foods and promote them. Focus on the enjoyment of sustainable meal offerings, and offer a free taste of new dishes
4. *Poor understanding of sustainable diets* Despite interest in the topic, most people show a very limited understanding and poor ability to grasp the guidelines	Simpler messaging but offering possibilities to learn more (e.g., through the GreenApes app)
5. *No awareness about learning experience* The fun learning experience that the challenges provided on the app was not communicated clearly, and people were not aware of it	More publicity about the learning experience Creating learning activities in the cafeteria that link to the app (e.g., guess the impact competition)
6. *Value for money* Many bring their own food to limit expenses especially in university cafeterias, whereas participation in the project is very much based on purchasing meals. Many also don't consider the meals to be value for money	Economic incentives for buying sustainable meals, e.g., by promotions and vouchers as rewards People who bring their own food should also be able to complete challenges and win rewards
7. *Not digitally savvy* Not everyone is a heavy app user or enjoys using it as a means for learning, whereas the use of the app is central to the learning experience	Create opportunities for reflection and learning outside the app environment, e.g., by providing more informative short displays and activities on site
8. *Positive, practical framing* A positive framing focusing on the benefits of a sustainable diet as well as practical tips on how to eat sustainably were generally appreciated	Stay positive and provide easy tips on small actions people can take to contribute and tell them how it makes a positive impact

- *Implementing toolkits and fostering ownership*: In order to make the implementation of project activities more resilient to external circumstances such as COVID-19 disruptions and the resources of the SU-EATABLE LIFE project team, participating cafeterias and caterers should be the main drivers of the project internally with support from the project team. We will develop an easy-to-use and comprehensive toolkit for stakeholders to adapt and implement the intervention according to their needs and also establish the key elements and commitments required to participate. The staff needs to be fully integrated in the intervention deployment and if possible, in its development. Educating populations about the risks associated with current diets has shown little success to date, and interventions nudging people toward more sustainable choices offer limited scope for lasting change (Guthrie et al., 2015). In contrast to such interventions that are generally "top-down" in nature, approaches in health promotion

TABLE 7.4   Key learnings from Stage I and focus areas for adaptation.

PRIORITIZE REPLICABILITY Create a toolbox and process that can be easily replicated in many cafeterias and are not heavily dependent on the project team and financial resources		
FOSTERING OWNERSHIP	COMMUNICATION STYLE	LEARNING EXPERIENCE
Cafeteria stakeholders share lead and commit resources	Positive, concrete, practical and simple	Positive sensory experiences and fun opportunities for learning
KNOW OUR AUDIENCES	DIGITAL EXPERIENCE (APP)	CAFETERIA EXPERIENCE
Tailored to audiences, time constraints and specific food motivations	Simplified toolbox style and inclusion of meaningful rewards	Inclusive, food-focused, increase tasty options and fit routines
SCIENTIFIC FOUNDATION Theoretical basis for the approach and understanding of the factors, mechanisms and barriers to engagement in sustainable diets		

are characterized by active participation of all actors involved in the change process and aim to empower people to achieve longer-term and wider-reaching behavioral and socioeconomic impact (Tengland, 2012). To facilitate a participatory, cocreation approach within the project, there needs to be dialog, integration, and exchange of views and ideas especially with the chef, kitchen and floor staff. Establishing such a dynamic may allow the staff to become part of the process toward a more sustainable food offering rather than passive observers. This would provide them with a sense of agency but also give the researchers the opportunity to set up intervention activities, which may be more relevant and understandable to them, thus ensuring a potentially more successful intervention and more engaged participants.

- *Increasing flexibility and reducing effort:* Research on intrinsic motivation for long-lasting change warns against putting deadlines and pressure (Steg, 2016). It instead emphasizes the need to provide flexibility and create optimally challenging contexts and opportunities for people to make their own choices. To translate this to the intervention, the experimental time frame will be shortened, and tasks and challenges will be open at all times. This translates into a toolbox-style digital experience, where users can pick and choose according to their interests and learning journey.
- *Know our audiences:* The principles of cocreation will also be applied to cafeteria consumers, increasing their role in the design of activities in order to better understand their time constraints, food motivations, and challenges to tailor content. A buffer of customization regarding the intervention activities will be created by using a toolbox model in which intervention implementers can pick the best-suited activities for their audiences.
- *Communication style, learning experience, cafeteria experience, digital experience:* Everything surrounding the direct and indirect promotion of sustainable diets needs to be done in a positive, fun, practical manner with very actionable tips. Furthermore, activities need to be conducted in a way that is embedded in the preexisting routines of consumers and take delicious food as an entry-way to instilling more sustainable food practices. Food is a prime entry point as eating is the ultimate daily routine that is followed by all and a moment of enjoyment.

Despite the interruption of the intervention due to the COVID-19 pandemic, the intermediate evaluation of project activities provided a number of learnings for the adaptation of the project. Overall, the original learning design was structured around monthly themes/challenges for customers and providing information to catering staff. This did not achieve high levels of engagement or withstand the pressures of time in a fast-paced cafeteria environment. A toolbox approach that provides a range of resources for greater flexibility of implementation and participation was identified as an important factor for the integration of activities into everyday life routines and practices. In addition, there were a number of interesting questions raised regarding people's perception of sustainable food practices and actual behavior, the role of budget, and the importance of social desirability when it comes to sustainable eating. It would be interesting to explore this more deeply in the interviews at the end of a complete run of experiments.

In conclusion, the current events highlighted the importance of prioritizing the replicability of experiments for different settings and the need to create a process and set of resources that can be easily adapted, applied, and validated by participating cafeterias with minimal support from the external project team. These findings have been applied to design and implement the second stage of project activities that started when cafeteria could reopen in October 2021. Four university cafeterias and one business cafeteria in the UK were recruited among members of the Sustainable Restaurant Association. Activities included on-site and online tools for managers, kitchen- and floor staff, university students, and employees that aimed to enable engagement in sustainable diets. The overall outcomes of the project activities can be found at the SU-EATABLE website.[1]

Acknowledgments

This chapter is based on the efforts of all members of our interdisciplinary project team that consists of social and natural scientists, social media designers, food provisioning intermediaries, communication experts, nutritionists, and dieticians working at Wageningen University (NL), GreenApes (IT), the Sustainable Restaurant Association (UK), and the Barilla Center for Food and Nutrition (IT). The EU SU-EATABLE LIFE project is cofinanced by the European Commission.

References

Abete, I., Romaguera, D., Vieira, A. R., Lopez de Munain, A., & Norat, T. (2014). Association between total, processed, red and white meat consumption and all-cause, CVD and IHD mortality: A meta-analysis of cohort studies. *The British Journal of Nutrition, 112*(5), 762–775. Available from https://doi.org/10.1017/S000711451400124X.

Bacon, L., & Krpan, D. (2018). Not) Eating for the environment: The impact of restaurant menu design on vegetarian food choice. *Appetite, 125*, 190–200. Available from https://doi.org/10.1016/j.appet.2018.02.006.

Böhme, T., Stanszus, L., Geiger, S., Fischer, D., & Schrader, U. (2018). Mindfulness training at school: A way to Engage Adolescents with Sustainable Consumption? *Sustainability, 10*(10), 3557. Available from https://doi.org/10.3390/su10103557.

Botto, S., Niccolucci, B., Rugani, V., Bastianoni, C., & Gaggi, C. (2011). Towards lower carbon footprint patterns of consumption: The case of drinking water in Italy. *Environmental Science & Policy, 14*(4), 388–395.

---

[1] https://www.sueatablelife.eu/

British Dietetic Association. (n.d.-a). Plant-based diet. Retrieved September 2017, https://www.bda.uk.com/food-facts/plant-based_diet.

Bryngelsson, D., Wirsenius, S., Hedenus, F., & Sonesson, U. (2016). How can the EU climate targets be met? A combined analysis of technological and demand-side changes in food and agriculture. *Food Policy*, *59*, 152–164. Available from https://doi.org/10.1016/j.foodpol.2015.12.012.

Chen, G. C., Lv, D. B., Pang, Z., & Liu, Q. F. (2013). Red and processed meat consumption and risk of stroke: A meta-analysis of prospective cohort studies. *European Journal of Clinical Nutrition*, *67*(1), 91–95. Available from https://doi.org/10.1038/ejcn.2012.180.

Clune, S., Crossin, E., & Verghese, K. (2017). Systematic review of greenhouse gas emissions for different fresh food categories. *Journal of Cleaner Production*, *140*(2), 766–783.

Doria, M. F. (2006). Bottled water versus tap water: Understanding consumer's preferences. *Journal of Water and Health*, *4*(2), 271–276. Available from https://doi.org/10.2166/wh.2006.008.

Ecrin, A., Aldaya, M., & Hoekstra, A. (2011). Corporate water footprint accounting and impact assessment: The case of the water footprint of a sugar-containing carbonated beverage. *Water Resources Management*, *25*, 721–741.

European Commission. (2020). Communication from the Commission to the European Parliament, the Council, the European Economic and Social Committee and the Committee of the Regions: A Farm to Fork Strategy for a fair, healthy and environmentally-friendly food systemCOM/2020/381 final. European Commission.

Fanzo, J., & Davis, C. (2019). Can diets be healthy, sustainable, and equitable? *Current Obesity Reports*, *8*(4), 495–503. Available from https://doi.org/10.1007/s13679-019-00362-0.

Feskens, E. J. M., Sluik, D., & Van Woudenbergh, G. J. (2013). Meat consumption, diabetes, and its complications. *Current Diabetes Reports*, *13*(2), 298–306. Available from https://doi.org/10.1007/s11892-013-0365-0.

Filimonau, V., Lemmer, C., Marshall, D., & Bejjani, G. (2017). Restaurant menu re-design as a facilitator of more responsible consumer choice: An exploratory and preliminary study. *Journal of Hospitality and Tourism Management*, *33*, 73–81. Available from https://doi.org/10.1016/j.jhtm.2017.09.005.

Fischlin, A., Midgley, G., Price, J., Leemans, R., Gopal, B., Turley, C., Rounsevell, M., Dube, O., Tarazona, J., & Velichko, A. (2007). In M. Parry, J. Canziani, J. Palutikof, P. van der Linden, & C. Hanson (Eds.), *Ecosystems, their properties, goods, and services*. Cambridge University Press.

Foolmaun, R., & Ramjeawon, T. (2008). Life Cycle Assessment (LCA) of PET bottles and comparative LCA of three disposal options in Mauritius. *International Journal of Environment and Waste Management*, *2*(1/2), 125–138.

Guo, J., Astrup, A., Lovegrove, J. A., Gijsbers, L., Givens, D. I., & Soedamah-Muthu, S. S. (2017). Milk and dairy consumption and risk of cardiovascular diseases and all-cause mortality: Dose–response meta-analysis of prospective cohort studies. *European Journal of Epidemiology*, *32*(4), 269–287. Available from https://doi.org/10.1007/s10654-017-0243-1.

Hall, K. D., Ayuketah, A., Brychta, R., Cai, H., Cassimatis, T., Chen, K. Y., Chung, S. T., Costa, E., Courville, A., Darcey, V., Fletcher, L. A., Forde, C. G., Gharib, A. M., Guo, J., Howard, R., Joseph, P. V., McGehee, S., Ouwerkerk, R., Raisinger, K., ... Zhou, M. (2019). Ultra-processed diets cause excess calorie intake and weight gain: An inpatient randomized controlled trial of ad libitum food intake. *Cell Metabolism*, *30*(1), 67–77. e3. Available from https://doi.org/10.1016/j.cmet.2019.05.008.

Jambeck, J. R., Geyer, R., Wilcox, C., Siegler, T. R., Perryman, M., Andrady, A., Narayan, R., & Law, K. L. (2015). Marine pollution. Plastic waste inputs from land into the ocean. *Science (New York, N.Y.)*, *347*(6223), 768–771. Available from https://doi.org/10.1126/science.1260352.

John Hopkins Centre for a Liveable Future. (n.d.-b). Implementing meatless Monday in food service operators. Retrieved July 2021, https://clf.jhsph.edu/sites/default/files/2019-02/implementing-meatless-monday-in-food-service-operation-report.pdf.

Johns, T., & Eyzaguirre, P. B. (2006). Linking biodiversity, diet and health in policy and practice. *In Proceedings of the Nutrition Society*, *65*(2), 182–189. Available from https://doi.org/10.1079/PNS2006494.

Maher, J., & Burkhart, S. (2017). Experiential learning for engaging nutrition undergraduates with sustainability. *International Journal of Sustainability in Higher Education*, *18*(7), 1108–1122. Available from https://doi.org/10.1108/IJSHE-01-2016-0010.

Mekonnen, M., & Hoekstra, A. (2010). The green, blue and grey water footprint of farm animals and animal products, Value of Water Research Report Series No. 48. UNESCO-IHE.

Monteiro, C. A., Cannon, G., Moubarac, J. C., Levy, R. B., Louzada, M. L. C., & Jaime, P. C. (2018). The un Decade of Nutrition, the NOVA food classification and the trouble with ultra-processing. *Public Health Nutrition, 21*(1), 5–17. Available from https://doi.org/10.1017/S1368980017000234.

Niccolucci, V., Botto, S., Rugani, B., Nicolardi, V., Bastianoni, C., & Gaggi, C. (2011). The real water consumption behind drinking water: The case of Italy. *Journal of Environmental Management, 92*(10), 2611–2618.

Notarnicola, B., Serenella, S., Assumpció, A., McLaren, S. J., & Saouter, E. (2017). The role of life cycle assessment in supporting sustainable agri-food systems: A review of the challenges. *Journal of Cleaner Production, 140*(2), 399–409.

Oostindjer, M., Aschemann-Witzel, J., Wang, Q., Skuland, S. E., Egelandsdal, B., Amdam, G. V., Schjøll, A., Pachucki, M. C., Rozin, P., Stein, J., Lengard Almli, V., & Van Kleef, E. (2017). Are school meals a viable and sustainable tool to improve the healthiness and sustainability of children's diet and food consumption? A cross-national comparative perspective. *Critical Reviews in Food Science and Nutrition, 57*(18), 3942–3958. Available from https://doi.org/10.1080/10408398.2016.1197180.

Pan, A., Sun, Q., Bernstein, A. M., Schulze, M. B., Manson, J. A. E., Stampfer, M. J., Willett, W. C., & Hu, F. B. (2012). Red meat consumption and mortality: Results from 2 prospective cohort studies. *Archives of Internal Medicine, 172*(7), 555–563. Available from https://doi.org/10.1001/archinternmed.2011.2287.

Petersson, T., Secondi, L., Magnani, A., Antonelli, M., Dembska, K., Valentini, R., Varotto, A., & Castaldi, S. (2021). A multilevel carbon and water footprint dataset of food commodities. *Scientific Data, 8*(1). Available from https://doi.org/10.1038/s41597-021-00909-8.

PlasticsEurope. (2017). Plastics- the Facts 2017; An analysis of European plastics production, demand and waste data. Plastic Europe, Association of Plastics Manufacturers. Available from https://plasticseurope.org/wp-content/uploads/2021/10/2017-Plastics-the-facts.pdf.

Poore, J., & Nemecek, T. (2018). Reducing food's environmental impacts through producers and consumers. *Science (New York, N.Y.), 360*(6392), 987–992. Available from https://doi.org/10.1126/science.aaq0216.

Roberts, S. B., McCrory, M. A., & Saltzman, E. (2002). The influence of dietary composition on energy intake and body weight. *Journal of the American College of Nutrition, 21*(2), 140–145. Available from https://doi.org/10.1080/07315724.2002.10719211.

Ruel, M. T. (2003). Operationalizing dietary diversity: A review of measurement issues and research priorities. *The Journal of Nutrition, 133*(11), 3911S–3926S. Available from https://doi.org/10.1093/jn/133.11.3911s.

Sanyé, E., Oliver-Solà, J., Gasol, C. M., Farreny, R., Rieradevall, J., & Gabarell, X. (2012). Life cycle assessment of energy flow and packaging use in food purchasing. *Journal of Cleaner Production, 25*, 51–59.

Sardi, L., Idri, A., & Fernández-Alemán, J. L. (2017). A systematic review of gamification in e-Health. *Journal of Biomedical Informatics, 71*, 31–48. Available from https://doi.org/10.1016/j.jbi.2017.05.011.

Satija, A., Bhupathiraju, S. N., Spiegelman, D., Chiuve, S. E., Manson, J. A. E., Willett, W., Rexrode, K. M., Rimm, E. B., & Hu, F. B. (2017). Healthful and unhealthful plant-based diets and the risk of coronary heart disease in U.S. adults. *Journal of the American College of Cardiology, 70*(4), 411–422. Available from https://doi.org/10.1016/j.jacc.2017.05.047.

Satija, Ambika, Bhupathiraju, S. N., Rimm, E. B., Spiegelman, D., Chiuve, S. E., Borgi, L., Willett, W. C., Manson, J. E., Sun, Q., Hu, F. B., & Moore, S. C. (2016). Plant-based dietary patterns and incidence of type 2 diabetes in US men and women: Results from three prospective cohort studies. *PLoS Medicine, 13*(6)e1002039. Available from https://doi.org/10.1371/journal.pmed.1002039.

Springmann, M., Wiebe, K., Mason-D'Croz, D., Sulser, T. B., Rayner, M., & Scarborough, P. (2018). Health and nutritional aspects of sustainable diet strategies and their association with environmental impacts: A global modelling analysis with country-level detail. *The Lancet Planetary Health, 2*(10), e451–e461. Available from https://doi.org/10.1016/S2542-5196(18)30206-7.

Tilman, D., & Clark, M. (2014). Global diets link environmental sustainability and human health. *Nature, 515*(7528), 518–522. Available from https://doi.org/10.1038/nature13959.

Toumpanakis, A., Turnbull, T., & Alba-Barba, I. (2018). Effectiveness of plant-based diets in promoting well-being in the management of type 2 diabetes: A systematic review. *BMJ Open Diabetes Research & Care, 6*(1)e000534. Available from https://doi.org/10.1136/bmjdrc-2018-000534.

Tuso, P. J., Ismail, M. H., Ha, B. P., & Bartolotto, C. (2013). Nutritional update for physicians: Plant-based diets. *The Permanente Journal, 17*(2), 61–66. Available from https://doi.org/10.7812/TPP/12-085.

UN General Assembly. (2015). Transforming our world: The 2030 Agenda for Sustainable Development (A/RES/70/1). Available from https://www.refworld.org/docid/57b6e3e44.html.

Van Woerkum, C., & Bouwman, L. (2014). Getting things done": An everyday-life perspective towards bridging the gap between intentions and practices in health-related behavior. *Health Promotion International*, *29*(2), 278–286. Available from https://doi.org/10.1093/heapro/das059.

Vanham, D., & Bidoglio, G. (2013). A review on the indicator water footprint for the EU28. *Ecological Indicators*, *26*, 617–675.

Vanham, D., Comero, S., & Gawlik, B. (2018). The water footprint of different diets within European sub-national geographical entities. *Nature Sustainability*, *1*, 518–525. Available from https://doi.org/10.1038/s41893-018-0133-x.

Wakeland, W., Cholette, S., & Venkat, K. (2012). In J. Boye, & Y. Arcand (Eds.), *Food transportation issues and reducing carbon footprint*. Springer. Available from https://doi.org/10.1007/978-1-4614-1587-9_9.

Willett, W., Rockström, J., Loken, B., Springmann, M., Lang, T., Vermeulen, S., Garnett, T., Tilman, D., DeClerck, F., Wood, A., Jonell, M., Clark, M., Gordon, L. J., Fanzo, J., Hawkes, C., Zurayk, R., Rivera, J. A., De Vries, W., Majele Sibanda, L., ... Murray, C. J. L. (2019). Food in the Anthropocene: The EAT–Lancet Commission on healthy diets from sustainable food systems. *The Lancet*, *393*(10170), 447–492. Available from https://doi.org/10.1016/S0140-6736(18)31788-4.

Design thinking workshop: using experiential learning, creativity, and empathy to learn about the complexities of food insecurity and sustainability [☆]

Sarah M. Zehr[1], Deana McDonagh[1], Jennifer Vokoun[2], Francesca Allievi[3] and Sonia Massari[4]

[1]Department of Education Policy, Organization & Leadership, College of Education, University of Illinois Urbana-Champaign, Urbana-Champaign, IL, United States [2]School of Art & Design, College of Fine & Applied Arts, Walsh University, North Canton, OH, United States [3]Center for Sustainable Food Design, Jamk University of Applied Sciences, Jyväskylä, Finland [4]Tourism and Service Business, School of Business, Pisa University, Pisa, Italy

Introduction

In the past couple of decades, the number of sustainable agriculture projects has been steadily increasing in universities across the United States, and they have been linked with an enhanced critical thinking as well as a more significant relationship with the surrounding community (La Charite, 2016). On the other hand, formal education and nonformal education on food sustainability issues have not caught up yet (Massari, Allievi, et al., 2021) and fail to have a long-term impact on how students relate to sustainability on a general

---

[☆] This project adhered to Institutional Research Board approval (University of Illinois Urbana Champaign IRB Protocol Number 18636).

Food Sustainability and the Media
DOI: https://doi.org/10.1016/B978-0-323-91227-3.00002-0

level. This is also due to institutional and cultural barriers, which hinder more resources to be allocated to sustainability (Kioupi & Voulvoulis, 2019).

While higher education faces these difficulties in educating youth about (food) sustainability, the media and social media have the potential to reach this audience and bring about change. Higher education institutions could use social media as a leverage point in this sense, but often fail to fully exploit this potential (Carpenter et al., 2016). Some studies have identified how social media can help to communicate topics such as sustainable consumption and recycling; however, also negative effects are present and include a lack of critical thinking for the assessment of fake news, as well as the promotion of a homogeneous consumption style, which has severe impacts on both environmental resources and inequalities. In the current setting of a food system dominated by the power of big corporations, the role and effect of social media are twofold: on the one hand, it allows horizontal, self-organized movements to reach larger audiences and offers an alternative to the industrialized food system, on the other hand, it can generate forces to regain the power lost (Stevens et al., 2016).

Despite the increasing information available on the link between the current average dietary patterns in Western countries and a variety of serious environmental and social issues, the definition of "sustainable diet" remains obscure for most. In this sense, media do not support a more comprehensive information on these topics. Some studies have highlighted how young people perceive that they receive too little information on sustainable diets from both media and education, and this hinders their full understanding of these topics. For instance, in 2019, Barilla Foundation commissioned Ipsos Italia to investigate the relationship between the new generations and Sustainable Development Goals[1]. The resulting data highlighted the often partial vision that students (age 14−27) have concerning sustainability, frequently clear on environmental factors but underestimating the importance of food behavior[2]. Young Italians seem to be supportive of the fight to reduce the impact of human behavior on climate change, but they are not sufficiently informed on the strategies that can be implemented to achieve lasting results, they are not familiar with the Sustainable Development Goals promoted by the United Nations, and above all, they do not know the extent to which agricultural production and the food they eat have an impact on sustainability.

According to the data collected, 44% of the young people interviewed are uninformed on the issues of politics, current events, and the economy, while only 15% were attentive and constantly informed. In fact, as often occurs among older people as well, youth tend to relate sustainability only to environmental aspects, while the equally important issues of sustainability associated with the economy (13%), society (9%), and food and nutrition (9%) remain in the background. The lack of information is directly correlated with the general attitude and the various everyday behaviors in relation to waste and to the choices of food sustainability: the attentive youth and informed youth tend to prefer products from sustainable agriculture, to always read food labels carefully, and to try to avoid wasting water. Forty-five percent of the sample who are knowledgeable or at least

[1] Available from https://sdgs.un.org/goals

[2] Available from https://www.barillacfn.com/en/magazine/food-and-sustainability/youth-and-sdgs-few-know-the-role-of-food-and-nutrition/

superficially knowledgeable about the issue received their information from the school or the university, whereas only in the age 24–27 segment was the media the main source of information, in particular Internet and newspapers. In fact, being familiar with the Sustainable Development Goals is not a sufficient condition for the youth to feel an urgent duty to act immediately. What makes the difference, instead, is the sense of engagement in taking charge of the problem, regardless of the qualified knowledge about the issue.

In the middle of these contradictions and contrasting forces, there seems to be a gap in both formal education and information from the media that hinders these from being effective to promote a comprehensive understanding of food sustainability. For this reason, the case presented here shows a different approach, based on experiential learning. In the workshop described, participants experience firsthand key topics of food sustainability, allowing for the activation of empathy and for a deeper understanding.

In this chapter, an example will be presented to put forward the case of how, by using design thinking and therefore acting on mechanisms of empathy, the understanding of sustainable diets can be transferred more quickly and effectively.

Presidents United to Solve Hunger Leaders Forum and the Universities Fighting World Hunger: multidisciplinary and multigenerational experiment to experience sustainability through food

The case presented here took place during the Presidents United to Solve Hunger (PUSH) Leaders Forum and the Universities Fighting World Hunger (UFWH) Summit organized in March 2018. PUSH is an organization hosted at Auburn University (US) that brings together university leaders across the United States (and a few institutions in Canada and other countries) to discuss and collaborate on initiatives to address world hunger (PUSH website). UFWH is a sister organization for students who are interested in fighting hunger (UFWH website).

Though the PUSH and UFWH events are separate and target very distinct audiences, the conference organizers wanted to propose one event that brought the two groups together. They decided to include a design thinking workshop to transition from the PUSH meeting for university presidents and their designees to the UFWH event for students. Three professors from three universities (University of Illinois Urbana-Champaign, Walsh University, and Roma Tre University in Rome, Italy) who specialized in design agreed to collaborate to develop a design thinking workshop focused on food sustainability (the three-professor team along with an additional university administrator who assisted with the process will be referred to as the workshop leaders throughout the rest of the article). Approximately 125 university leaders and students participated in the workshop. The workshop had two main goals. First, the workshop leaders wanted to introduce the participants to the design thinking process and to provide them with a unique and impactful experience. Second, they aimed to enhance participants' understanding of the food system and sustainability and to encourage them to brainstorm and discuss potential solutions.

Participants were randomly assigned to tables of 10 so most did not sit with people with whom they were well acquainted. The workshop leaders led the participants through a series of 10-minute presentations focused on various aspects of food sustainability. At the end of each presentation, a thought-provoking question was posed to participants, and they were asked to think individually about the emotions they were feeling and to collaboratively reflect on the question. After a few minutes of individual time, they discussed the question with the others at their table. During the second half of the workshop, participants received one of four specially designed boxed lunches. Meals ranged from meager, representing malnutrition and an insecure diet, to a healthy meal (based on human health), to a sustainability balanced meal (planet health), to a high calorie and unhealthy meal (more than one person could possibly eat) that represented wealth and overabundance. Each boxed lunch appeared to be identical from the outside and the white boxes were unlabeled, so participants randomly chose the meals from the selection on the table. The boxed lunches were a critical part of the workshop design, and the groups at each table continued to discuss and generate ideas as they ate. Fig. 8.1 provides details about what was included in each boxed lunch category, including pictures of the meals.

The following questions were posed to elicit the impact of the design thinking workshop:

- How does participation in an experiential learning activity impact the participant's understanding of the complex concept of food sustainability?
- How does participation in an experiential learning activity impact the participant's motivation and perceived ability to address a complex societal problem?

Theoretical background and definition of key concepts

Since the 2008 financial crisis in the United States, 14% or more of U.S. households reported being food insecure (Zepeda, 2018). The U.S. Department of Agriculture defines food insecurity as reduced quality, variety, or desirability of diet. In 2019, before the COVID-19 pandemic, 690 million people worldwide (about 9% of global population) suffered from undernourishment, and projections estimate that the pandemic will add another 83–120 million people to this amount (FAO, IFAD, UNICEF, WFP and WHO, 2020).

While most people may assume that people who report being food insecure are unemployed, in actuality, many individuals who are employed full time do not generate enough income to be considered food secure and may not have access to healthy, nutritious food consistently (Zepeda, 2018).

Zepeda (2018) conducted a case study of 20 food insecure individuals who were categorized as middle class but did not use food pantries. She used the Asset Vulnerability Framework (AVF) to classify the causes of nonhealthy and nonsustainable food habits among the participants. The AVF identifies five potential causes for food insecurity: labor (unemployment or underemployment), productive (using available money for expenses such as housing or transportation that provide a means to make money, such as going to work versus spending it on food), human capital (health or education expenses), household relations (who/how many live/lives in the household), and social capital (avoiding

Meal Type	Boxed	Plated	Contents
Healthy ~20%		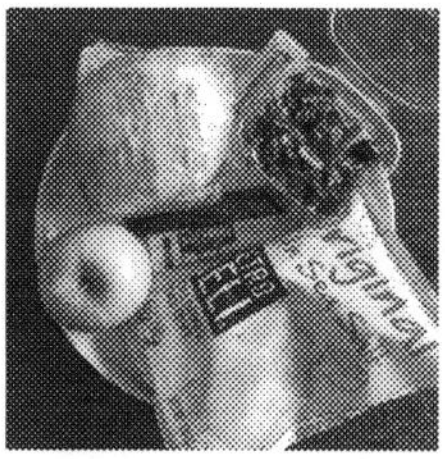	Apple Mixed Greens Crisps/chips Turkey/Ham Sandwich Mayonnaise/Mustard Salt/Pepper/Cutlery 520 calories
Sustainable ~20%			Antioxidant Salad Salad dressing Mayonnaise/Mustard Salt/Pepper/Cutlery 220 calories
Obesity/ Abundance ~10%		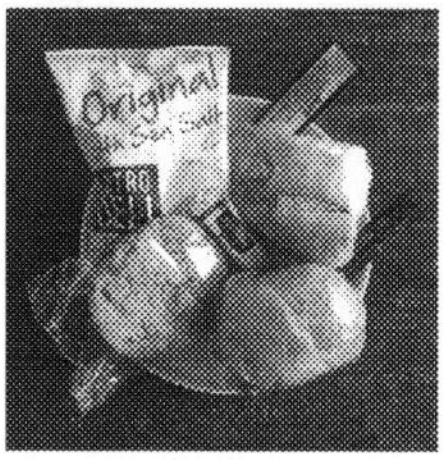	Biscuits/cookies (x2) Crisps/chips Ham Sandwich Pesto Chicken Wrap Mayonnaise/Mustard Salt/Pepper/Cutlery 2,460 calories
Food Insecure ~50%			Crisps/chips Salt/Pepper/Cutlery 360 calories

FIGURE 8.1   Randomly assigned boxed lunches (certified dietitian developed).

shame or blame). Several of the research subjects reported feeling ashamed of using a food pantry or that they did not want to take food if others might "really" need it, more so than they did.

College and university students are not immune to unsustainable food choices and habits. A study at North Carolina University revealed that 14% of students had been food insecure in the past 30 days (Wright et al., 2020). Food insecurity and food sustainability issues impact people in multiple ways, including their mental health, and can impact students' education and grades. The researchers offered several recommendations to faculty who may have students who are food insecure in their courses. Examples include reducing course costs, such as using open educational materials and including resources available to students in the course syllabus (Wright et al., 2020).

Experiential learning

Kolb et al. (2000) claimed that experience represented a critical piece of the learning process. Kolb (1984) derived his theory of experiential learning based on previous work by Dewey, Lewin, and Piaget. He believed that learners developed concepts by comparing and contrasting what they experience to existing mental concepts. The learning process involves four stages:

- The learner undergoes a *concrete experience*.
- Through *reflective observation*, the learner compares and contrasts the experience with existing mental concepts.
- The learner considers and analyzes the experience through *abstract conceptualization*.
- The learner then participates in *active experimentation* to either assimilate the new data or reconstruct/adapt existing concepts to incorporate any nonconforming data.

Kolb et al. (2000) also identified four learning styles. While one individual can exhibit more than one learning style, most have a preferred style through which they excel.

Learning style	Preference/tendency	Learning patterns/strengths
Diverging	CE/RO	Considers various perspectives; good for idea generation or brainstorming
Assimilating	AC/RO	Focuses on logic and abstract concepts; good for summarizing a significant amount of information
Converging	AC/AE	Tends to prefer a practical approach; good for solving problems and technical issues
Accommodating	CE/AE	Relies on intuition over logic and learns from hands on experience; good for implementing plans

CE, concrete experience; *RO*, reflective observation; *AC*, abstract conceptualization; *AE*, active experimentation.

Mezirow's (1997) transformational learning theory aligns well with Kolb's theory of experiential learning. Mezirow also frames learning as a critical reflection of existing assumptions or concepts. He outlined four processes of learning: elaborating or reinforcing an existing point of view, establishing a new point of view (no current point of view or assumption), transforming a point of view (similar to Kolb's idea of adapting an existing mental concept), and becoming aware of generalized bias. Mezirow described transformational learning as learner-centered, participatory, and interactive.

Experiential learning tends to be "messy" and unstructured, in contrast to typical problems students encounter in a classroom (Kerka, 1997). Ambrose and Poklop (2015) described a disconnect between the classroom and the real world, proposing that problems students work on in most classes tend to be more defined and straightforward compared with what student encounter once they enter the workplace. Based on the constructivist theories, Piaget et al. (2008) claimed that students were more likely to develop knowledge when it is accessible, plausible, and useful. Vygotsky's zone of proximal development placed an upper limit on what an individual could learn; however, carefully designed activities can facilitate learning by using simulations, familiar concepts, and other techniques to improve the accessibility of certain ideas or concepts.

Design thinking as an experiential learning activity to solve complex problems

Several researchers discussed the use of design thinking in experiential learning. Zupan et al. (2018) discussed the use of design thinking in education as early as the primary grades. Students in the study identified problems in the community, brainstormed solutions, developed prototypes, and some even presented their ideas to the community at the conclusion of the activity. Kremel and Edman (2019) used design thinking to teach entrepreneurship in a study where students were asked to come up with a solution that enabled older individuals to continue to live in their home. Students worked with aging individuals and demonstrated significant engagement throughout the process, though they did report that the reflection logs they were asked to keep were not valuable. In a third example, McGann et al. (2018) explored how involvement of the community in design thinking impacts the process. They found that the public sector tends to favor incremental change over radical transformation and values evidence-based policymaking.

Despite the growing acceptance and use of design thinking in education, no standard pedagogy has emerged. McLuskie and Dewitt (2019) surveyed 39 design thinking educators and could not identify a standard model, though there were some common themes. Most models incorporated some form of identification and definition of a problem, reflection, and prototyping or experimentation, though terminology tended to differ. Clemmensen et al. (2018) explored how culture impacts the design thinking process. Their findings supported the dynamic constructivist theory of culture, or that culture influences cognitive structures. However, they also endorsed Dorst's claim that abduction tends to be the most common method of reasoning in design thinking across cultures.

Beckman and Barry (2007) recommended incorporating cross-disciplinary teams with divergent views and assigning roles based on learning styles defined by Kolb and his colleagues. Interestingly, another study found that teams with members from cross-disciplinary and diverse backgrounds and fewer designers tended to perform better (Luccarelli et al., 2019). Another study by Brooks and Hehn (2020) supported cross-disciplinary teams. In addition, they suggested opening discussions with questions such as "How might we...?" They concluded that design thinking opens discussion and expands the number of potential solutions by incorporating empathy and ideation into the process.

A recent research study explored the role of design thinking—mainly centered on consumer engagement in tackling food waste, while at the same time achieving other sustainability goals in line with the UN Agenda 2030, from healthy eating to well-being and food security. A new framework (CEASE) was proposed for the design of new food experiences aimed at reducing food waste, while simultaneously promoting the well-being of individuals and society (Massari, Allievi, et al., 2021). A shift in design thinking mindset is underway, from regarding a product simply as a physical object to considering it part of a set of relationships that fulfill various purposes for different people and can create a tangible impact and improve the lives of individuals and the entire community.

The impact of experiential learning on motivation and perceived ability to address a complex problem

Several studies have investigated how experiential learning can influence motivation and perceived ability to make an impact on a complex problem. Kenney and Young (2019) used an experiential learning simulation in a social work class to help students understand food insecurity. Students were required to limit their expenditures on food for 1 week to $6.10 (approximately £4.93), equivalent to the amount provided by the Supplemental Nutrition Assistance Program (SNAP) in the United States, which provides financial assistance to purchase food for low-income individuals. Students reported that they often felt hungry and that it impacted their mood, performance, and state of mind. They also significantly enhanced their knowledge of SNAP and felt empathy for those who were forced to participate in the program.

In another example, Northrup et al. (2020) evaluated two experiential learning activities to determine which was more effective in helping students to understand food insecurity. The first was a poverty simulation where students role-played a family and experienced four rounds representing weeks where they had to seek jobs, deal with unexpected expenses, and figure out how to make ends meet with very limited funds. The second was a hunger banquet where students received a meal designed to represent poverty, middle class, or wealth. While both simulations improved students' understanding of poverty, the poverty simulation was more effective in increasing students' empathy for those living in poverty. While both simulations were experiential in nature, the poverty simulation required students to role play and make difficult choices using the limited resources available to them.

A third study by Hendricks and Drysdale (2015) asked students to participate in a survey that asked if they would be willing to support a tax increase of $5 that would increase U.S. aid to address poverty in Africa and/or donate time to such a cause. Students were randomly assigned to three groups; one received just the questions about the tax increase and volunteering time while a second group received the questions along with some statistics about poverty and hunger in Africa, and a third group received the questions along with a picture of a young African girl who lived in poverty and information about her specific situation. The students who received the picture of the girl and read her story were willing to give 30% more than the other two groups, indicating that the personal story and photo fostered more empathy. They were also more willing to volunteer their time to address the issue.

The literature suggests that experiential learning can positively impact learning outcomes of complex topics and that incorporating empathy into a learning experience may also improve learning. The workshop leaders used Kolb's (1984) and Mezirow's

(1997) theories as a framework to create an experience that would encourage transformative learning of the complex topic of food insecurity and sustainability as well as how media impacts food consumption.

PUSH-UFWH design thinking workshop: everyone should be an agent of change

The food system and its sustainability are complex, being characterized by *interconnectedness* (humans and other organisms are all connected to each other), *emergency* (each action and interaction may affect the system), and *modularity* (humans are both connected and disconnected in multiple subgroups, allowing different view and values on the definition of sustainability) (Dentoni et al., 2017). Furthermore, the link between food choices and environmental (from water pollution to land degradation) and health problems (such as diabetes and undernutrition) has called for a transition to more sustainable diets (Willett et al., 2019).

While such complexity might seem discouraging, it can instead serve as the starting point to reflect on the key role that human actions connected to food production and consumption play on the use of natural resources and on the generation of inequalities. It thus becomes evident how a collective culture geared toward competition does not serve the purpose of sustainability. The discussion on sustainability is often linked with that of Sustainable Development Goals (SDGs); however, these 17 goals present a mostly anthropocentric point of the relationship between nature and humans; by accounting for an ecocentric perspective, the empathic recognition of nature's inherent value would be enabled, easing the paradigm change necessary to tackle sustainability (Koiupi & Voulvoulis, 2019). In this sense, as food can be considered to be intrinsically connected, one way or another, with all the SDGs (SRC, 2016), keeping the common good, as environmental resources and health, at the center of the discussion is key.

With all the above in mind, each person is called to be in his/her personal and professional life, an agent of change for a more sustainable food system.

Workshop theoretical framework

The workshop leaders structured the design thinking workshop based on Kolb's (1984) experiential learning theory and Mezirow's (1997) theory of transformational learning with the intent of enhancing participants' understanding of food sustainability and encouraging them to consider opportunities to address the issue in their communities or in a broader context. They also incorporated the principles of design thinking to encourage empathy and engagement throughout the process by asking them to focus on their reactions and emotions to the questions and prompts.

Kolb's (1984) theory begins with an experience, which the individual then reflects on and analyzes to conceptualize it within existing cognitive structures. Then the individual actively experiments to fully flesh out the concept and to reinforce, add to, or replace existing concepts. Mezirow's (1997) transformational learning theory focuses on a situation where an existing cognitive structure must be replaced or adapted significantly. Successful transformative learning requires critical reflection of assumptions and willingness to adapt or replace existing concepts with new realizations or discoveries. In addition, the workshop leaders believed that there is a connection between constructive discomfort and transformational learning that "create[s] disorienting contradictions that prompt critical self-assessment of values and beliefs" (Nolan & Molla, 2018, p. 5) that lead to thinking and action. Mezirow (1997) states that "[e]ducation that fosters critically reflective thought, imaginative problem posing, and discourse is learner-centered, participatory, and interactive, and it involves group deliberation and group problem solving" (p. 10).

In the case of the design thinking workshop, participants attended the PUSH and UFWH events because they had a prior interest in food issues. Many of them had studied the concept and had significant prior knowledge of food sustainability issues, including food insecurity. The workshop leaders began the workshop with a series of short presentations focused on common misconceptions about the relationship between sustainability and food. After each presentation, they invited participants to first think about their emotions or feelings and then to consider a question or prompt individually. Once they had time to reflect individually, they shared their thoughts and ideas with others at their table. The focus on feelings and emotions was intended to encourage empathy. About halfway through the workshop, after the workshop leaders set the stage, the students then opened their boxed lunches, which was part of the experience. Their responses were observed as they reacted to what they received in the lunches. There was also an intentional element of discomfort in the process. Those who received a meager meal may have felt "cheated" while those who received a balanced or abundant meal may have experienced guilt—but in both cases, they were uncertain how to appropriately react and felt discomfort. Incorporation of empathy and discomfort led to vulnerability, which facilitates a willingness to consider new ideas or perspectives. Then they were asked to discuss challenges related to food system and sustainability and how it could be addressed, representing conceptualization and experimentation.

Workshop methodology

The purpose of this study was to understand how the design thinking workshop impacted participants' understanding and views of food choices and food systems as well as their motivation to take sustainable actions and their perceived ability to make an impact. Throughout the activity, the workshop leaders observed the various groups at the tables and how they worked together. Participants were seated at tables with 10 seats, but not all tables were completely filled. The workshop leaders each observed approximately three to four tables. They observed reactions and listened to the discussions that occurred at the various tables throughout the duration of the workshop and answered questions from participants. The workshop included two parts. The first provided a series of mini-presentations about various aspects of food insecurity and sustainability and the impact of media on food consumption and systems. The second half focused on the distribution of boxed lunches, which were designed to loosely mimic the distribution of food in the world. After the workshop, the workshop leaders asked each participant to complete a written survey about his or her experience, and they collected the sticky notes that participants had used to record their emotions and ideas. See Figs. 8.2—8.6 for photos of the workshop setup and aftermath.

FIGURE 8.2 Workshop room set up prior to participants arriving.

FIGURE 8.3 Table 5 set up.

FIGURE 8.4 Participants during workshop (note that faces have been obscured to protect identity of participants).

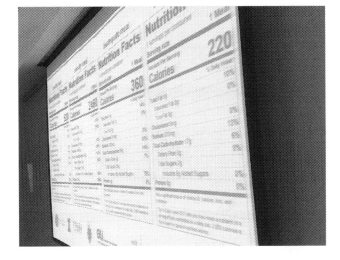

FIGURE 8.5 Slide with nutrition facts for the four meals.

Participants

Participants in the study included PUSH Leaders Forum and UFWH Summit attendees. Approximately 125 university students (n = 95) and leaders (university presidents or their designees; n = 30) participated in the workshop (refer to Table 8.1 for further details). Due to the nature of the two events, the participants were very likely knowledgeable about and had a specific interest in food sustainability since they had elected to attend one of the conferences. The workshop leaders did not recruit participants; instead, they used the convenience sample of students and university leaders who participated in the workshop. Participants were randomly assigned to tables so the likelihood that they were seated with several people they were not acquainted with was high.

FIGURE 8.6   After the presentation.

Data collection

All attendees were asked to read and sign a consent form at the beginning of the workshop (approved by the University of Illinois Urbana-Champaign Ethics/Institutional Research Board). The workshop incorporated a quasi-pre/post design by incorporating survey questions that asked participants to compare their knowledge level prior to and after the completion of the workshop. The workshop leaders observed the groups at their tables throughout the activity. Each workshop leader observed a group of three to four tables and took notes on the reactions of participants. The initial half of the workshop consisted of a series of mini-presentations on topics related to the complexity of food insecurity and sustainability and the impact of media on food choices and systems. During each round, one of the faculty members shared a brief presentation and then asked the groups at each table to write down their current emotions before starting a discussion or sharing ideas with the others at their table. Then participants were asked to write down some ideas related to a question about the topic on a sticky note to foster individual reflection among participants. Next, participants discussed their responses with others at their

TABLE 8.1   Summary of responses to closed response questions.

Question	Responses
Age	17–24 years: 69%
	25–30 years: 11%
	30–55 years: 13%
	55+ years: 7%
Gender	Female: 71%
	Male: 29%
	Other: 0%
Role	Student: 78%
	Higher education professional: 14%
	Other professional: 7%
	Other: 2%
How well did you understand food insecurity prior to the workshop?	Very well: 55%
	Somewhat: 38%
	Not much: 6%
	Not at all: 1%
Meal received	Obesity/wealthy meal: 11%
	Healthy/balanced meal: 20%
	Sustainable meal: 21%
	Food insecure meal: 48%
Did the workshop help you to better understand issues related to food insecurity?	A great deal: 43%
	Somewhat: 40%
	Not much: 12%
	Not at all: 5%
Did the workshop help you to identify new opportunities to address food insecurity?	A great deal: 53%
	Somewhat: 33%
	Not much: 11%
	Not at all: 3%

tables. The workshop leaders wandered around the room while groups were engaging in discussion to answer questions, but also to listen in to some of participants' ideas and thoughts as well as to observe how they interacted with one another. Other than responding to questions, they did not interact or engage in the discussions at the tables. After each mini-presentation and discussion, they asked two to three groups to share an insight from their discussion.

During their observations, the workshop leaders agreed to consider the following questions:

- How did the groups engage in discussion? Did one individual tend to dominate or did several individuals interact with each other?
- Did each participant at the table share their ideas and then the group started discussing? Or did the group tend to engage in discussion after everyone shared a thought or idea?
- Did most individuals use the sticky notes to record their emotions and ideas?
- Did the group come to consensus? Were they able to share a couple of ideas during the debrief period?

Once the boxed lunches were opened, the workshop leaders considered additional questions in their observations:

- What were the initial reactions of individuals at each table once participants opened their lunches? What emotions did you observe?
- How did the individuals at each table engage with each other after opening their lunches?
- Did participants at the tables you observed end up sharing food with each other? Or did they each keep their own lunches?
- How did the sentiment at each table shift over time, if at all?

At the conclusion of the presentations and lunch, participants completed a written survey as their last activity of the workshop (available in the Appendix). The survey elicited a few demographic questions and what type of meal they received during the luncheon. As part of the quasi-pre/post design, closed-ended questions were included addressing whether the workshop increased their knowledge about food security and whether it helped them to understand food sustainability and food sustainability issues better. Participants also responded to four open-ended questions: their thoughts about the workshop and how they would describe it; emotions they experienced during the workshop; anything about the workshop that surprised them; and the most important thing they took away from experience. The surveys, consent forms, written notes, and all the sticky notes with comments on emotions and individual ideas were collected from each group at the end of the event.

Data analysis

The workshop leaders shared notes on their observations of participant behaviors and discussions during the workshop. The notes were combined and reviewed to extract general themes. Each of the workshop leaders participated in the consolidation process and discussed their conclusions about the workshop outcomes. Discussions took place via a series of conference calls among workshop leaders.

General statistics about the closed-end questions to the survey were calculated (see Table 8.1) to understand the demographics of the participants, what meals they received, etc. The survey results also provided insight about participants' knowledge of food insecurity and sustainability before and after the workshop. For the open-ended questions, each response was coded, categories formed, and key themes for each open-ended question identified for each group. Each of the workshop leaders participated in the coding process at some level to support consistency in response coding. The themes and some example quotes are captured in Tables 8.2—8.5.

The information from surveys and the observations during the workshop were combined to address the research questions. The notes and potential solutions that each group offered were recorded and captured. All workshop leaders also read through the solutions offered to gain a sense of the level of ideas that groups generated during their discussions.

Results

Meal distribution during the workshop was intended to roughly mimic sustainable development issues in the world. Based on the survey, 11% received a meal with too much food

TABLE 8.2   Summary of responses to "Please provide your thoughts on your workshop experience. How would you describe it to others?

Theme	Percentage	Example quote
Educational experience	29	The experience was amazing and educating. I learned a lot about food insecurity from different perspectives.
Collaborative and/or interactive experience	22	An innovative way to bridge students and professionals and share ideas in an equitable way.
Unique experience	17	I thought this experience was very unique, and really helped individuals be creative and form ideas.
Encouraged empathy	12	Great brainstorming, creates empathetic environment, understanding others' situations.
Activity has limitations	11	I have seen this type of activity before, I feel like it is just "trying on" someone else's lived experience.
Encouraged discussion	10	What a great activity for students. I was inspired by the thoughtful discussion at my table.

Other examples of responses:
- I learned a lot more about food insecurity and now I am even more motivated to make a difference!
- Thought provoking.
- I would describe the workshop as an interactive way to see food insecurity.

for one person, which was labeled a wealthy or abundant meal. Twenty percent received a meal with balanced nutrition and adequate food, designed to represent the middle class. Twenty-one percent received a sustainable meal, which was made with sustainable ingredients that have a lower impact on earth's resources. Finally, 48% reported receiving a food insecure meal of a single bag of potato chips. Not only did this meal contain inadequate calories, it also offered very little nutritional value. The meal distribution created an opportunity to more clearly see how many people are impacted by food insecurity.

During the mini-presentations in the first part of the workshop, the workshop leaders observed their assigned tables during the individual reflection and group discussion breaks. Participants were first asked to focus on and write down their emotions and then to write down their thoughts and ideas. This provided each participant with an opportunity to reflect on what they were feeling and then respond to the question individually before discussing as a group. Workshop leaders compared observation notes and found that at most tables, the vast majority of participants engaged once group discussions began.

The workshop leaders observed the participants at their assigned tables as they opened their meals. The workshop leaders noted and recorded reactions of individuals and also observed behavior and interactions of participants at each table. After the workshop, the leaders compared their notes and identified key themes across their observations. Most participants expressed surprise, as they expected standard conference fare, and did not anticipate the meals were part of the workshop. Many conveyed clear disappointment, particularly those who received the food insecure meal, as they worried they were going to leave the workshop hungry. Others were upset or angry because they had food allergies

TABLE 8.3 Summary of responses to "Please explain how you felt during the workshop. What emotions did you experience?

Theme	Percentage[a]	Notes
Frustrated	25	Includes annoyed
Hope	22	Includes optimistic
Empowered	14	Includes encouraged, energized, helpful, creative, unity
Inspired	14	
Despair	10	
Sad	9	Includes upset
Motivated	8	Includes curious
Anger	8	
Guilty	8	Includes shame
Confused	6	
Nervous	5	
Happy/joy	5	
Excited	5	
Empathic	5	
Disappointed	5	Includes discouraged, helpless, overwhelmed, powerless, raw
Other	19	For example, equal, included, grateful, hungry, mixed, satisfied, understood

[a]Note that the percentages add up to more than 100% because survey respondents could report multiple emotions.
Examples of responses:
- I was disappointed when I was handed chips and others received more. I felt that I can empathize with people who have to be food insecure every day.
- At first, thinking was more negative, but as the workshop continued, it turned positive.
- I envied the person who received two sandwiches, chips, and two cookies.
- Empowered by my peers at the table.

or intolerances or other dietary restrictions that did not align with their meal. Interestingly, many of those who received the abundant or wealthy meal appeared to be embarrassed or ashamed to have received so much food when others had so little. Clearly, many of the participants experienced discomfort during this part of the experience.

The workshop leaders continued to observe behavior at the tables as participants ate their meals. Reactions ranged from nervous laughter to anger at the inequality of the meals to shock or disbelief that some received so little food. One peculiar result was the different outcomes at different tables. At some of the tables, everyone put all the food in the middle and shared it among the members of the group. At other tables, participants did not share, potentially because they felt it was "breaking the rules" or were unsure if it was "okay" or acceptable to share food with others. (Note that the workshop leaders intentionally did not establish or imply any rules about the meals.) The workshop leaders noticed that those at the tables where food was shared became much more relaxed and

TABLE 8.4   Summary of responses to "Did anything about the workshop surprise you? Please explain.

Theme	Percentage	Example quote
The lunches	36	The different lunches, making us feel what some students go through daily.
Team member reactions	17	The people at our table did not eat and wanted to share.
Process	16	I was surprised about what the design thinking process was, I come from an engineering background which explains the design process but was a different application of that.
Ideas generated	14	The diversity of proposed solutions.
Information provided	10	The amount of food insecurity on American campuses!
Other	6	Talking to other students and understanding their level of knowledge was surprising because it was a great spectrum of perspectives.

Other examples of responses:
- The passion and empathy, and clear talent of the people around me.
- The different lunches. Randomness was a good experience on food insecurity.
- How successful the process was in changing feelings from beginning to end. Many of us started with feelings of being overwhelmed or frustrated and were hopeful by the end.
- The amount of new ideas achievable.
- The meal interactive portion because it felt real and a simulation of what people are really feeling.

TABLE 8.5   Summary of responses to "Describe the most important insight you took away from your experience today.

Theme	Percentage	Example quote
Collaboration is key to resolution	36	Community is key. Interdisciplinary work is essential.
Learned from experience	22	I learned more about how prevalent the issue of food insecurity is and how solutions can be implemented on campus.
Resolution is possible	12	There is a universal understanding and desire for food insecurity solutions among young people.
What others are doing	12	How to make impact and new ideas to implement with my university food program.
Empathy is important	10	You have to experience in order to truly feel and understand.
Ideas generated	9	There is much more than I can be doing.

Other examples of responses:
- The most important insight I will take away is that we can make a change. Thank you!
- Food insecurity is unfair and unjust, and we can solve it.
- Food insecurity is not just having any food but also not having enough healthy foods.
- Feelings are so valuable in development! Process we use to come up with ideas is not to be overlooked.

comfortable compared with those where food was not shared. The participants at the tables that did not share continued to exhibit uneasy behaviors, with some appearing nervous or ashamed, and there was much less conversation and laughing at those tables as well. This was consistent across the observations by the four workshop leaders.

A post-workshop survey asked participants about the impact of the experience. (Note that the survey instrument is available in the Appendix and the survey results are included in Tables 8.1—8.5). Prior to the workshop, 55% reported that they understood food security very well and 38% understood it somewhat. Despite the significant level of prior knowledge, 43% responded that the workshop helped them to understand food sustainability issues a great deal, and an additional 40% said that it helped somewhat. Therefore, more than 80% of participants reported that the workshop improved their understanding of food insecurity and sustainability. Along similar lines, 53% said that the workshop helped them a great deal in identifying new opportunities to address food security, and another 33% said it helped somewhat, meaning more than 85% left with a new opportunity they could pursue.

The survey also included four open-ended questions. The first asked how participants would describe the experience. The two most common responses were: an educational experience (29%) and an interactive or collaborative experience (22%). One participant responded: "An innovative way to bridge students and professionals and share ideas in an equitable way." The second question asked about the emotions participants experienced. Twenty-five percent cited frustration and 22% responded hope, but note that participants could provide more than one response. The range of responses are provided in Table 8.3. One participant stated: "I felt that I can empathise with people who have to be food insecure every day." Next, participants were asked what surprised them the most about the experience and the most common response was the lunches (36%). Three other common responses included: team member reactions (17%), the process (16%), and ideas generated (14%). One respondent was an engineering student who was surprised by the design thinking process as it was different than his or her experience of the design process in engineering. Another participant talked about the change in emotions from at different points: "Many of us started with feelings of being overwhelmed or frustrated and were hopeful by the end." Finally, the survey asked about insights from the experience. The most common response was that collaboration is critical to addressing food sustainability, reported by 36%. Twenty-two percent reported that they learned from the experience. One participant discussed the importance of empathy by experiencing emotions that helped him or her to better understand food insecurity. Another reported that the most important takeaway was "that we can make a change."

Discussion and reflections

First, the authors will revisit the research questions posed at the beginning of the chapter. Then they will share additional insights from the workshop observations and survey.

How does participation in an experiential learning activity impact the participant's understanding of the complex concept of food sustainability?

In the survey after the workshop, more than 90% of the participants reported that they understood food systems very well or somewhat. This was not surprising given participants opted to attend the PUSH or UFWH events and therefore likely had a vested interest in addressing world hunger. Even though so many came with prior knowledge, more than 80% of participants responded that the workshop helped them to understand food sustainability issues a great deal or somewhat.

The experience of receiving a nutritionally unbalanced meal—or observing someone else receive one—made an impact on the participants. It required them to reevaluate what it really meant that approximately 50% of the world population is food insecure. Seeing the number of nutritionally inadequate lunches made the statistic much more real and accessible, thus encouraging participants to challenge their assumptions and revise or replace their current mental schemas about food sustainability. The boxed lunches also demonstrated that nutritionally unbalanced meals are just one aspect of food sustainability, as food may be accessible and affordable but not healthy and nutritious.

Another aspect of the experience that caused participants to critically reassess their assumptions was having limited choice in what they received. While they were able to pick their own boxed lunch, they did not know what was inside, and some expressed disappointment or anger that their food intolerances or dietary restrictions did not allow them to enjoy the meal they selected. The experience mimicked what those who face food insecurity may face every day, especially for those who also deal with diseases such as diabetes or special diets such as vegetarianism. The lack of choice represented an aspect of food choices that may not have occurred even to those who are well acquainted with the concept.

> How does participation in an experiential learning activity impact the participant's motivation and perceived ability to address a complex societal problem?

The structure of the workshop created an occurrence where participants experienced unequal food distribution, something they may not have truly experienced in the past. Most of the participants likely came from universities or colleges where they have a dining hall and have access to many choices and adequate nutrition. While a brief encounter, experiencing the unequal distribution resulted in a memorable experience that motivated them to address the problem of food sustainability.

The emphasis on emotions and how participants were feeling positioned them to experience empathy for those who are food insecure. Several complained that it was unfair that they received less food or just a bag of potato chips, but they then recognized that those who suffer from the consequences of malnutrition and food insecurity rarely make that choice directly. Empathy is fundamental to bring about an understanding of food sustainability that is holistic and effective (Massari, Allievi, et al., 2021) and may need to be activated through specific educational activities such as this one.

Once the shock of the lunches had resided, the workshop leaders asked participants to think about potential solutions or actions they could take to address food sustainability issues. The workshop leaders encouraged participants to share successful initiatives or programs from their home universities or communities, thus fostering conversations about smaller-scale solutions. The intent of the focus on smaller-scale actions was to reduce the feeling of inadequacy when a complex problem seems completely overwhelming. By initially discussing smaller steps that could be taken, participants believed that they had the capacity to make a difference at some level, even if it was local or centered on their home institution. The perception that they could make any impact could lead to larger-scale solutions in the future.

Additional insights

The workshop leaders observed that the discussions about solutions and actions certainly resulted in more animated and hopeful tones and conversations compared with the frustration and negative emotions that were more prevalent in the first part of the workshop and immediately after the lunch distribution. This mirrored the comments in the survey from many participants that their initial emotions were negative (frustration, despair, overwhelmed) but by the end of the workshop they felt much more positive (hope, inspired, empowered).

The workshop was structured to support this transition from negative to positive. The first part of the workshop focused on the stark reality of food sustainability, including statistics and numbers. Then the lunches simulated the distribution of wealth and food systems in the world, making it very clear that food sustainability is a widespread issue. The situation encouraged participants to feel empathy, motivating and priming them to think about what concrete actions they could take to address it. As a result, they left with one or more ideas that they could implement quickly in their own communities or at their institutions back home.

The design thinking process proved to be uncomfortable for some during the workshop. Many people are uncomfortable with the ambiguous and unstructured nature of the experience. Several reported feeling frustrated and uncertain of the purpose of the workshop during the earlier stages, which was likely amplified by the fact that they were working with several others they did not know. However, such discomfort can increase the likelihood that individuals may truly reassess their assumptions and/or existing conceptual frames.

The workshop leaders anticipated that there would be some frustration and discomfort early in the process and were prepared to encourage participants to work through it, including ensuring everyone that they would be fed before they left. Transformational learning experiences require more effort and investment of resources than traditional and informative learning, but often create a lasting impact of a higher magnitude. The survey responses supported past research in demonstrating that participants did learn from the experience and that they would likely leave with action steps they were ready to implement.

Conclusions

As presented above, traditional media and social media have contradictory messages and forces moving them, making them insufficient to provide young people to form an informed opinion about food sustainability issues and to act upon them. The current formal education (food studies related curricula and food education in general) is also facing some issues in this sense, lacking the adequate spaces, funds, and culture to include (food) sustainability in the curricula and practices of the courses and the campuses.

New forms of information and education are needed in order to enable young people to experience food sustainability issues, such as sustainable development goals and food insecurity, firsthand, so to enable empathic mechanisms, which in turn result in proactive approaches to tackle the current challenges of our food system.

The workshop in this research study provides an example of how experiential learning can be used in generating transformational learning outcomes. Design thinking represents one instance of experiential learning that educators can use to engage students and to help them to understand and integrate what they experience into their existing knowledge or meaning. As a result, experiential learning helps people to continue to move forward on the path to autonomous learning.

In this case, the workshop effectively engaged participants and helped even those who were already knowledgeable about food habits and systems to enhance their understanding and to generate new opportunities to address the issue. Not only did participants learn about what others were doing and think about how they could implement new ideas or programs on their own campuses, they also left feeling empowered and at least partly responsible for taking action. And the majority of them appeared to be excited about going back home and trying new things.

Though many of the participants started the experience feeling frustrated and uncomfortable, the majority of them left with an improved understanding of a complex issue and a new set of resources and tools to address food sustainability in their local communities. The structure of the experience encouraged empathy and engagement in the design thinking process. "A defining condition of being human is that we have to understand the meaning of our experience" (Mezirow, 1997, p. 5).

Finally, this chapter can provide some suggestions for the use of social media to educate people on food sustainability and to foster the agency of young adults around these issues. Through social media, people can create new conversations, ask and answer questions, and discover new possibilities to help refine their solutions. If food and sustainability education aims to teach students how to think and act in a sustainable manner, then the design thinking process is undoubtedly an important part for resetting and creating a new information experience in this field.

Incorporating social media in food and sustainability education could allow more collaboration and conversation, for the development of empathy and stronger leadership skills in individuals.

As part of the process of "learning how to think," the concept of design thinking has been gaining traction in classrooms, from primary education through college and beyond. The concept of design thinking is illusorily simple: there is a problem and potential solutions are identified, a prototype solution is created, and then refined based on feedback. However, the idea behind this type of thinking in education is not only to encourage creativity and find new ways to approach challenges, but instead to encourage transdisciplinary and system thinking and more collaborative in-depth research among students. Along these lines, the dissemination of information on food sustainability to young people should be made more accessible (also through the use of social media), using an approachable, understandable, and accessible language, which can give them the opportunity to both discuss together and broaden their knowledge and then move from theory to practical actions.

Another key aspect of design thinking is the concept of empathy. As shown in the example presented in this chapter, the process of design thinking requires problem-solvers to consider the end users of their solutions throughout the entire process. Rather than focusing on problem-solving for its own sake, design thinking instead focuses on people who need solutions and keeps their needs and desires in the forefront of the design process. As described in the case of the workshop analyzed here, design thinking encouraged participants to learn to listen to others and fully understand their needs to design solutions that truly work for them and the planet. As highlighted also in the EOE model (Massari, Allievi, et al., 2021), which presents a three-level empathy process (empathy toward yourself, others, and the environment), media should account for these three different levels to improve the activation of empathic mechanisms in their audience. This approach would ensure the shift toward a more systemic approach to sustainability issues.

Appendix

Survey instrument

A written survey was distributed to conference attendees with the consent form (two copies) attached to the first page.

- Age (options: 18–24; 25–30; 30–55; 55+)
- Gender (options: Male; Female; Other)
- Profession/affiliation (options: Higher education professional; Other professional; Student; Other/please specify)
- How well did you understand the issue of food insecurity before today's workshop? (options: Very well; Somewhat; Not much; Not at all)
- What meal did you receive at the workshop? (options: meal descriptions)
- How did your meal compare with the meals of others? (options/check all that apply: Less food; more food; healthier food; more sustainable food; other/please specify)
- Did the workshop help you to better understand issues related to food insecurity? (options: A great deal; Somewhat; Not much; Not at all)
- Did the workshop help you to identify new opportunities to address food insecurity? (options: A great deal; Somewhat; Not much; Not at all)
- Please provide your thoughts on your workshop experience. How would you describe it to someone else? (open response)
- Please explain how you felt during the workshop. What emotions did you experience? (open response)
- Did anything about the workshop surprise you? Please explain. (open response)
- Describe the most important insight you took away from your experience today. (open response).

References

Ambrose, S. A., & Poklop, L. (2015). Do students really learn from experience? *Change: The Magazine of Higher Learning, 47*(1), 54–61.

Beckman, S. L., & Barry, M. (2007). Innovation as a learning process: Embedded design thinking. *California Management Review, 50*(1), 25−56.

Brooks, R., & Hehn, J. (2020). Design thinking complex problems. *Developing Leaders, 34*, 67−70.

Carpenter, S., Takahashi, B., Cunningham, C., & Lertpratchya, A. (2016). Climate and sustainability the roles of social media in promoting sustainability in higher education. *International Journal Of Communication, 10*, 19, Retrieved from. Available from https://ijoc.org/index.php/ijoc/article/view/4694.

Clemmensen, T., Ranjan, A., & Bodker, M. (2018). How cultural knowledge shapes core design thinking: A situation specific analysis. *Codesign, 14*(2), 115−132.

Dentoni, D., Waddell, S., & Waddock, S. (2017). Pathways of transformation in global food and agricultural systems: Implications from a large-systems change theory perspective. *Current Opinion in Environmental Sustainability, 29*, 8−13.

FAO, IFAD, UNICEF, WFP and WHO. (2020). *The state of food security and nutrition in the* World 2020. *Transforming food systems for affordable healthy diets.* Rome: FAO.

Hendricks, N., & Drysdale, K. (2015). In-class experiment assesses empathy for international hunger and poverty. *NACTA (North American Colleges and Teachers of Agriculture) Journal, 59*(3), 253−258.

Kenney, J. L., & Young, S. R. (2019). Using experiential learning to help students understand the impact of food insecurity. *Journal of Social Work Education, 55*(1), 64−74.

Kerka, S. (1997). Constructivism, workplace learning, and vocational education. *ERIC Digest*, No. 181.

Kioupi, V., & Voulvoulis, N. (2019). Education for sustainable development: A systemic framework for connecting the SDGs to educational outcomes. *Sustainability, 11*(21), 6104. Available from https://doi.org/10.3390/su11216104.

Kolb, D. A. (1984). *Experiential learning: Experience as the source of learning and development.* Englewood Cliffs, NJ: Prentice Hall.

Kolb, D. A., Boyatzis, R. E., & Mainemelis, C. (2000). Experiential learning theory: Previous research and new directions. In R. L. Sternberg, & L. F. Zhang (Eds.), *Perspectives on cognitive, learning, and thinking styles.* New Jersey: Lawrence Erlbaum.

Kremel, A., & Edman, K. W. (2019). Implementing design thinking as didactic method in entrepreneurship education: The importance of through. *The Design Journal, 22*, 163−175.

La Charite, K. (2016). Re-visioning agriculture in higher education: The role of campus agriculture initiatives in sustainability education. *Agriculture and Human Values, 33*, 521−535.

Luccarelli, M., Tillman, S., Lay, R., Grundmeier, A., & Hogsdal, S. (2019). Sustainable design education for elementary schools: Interdisciplinary development of new educational models through design thinking. *International Journal of Design Education, 13*(4), 1−25.

Massari, S., Allievi, F., & Recanati, F. (2021). Fostering empathy towards effective sustainability teaching: From the Food Sustainability Index educational toolkit to a new pedagogical model. In W. Leal Filho, A. L. Salvia, & F. Frankenberger (Eds.), *Handbook on teaching and learning for sustainable development.* Cheltenham: Edward Elgar Publishing.

McGann, M., Blomkamp, E., & Lewis, J. M. (2018). The rise of public sector innovation labs: Experiments in design thinking for policy. *Policy Science, 51*, 249−267.

McLuskie, P. & Dewitt, S. (2019). Design thinking pedagogy and enterprise education. *Proceedings of the 14th European Conference on Innovation and Entrepreneurship.* ECIE, 648−656.

Mezirow, J. (1997). Transformative learning: Theory to practice. *New Directions for Adult and Continuing Education, 74*, 5−12.

Nolan, A., & Molla, T. (2018). Teacher professional learning through pedagogy of discomfort. *Reflective Practice: International and Multidisciplinary Perspectives, 19*(6), 721−735.

Northrup, A., Berro, E., Spang, C., & Brown, M. (2020). Teaching poverty: Evaluation of two simulated poverty teaching interventions with undergraduate nursing students. *Journal of Nursing Education, 59*(2), 83−87.

Stevens, T., Aarts, T., Termeer, C. J. A. M., & Dewulf, A. (2016). Social media as a new playing field for the governance of agro-food sustainability. *Current Opinion in Environmental Sustainability* (18), 99−106. Available from https://doi.org/10.1016/j.cosust.2015.11.010.

Willett, W., Rockström, J., Loken, B., Springmann, M., & Murray, C. (2019). Food in the Anthropocene: The EAT−Lancet Commission on healthy diets from sustainable food systems. *The Lancet, 393*, 447−492.

Wright, S., Haskett, M. E., & Anderson, J. (2020). When your students are hungry and homeless: The crucial role of faculty. *Communication Education, 69*(2), 260−267.

Zepeda, L. (2018). Hiding hunger: Food insecurity in middle America. *Agriculture & Human Values, 35,* 243–254.

Zupan, B., Nabergoj, A.S., & Cankar, R. (2018). Design thinking as a methodology for teaching entrepreneurial mindset in primary education. *Proceedings of the 13th European Conference on Innovation and Entrepreneurship.* ECIE, 893–899. ISBN: 978–191121897-5.

Further reading

Cakir, M. (2008). Constructivist approaches to learning in science and their implications for science pedagogy: A literature review. *International Journal of Environmental and Science Education, 3*(4), 193–206.

Massari, S., Antonelli, M., Principato, L., & Pratesi, C. A. (2021). Design thinking to engage consumers in achieving zero waste food experiences: The CEASE framework. In W. Batat (Ed.), *Design thinking for food well-Being.* Cham: Springer.

Nadan, Y., & Stark, M. (2017). The pedagogy of discomfort: Enhancing reflectivity on stereotypes and bias. *British Journal of Social Work, 47,* 683–700. Available from https://doi.org/10.1109/bjsw/bcw023.

President United to Solve Hunger. (2020). Available from https://wp.auburn.edu/push/?page_id = 63. Accessed 8 April 2020.

Universities Fighting World Hunger. (2020). Available from http://www.universitiesfightingworldhunger.org/. Accessed 8 April 2020.

Global food ecosystems: new models to cover messages about food systems

Sara Roversi

Future Food Institute, Bologna, Emilia-Romagna, Italy

Intro

As human beings, we create our world and our belief systems through communication: having a developed precortex permits us to imagine and create highly complex communication systems. However, the narrative we followed until now regarding the food sector has been unsustainable and disruptive for our planet. Unsustainable systems are able to maintain themselves only if they are accompanied by a strong communication and belief system. So the way we communicate around the food sector is crucial to develop better and more sustainable food systems. Studies show the unsustainability of our food systems: high water use, widespread pesticide use, and resistance, mono-variety crops, increase in diet-related disease. How have we got here? Which communication and which belief system has led us here? New values are needed to drive our actions as consumers, businesses, and social actors.

What climate change, the global pandemic and the agri-food system have in common are the cumulative impacts that they all generate, far beyond their specific focus area. The evident paradoxes of the current system and the silos thinking that is at the basis of the "old world" are now leading the global society toward a wake-up call, where inclusion and global interconnection are the keywords. Additionally, food converges with many different sectors. To achieve food security at a global level, it is necessary not only to rethink, through the lenses of sustainability along the whole food supply chain but also to consider the role food plays at a social, environmental, institutional, individual, economic, and cultural level: from farm to fork and beyond.

Since 2014, Future Food Institute has represented a benchmark in delivering new models to reshape and regenerate the global food ecosystem to protect the Planet, empower people, and enable prosperity. Starting from the premise that "short-termism" has pervaded the public and entrepreneurial sphere and has, in turn, led us to the global challenges we now face. Future Food embraces the unconventional strategy to start from the

DOI: https://doi.org/10.1016/B978-0-323-91227-3.00009-3

bottom, specifically from **education**, as a means to raise awareness and shift the mindset toward global issues. It has done this by investing in designing learning experiences that involve the global community of food system experts and innovators and, at the same time, local communities rooted in ancient food traditions.

Future Food focuses its efforts on learning together to care about relationships, training listening skills, stimulating creativity, improving critical thinking, developing abilities to codesign for prosperity, and being different but united. This is an approach that embraces inclusion for diversity as an added value of learning. Over time, it has allowed Future Food to develop its second principal pillar of action: **community**. Inspiring the broader and multistakeholder community to take action on SDGs and, as a result, nudging a behavioral shift is a process that requires long-term planning. Equally, only once the community is cohesive, it is possible to enable prosperity by implementing tangible innovations to help start-ups, corporates, or institutional partners make meaningful impacts.

The third pillar of action of the Future Food is both a driver for global economic development and a catalyst for change: **innovation**. Future Food has developed a theory of change, based on three phases (inspiration, aspiration, and action) to lead food players to learn, support, regenerate, innovate, and impact. The peculiarity of this model is to act at different levels, create different branches, and involve different collaborations, around a structure that includes both for-profit and nonprofit companies. Future Food takes advantage of the potential for innovation and the highly agile and fast structure of for-profits while it is the not-for-profit core that develops guidelines, content, and research that will inspire companies, from a base of interlinked relationships between people, digital, and culture.

These three aspects (people, digital, and culture) represent the center of the Future Food compass, on which purpose-driven research is developed. This structure has the mission of breaking down the silos of food systems (using a systemic approach) and building a thriving society through food: the nexus of culture identities and values with environment and territories. All these elements compose a global ecosystem, which is creating an exponential positive change to sustainably improve life on Earth through food.

In this pursuit, the Future Food Institute is deeply committed to the principle of **integral ecological development**, recognizing that true sustainability requires a systemic approach. This encompasses not only environmental stewardship but also social equity, economic viability, and a profound respect for cultural and biological diversity. By embedding this principle in our initiatives, the Future Food Institute aims to nurture ecosystems that thrive on interconnectivity and mutual support, laying the groundwork for a future where humanity and nature coexist in harmony.

The global framework in which the ecosystem works is the UN Agenda 2030. Food systems and agriculture are transversal to the whole of the Sustainable Development Goals (SDGs) framework. Where speaking about SDGs means speaking about food and agriculture. Equally, considering food and agriculture means scaling up SDGs, as eating is an essential act of human beings. Through its three main pillars of activities (education, community, and innovation), Future Food works toward achieving all 17 SDGs. This translates to work in studying the changes and challenges of food systems, prototyping innovative projects, and empowering and bringing together a multistakeholder community, composed of start-ups, innovators, policymakers, academics, corporations, media, and young talents.

The ingredients of innovation for this model are represented by mindset, platforms, and community. In this article, we explain the definitions and transformational concepts for all these ingredients. We additionally describe some tools that this model is using to map the transformation of the role of food in achieving SDGs, such as the Food for Earth Regeneration Toolbox, composed of five innovative areas (food diplomacy; circular living; prosperity; climate-smart ecosystems; food identity) and its evolution throughout the recent global disruption (such as the global pandemic and the war in Ukraine) to restore the regenerative power of food in the whole ecosystem.

Value-based communication

The world is not suffering from a lack of intelligence or leadership, it is suffering greatly from a lack of wisdom —*Robert J. Stanberg (Cornell University).*

In a world overexposed to information and knowledge, up to the point of considering infodemic as one of the major causes of mistrust and confusion, there is a need for new models, systems, and mindsets coming from new forms of communication. Language is a powerful tool and, in this case, can be used to address behavioral change.

Value-based communication starts from the premise that we, as human beings regardless of geographic, religious, political, gender, age, or cultural differences, all share the same ancestral needs: to be happy, healthy, and have access to basic needs.

Value-based communication speaks directly to people's hearts and instantly engages individuals and communities, especially now, knowing that isolation, instability, and uncertainty accelerate depression and do damage to people's well-being.

After decades of hyperspecialization, hypercompetition, and a silo approach to global issues, the pandemic has reminded the world of the level of complexity and deep interconnection of the ecosystem we live with. Restoring common interests, and common goods and prioritizing balanced and inclusive well-being are only possible by restoring shared universal values. This is the starting point to build a fairer and more sustainable global food ecosystem.

For this reason, and the basis of the Good After COVID-19 movement (composed of a group of free thinkers, international experts, top leaders, and change-makers that have come together to identify the foundations of the seeds of change during and after the pandemic) values are central. This approach has been conveyed in the publication of a *Manifesto of Values for a New Green Marshall-type Plan*, now the book "Values for a Life Economy." Compassion and empathy, love for our Earth, love for each other, balance, stewardship of the Earth, youth empowerment, education, new citizenship, good ancestors, the life economy, and partnership are the basic principles for the new needed world. Principles that can be summarized in three simple words: The Golden Rule (the ethic of reciprocity): treat others the way you want to be treated.

Food for Climate League

The pandemic is just one of the crises humanity is facing. "We have deluded ourselves into thinking we can stay healthy on a sick planet," Pope Francis said in the midst of the first wave of COVID-19, in reference to the environmental and climate crisis in which we are immersed and given only 8 years left to heal, according to scientists at the Intergovernmental Panel for Climate Change. This translates into eight agricultural seasons to keep the global average temperature increase below the threshold of 1.5°, over which we would face even more devastating effects. Therefore, it is crucial to regenerate food systems, as they are both responsible for one-third of all GHG emissions and the victims of the climatic changes they contribute to.

The way we produce food, transform it, market it, discard it, communicate it, consume it, and lose it along the supply chain are closely connected to the current climate crisis, in a two-way relationship that must be broken. And it must be broken both through innovation throughout the system and through a new narrative able to generate positive behavior.

Nonetheless, change is already happening, and more and more players of various sizes and sectors are moving in this direction. Large institutions and private companies, such as FAO and Google, are interested and involved in constant work, also supporting Future Food in their work. Climate change can be fought through training programs, innovation projects, and territorial regeneration plans, but also by understanding and changing the way we express ourselves and talk about things.

These are the reasons that led us to enthusiastically support the inception of "Food for Climate League," which was also cited in the Washington Post as the nonprofit organization aimed at redefining "sustainable eating and (to) help businesses, nonprofits and governments promote food that's good for both humans and the planet." It is a project whose scope is the creation of a new narrative on the food−climate relationship, which democratizes food systems sustainability and allows us to face the climate crisis, one bite at a time. Food for Climate League (FCL) provides a sort of instruction manual for those who approach the world of food, both as professionals and consumers, to identify the languages with which to promote sustainability campaigns starting from food.

This project has its roots in the Google Food Lab and the mind of Eve Turow-Paul—writer, expert in food communication and Millenial and Gen. Z food culture—as well as other Lab members like myself.

As Dorothy Shaver—Food for Climate League Board of Directors and Global Sustainability Lead Knorr at Unilever and author of Future 50 Foods—wrote, "we believe that food is the answer and we all work hard to make that answer understandable, accessible, and delicious."

The aim of the project is to provide up-to-date tools to promote climate-virtuous products and consumption patterns, encouraging people to look for greater biodiversity on their plates, not to waste food, and to look for plant-based products.

We can fight the climate crisis and "climate-smart" solutions can be tasty, nutritious, and also Instagrammable, to reach Millennials and Generation Z more easily.

The guiding principle is one of communal and shared prosperity, both for the people and the planet, which arises from the acknowledgment that we influence one another, and that we can do it positively and sustainably. Science tells us that the way we consume jeopardizes the functions of carbon absorption and ecosystem support of natural ecosystems.

Nonetheless, through more balanced food choices and more sustainable diets, we could directly support regenerative agriculture, those farming techniques that prioritize soil health as part of crop production, improve crop yields and turn farmland and pastures into carbon sinks.

There is a direct nexus between food and climate: food production is a major contributor of GHG emission but yet highly depends on climate to survive. Hence, there is a clear and explicit lack of public awareness and knowledge of this nexus, because the global society still pollutes by eating, as we eat by overexploiting and wasting, and we eat what we pollute, as we eat from soils increasingly depleted of nutrients, filled with pesticides.

We have to act faster and reach as many people as possible to generate a real global impact with our diets.

It is therefore time to rethink the food and climate narrative so that we can create a just, tastier, and more nutritious food system that empowers all of humanity to address the climate crisis one bite at a time.

But how?

We need to redefine the concept of "climate-smart eating," making it more accessible and practically helpful for the whole population. Indeed, a global culture obsessed with food, passionate about taste, textures, and nutrient-rich foods can save humanity, promoting a positive impact on biodiversity and our psyche. As an example, it is proven that appreciating diverse flavors is an indicator of good psychophysical health, as stated by Eve Turow-Paul in an article that appeared in Forbes Magazine.

The climate crisis is a particularly difficult topic to talk about, as it can trigger negative emotional responses such as anxiety or guilt. Stimulating reflections on the food–climate nexus encapsulates many challenges, including understanding the complexity of agri-food systems and the need to address a profound change in our behavior.

Currently, there is no comprehensive, universally accepted, digestible, and empowering communication framework around food and the climate crisis.

Conversely, the general public is left to face a variety of terms, definitions, and recommendations that leave them disengaged and helpless.

The current dominant storytelling strategies revolve around negative and punitive narratives: "We have to eat less meat" and "We have to generate less waste," leading to a loss of communication efficacy and not reaching the general public. It is too big a task to be addressed by the single agribusiness companies.

By working together, we can jumpstart new inclusive food trends, able to catalyze a bigger impact by sustaining products and campaigns centered around sustainability.

Food for Climate League (FCL) has studied this issue in depth.

FCL has interviewed climate communications experts and spent countless hours researching and discussing the role of food in the digital age, particularly among the lifestyles of Millennials and Gen Zers. From our studies, we found that "climate-smart" eating is becoming a cultural norm, meeting people's most basic human needs. By identifying the optimal communication tactics, we can catalyze a global movement toward "good-for-you and good-for-the-planet" nutrition. This relationship outlines a communication framework that, if upscaled, can:

1. give a sense of control, and therefore security, to people when eating;
2. democratize good-for-the people and good-for-the-planet food;
3. create awareness-stimulating foods: what we eat must be sustainable, desirable, and culturally rooted.

About our methodology: we started from the analysis of the food and climate-related initiatives realized by 15 of the biggest multinationals in the food and beverage sector and reports coming from the most influential research centers, think tanks, and NGOs, among which the World Resources Institute, the FAO, and the One Planet Business for Biodiversity (OP2B) coalition.

For each firm, we analyzed food and climate-related messages, semiotics, and images. Moreover, we analyzed materials from Google Food, the Menus of Change University Research Collaborative (a joint initiative of the Culinary Institute of America and Stanford University), and our own Future Food Institute, which was added by Food Tank as one of the best 100 organizations concerning food and sustainability.

We therefore classified common attributes to identify the dominant food and climate narrative in the whole world on climate-related topics and more specifically on:

- Biodiversity in food supply.
- Vegetarian diets.
- Regenerative agriculture.
- Social justice in the food system.
- Food loss and waste.
- Water use.

In addition, in-depth one-on-one interviews were conducted with more than 20 leaders in the fields of climate communication, climate psychology, nutrition communication, lifestyles, consumer goods, food systems and social justice, agriculture, and food service to discuss the challenges and opportunities identified by the framework provided by the Food for Climate League.

What did we discover?

This research confirmed that, while nutrition and agriculture are powerful tools to mitigate the climate crisis, the current narrative of sustainable nutrition is complex and flawed and, as such, limits the development of a right food awareness, able to reinforce and move the food culture and highlight the effects of our choices on natural ecosystems. Experts agree that there is much to do in order to connect the dots between food choices, environmental health, human health, and social justice.

Care must be taken to avoid unintended consequences in recommending a rocket solution to such a complex problem.

Repeatedly, the Food for Climate League has been urged to develop simple, but not simplistic communication.

What brings many organizations together today is a sense of responsibility and common purpose on several issues such as protecting biodiversity, stopping deforestation, improving soil health, restoring aquifers and protecting watersheds, and improving social justice throughout the supply chain, with a focus on women, children, and small farmers.

Public and private partnerships are abundant, and new consortia are being created, but all of them are still using language that is too confusing and inconsistent. We heard repeatedly from interviewees the importance of converging on joint and more consistent messages.

The Report that Food for Climate League drafted deepens three action areas:

1. Education is not sufficient. The mistake of assuming that knowledge alone will lead to behavior change should not be made.

2. **Emotional factors are triggering.** For many, climate keywords can evoke difficult emotions such as guilt or anxiety and cause people to shut down; we need to find ways to engage people without triggering defensive emotions.
3. **Exclusive and impersonal.** The entire conversation about climate—and the link between food and climate in particular—needs to move away from elitist paradigms and become much more inclusive.

What would communication messages look like if we instead assumed that all people are aware of and concerned about the ongoing climate crisis, but do not engage in sustainable behaviors because they are inhibited by feelings such as anxiety or guilt, and because the dominant messages are too elitist and create marginalization, or because terms such a "biodiversity," "regenerative," and "circular" are too sophisticated?

We can turn the narrative upside-down, help companies and organizations illustrate how "climate-smart" food speaks to essential human needs, and fuel change at the systemic level.

Finding a solution is therefore critical. The research project involved numerous experts with expertise in psychology, Millennial/Gen Z behavioral sciences, food choice architecture, food service, storytelling, nutrition, and sustainability communications who were critical in developing the first set of best practices for a communications framework that recommends using the following in consumer marketing campaigns, employee education materials, packaging, and menus, etc:

1. **Speak using "human" truths.** Artisanal products and messages that respond to our innate desire to have situations under control, feel part of a community, and be proactive.
2. **Build upon current food trends.** Work within the existing culture to ease the transition toward new habits and generate a real shift of habits.
3. **Highlight positive sides.** Promote the collection of several new, exciting, and appetizing culinary experiences.
4. **Be consistent and repetitive.** Approach new languages and new behaviors in a consistent and repetitive attitude.
5. **Democratize.** Speak to as broad an audience as possible, and always discuss your hypothesis.
6. **Connect the dots.** Clarify the environmental impact of food choices with a clear and accessible storytelling approach.
7. **Empower.** Invite people to get involved as stakeholders and active participants, with the power to make a difference. Give space to imperfections.
8. **Depoliticize.** Focus on shared values and avoid common keywords that can fuel political identities.
9. **Recognize evolution and change as fruits of experience.** Consider diversity in people's view of the climate crisis.
10. **Be honest.** There is no need to avoid the obvious: the climate crisis is scary. But everything we do is significant.

If we have learned something from the COVID-19 pandemic and the disruption generated by the war in Ukraine, it is that humanity can and will join forces to minimize catastrophe. People can learn new words and phrases, such as "flatten the curve," and even personally adopt uncomfortable behaviors for the common good. We can do the same with food and the climate crisis. The first step is to come together and begin to share and embrace the same story. A story where every "bite" turns into a heroic act to save the planet.

This has been precisely the aim behind the realization of "Climate Menus" realized from the collaboration between the Future Food Institute, True Italian Taste (promoted and funded by the Italian Ministry of Foreign Affairs and International Cooperation), the General Consulate of Italy in New York City, I Love Italian Food, and some Italian Chefs based in New York, released on occasion of the 7th Edition of the Week of the Italian Cuisine in the World.

We must choose now whether to sit on the side of the problem or the side of the solution.

Food is life, health, and nourishment, it is a vehicle of values, culture, symbols, and identity, food is sociality.

Humanity will be able to adapt to the great changes we are experiencing only by putting the harmony of the ecosystem back at the center.

Boot Camps

One of the most effective ways to empower people is by activating them.

The Food and Climate Shaper Boot Camps, codesigned by the Future Food Institute (FFI) and the Food and Agriculture Organization of the UN (FAO), are indeed examples of how education can become a driver for reawakening a sense of agency and responsibility. Conceived as hybrid experiences where theory and practice merge together aimed at valorizing intergenerational, interdisciplinary, multicultural, multi geographical learning experiences, the FFI and FAO Boot Camps are characterized by three different phases:

- **Inspiration:** to combine learning and innovation to get a view of the future, through the eyes of big visionaries and experts. This first phase of the Future Food learning approach relies on masterclasses, conversations, and theoretical deep dives into specific topics that are possible thanks to the collaboration with the FAO E-learning Academy, which provides learning opportunities and multilingual e-learning courses for professionals working in food and nutrition security, social, and economic development and sustainable natural resource management but also to the heterogeneity of the Future Food faculty. The Future Food faculty is composed of over 150 ecosystem members currently shaping our food system, including FAO officers, university professors, startuppers and entrepreneurs, grassroots leaders, researchers, chefs and food scientists, industry executives, regenerative agriculture leaders, and policy and institutional representatives.
- **Aspiration:** to unlock participants' potential by experiencing the food system alongside the "doers" and "makers." To achieve this objective, the second step of the Future Food learning approach consists of creating direct and hands-on experiences to discover food heroes and innovative realities. Through **local discoveries**, with participants visiting

regenerative farms, fisheries, food supplies of diverse realities or through **climate suppers**, in-person or hybrid moments of conviviality to debate and investigate the challenges behind the agri-food system, participants have the chance to approach, interview, and directly experience firsthand how challenges can be turned into solutions. This activity is a tool to learn on the ground from true role models, but also to strengthen the relationship between the participants and their communities. Learning by doing and engaging with local food heroes is part of the process of thriving together and empowering people to raise a collective voice, strengthening the communication of sustainable topics.

- **Action:** Convert ideas into action and make a transformation in the communities in which participants are part. In this sense, the last phase of the experiential learning platform allows the participants to transform the current challenges and live experiences into possible solutions. Prototyping, teamwork-based innovation challenges, entrepreneurial mentoring, and inspirational talks are all part of the Action phase exemplified in a two-day hackathon. Only in this way, it is possible to create awareness about the crucial correlation between what we eat and climate change, making people conscious of the complexity of the main global issues and the necessity to have a systemic approach to solve them. The participants of the Boot Camps become agents of that positive change who can lead us into a better future: people capable of understanding the complexity that surrounds them, who look at the system with critical and holistic thinking, with the lens of integral ecology; citizens of this mad and beautiful world determined to devise new solutions for the sustainable development of their territories, to be able to enhance them while respecting local traditions, but capable of dialoguing and making the best use of latest-generation innovations; good communicators and communicators for (the) good, able to translate complex concepts in something more simple without being superficial. The Future Food Institute calls these kinds of change-makers **"Climate Shapers."**

Value-based communication is crucial in these activities, and it is one of the magic ingredients that the participants bring home after the experiential learning. All the communication used during these courses helps future Climate Shapers understand the urgency of the global emergencies we are facing solve problems and switch on lights rather than focusing only on the dark to better the climate and the world using ecosystemic and prosperity thinking approaches.

The Future Food Institute has been working on this innovative learning model since its birth. On June 10, 2019, Future Food had the honor of signing together with FAO Director-General, José Graziano Da Silva, a Memorandum of Understanding to work toward the improvement and development of education, innovation, and community-building activities of the FFI through continued integration of the work of the FAO e-learning Academy:

"Sustainability is the greatest goal of Humanity and the only way forward. For this great goal, we need competent, capable professionals who are able to make appropriate decisions, formulate targeted and sustainable policies and strategies, adopt innovative 'green' methodologies and technologies, we need climate shapers. We need climate shapers to redesign our agri-food systems, putting the nutritional and health status of citizens and sustainability at the centre. In other words, we can only achieve sustainability through capacity building and transfer of skills and competences."—Cristina Petracchi, Leader of FAO's eLearning Academy

Since July 2019, Future Food Institute and FAO have organized in-person Boot Camps in New York (United States), Tokyo (Japan), Thyngeri (Iceland), Pollica (Italy), Marettimo (Italy), thirteen editions of Boot Camps in a digital format in English, and four digital edition in the Japanese language.

Embracing the same three learning phases of inspiration, aspiration, and action, the **Food and Climate Shaper Digital Boot Camp**, which was awarded as "Most Innovative New Program" in June 2021 by GoAbroad and shortlisted by Reimagine Education jury as one of the 10 best online programs in the world in December 2021, has been designed to adapt education toward new communication tools and the changes of our *Humana Communicatas* triggered by the global pandemic. COVID-19 has resulted in schools shut down all across the world, changing education dramatically, with the distinctive rise of e-learning, whereby teaching is undertaken remotely and on digital platforms. In this context, our Digital Boot Camps have represented the answer of reshaping and evolving our learning initiative in the time of the pandemic. It is possible to keep on educating, not simply shifting the normal formal lessons into something digital, but creating new innovative ways of interacting with the students, making the lessons more interesting, involving, and creative. The digital edition, moreover, gives the possibility for everyone to have access to this kind of learning process: making it possible to connect to the lessons simply through a device with an internet connection. All the lessons are recorded, solving also the potential issue of time zones. In terms of program fees, FFI and FAO guarantee some scholarships assuring the right to education to everyone interested in the program, and increasing the democratization of knowledge. Being aware that change is something natural and typical of living organisms, Future Food is used to adapt to this world that is changing often and fast. Changing methods but keeping the same mission: training a new generation of climate-conscious leaders who are challenged to think from a foundation of values, embracing complexity, and oriented towards opportunity; learning through concrete projects and real-life challenges ensures understanding of essential needs, but also the power of codesigning and learning with and from others; applying the design thinking, mentorship, challenge-based work, and creativity sprints (aka Hackathons), all underpinned by a philosophy of "prosperity"—making things the planet wants.

Climate-Smart Cities, Climate-Smart Farms, Climate-Smart Kitchens, and Climate-Smart Oceans are the four main thematic tracks embraced during the Boot Camps to tackle the agri-food system in its complexity, which in time have evolved into four **regenerative tracks**. In fact, regenerative means much more than simply sustainable (do less or no harm); regeneration is more than restoration (going back to a previous state). Regeneration is the reenlivenment of degraded relationships at large, touching natural ecosystems, social fabric, and individuals. In this sense, the need to accelerate the path toward an **integral ecological regeneration** has ensured both digital and in-person Boot Camps serve two main goals:

1. Empower individuals to build community: Education leads to a climate-literate society where many people are able to take concrete and effective actions to regenerate the natural environment.

2. Empower organizations: Whether at the executive level or an entry-level employee, Boot Camps enhance independent thinking, inspiration, and teamwork. It is a way to develop professionals who will incorporate climate change solutions into their roles and responsibilities and have a larger positive impact on the organization as a whole. This is a platform for employees to break out of the box, learn, experiment, and come back with impactful projects to incorporate into their organization.

Until today, more than 500 Climate Shapers, from 60 different countries, have been trained, while more than 80 local discoveries have been created as forms of peer-to-peer learning. The Food and Climate Shaper Boot Camp is our opportunity to redesign the learning ecosystem; to transform the concept of education into a values-based process with the ultimate goal of the betterment of humanity; to demonstrate that an alternative narrative is possible and that we need communicators capable of spreading new positive and life-changing messages. In fact, being part of this learning pathway is not an end in itself: as soon as participants become Climate Shapers, they officially join the alumni group and become part of the Future Food Ecosystem. To date, the group is composed of more than 300 people who share the same mission and offer their expertise and diversity for a common purpose: restoring the agri-food system. Food innovation and regeneration are indeed **a glocal mission** and require the spirit of the explorers, driven by purpose and curiosity.

Hackathon

We are all experiencing the effects of decades of linear approaches, as a global community, as individuals, and as hosts of the natural ecosystem. Whether it is applied to usual production and consumption systems or in traditional learning methods, the result still remains the same: the impracticability to continue this way in the long term. To turn students into conscious, prepared, active citizens of tomorrow, to value individual expertise for the benefit of the whole community, and to face the global challenges we are facing we need to overcome **one-way forms of communication** (from teachers to students, from adults to youths, from employers to employees, from experts to the masses) and embrace multidimensional, intergenerational, multilevel, glocal exchanges. Building bridges of knowledge, cross-pollination of expertise, dialog between stakeholders, and occasions of mutual contamination of ideas and perspectives is pivotal to propose long-lasting solutions to multifaceted problems.

The principle at the basis of the hackathon is precisely that of cocreating with all the participants working on innovative solutions to real problems. Hacking a problem requires a creative, collaborative process and to be strategically focused on a goal: designing a concrete solution to the needs of a specific target. Therefore, rather than an application of a linear approach, the nature of the hackathon is grounded on a reverse **mindset:** starting from the insinuation of a doubt, not a given information, from a collective challenge or open question, not a solution.

There are thousands of hackathons happening all around the world. However, not all of them are able to generate a long-lasting impact, especially in the food ecosystem.

Since 2014, the Future Food Institute has organized **more than 200 hackathons**, touching the agri-food systems in all its phases and involving the widest plethora of local, national, and international actors on food: food and beverage companies, startups, UN agencies, universities, students, innovators, policymakers, entrepreneurs. Ranging from specific segments of the food supply chain, such as sustainable cocoa and organic food, to broader problems such as food services, malnutrition, and food waste, until considering the agri-food system in its complexity through hackathons on SDGs and the rights of the Earth, the power and scope of the solutions generated by these hackathons are empowered by the values underlying our activities. The *why* addressing our actions (before the what and how) are the common, universal values that bond together every human being despite their differences.

Shifting from "me" to "we" is the basis of the transition in our hackathons from **design thinking to prosperity thinking**, the methodology that takes into account the balance between man and planet, through a **human and Planet-centered approach**. This is an approach that is scientifically validated and coimplemented with Prof. Matteo Vignoli, cofounder of the Future Food Institute, and improved on a daily basis thanks to the scientific advice of Prof. Sonia Massari and our head of design Chhavi Jatwani. This is a way of learning, **challenge-based learning or problem-based learning**, which is grounded on pragmatism and can be applied within schools, as alternative hands-on education pathways. The partnership between Future Food and Cosmopolites, exemplified in the two editions of the "Hackathon in the School project," was aimed at achieving this exact purpose. More than 500 people involved, among students aged 16–18, teachers, and mentors, and eight challenge spaces with problems to be solved by Italian schools: culture, wood, business and territories, education, water, social innovation, food waste, and Mediterraneo. In the traditional education system, based on face-to-face lectures where students are asked to memorize data and information, hackathons are an innovative way of learning. As its etymology confirms, the hackathon is a marathon where team working is centered first on *understanding* what the people's and the planet's real needs are, investigating firsthand the causes of the problems before shaping the solutions. Through a process of cocreation, students are forced to become responsible for their own learning, to "show, don't tell!," to learn by doing: collaborating with others in the creation of a collective idea, dialoguing with peers and experts, prototyping, testing their ideas and solutions, organizing time and resources, presenting their results to a broader audience. It is a way of learning that embraces fun and creativity, but also teaches purpose and **soft skills**.

Hackathons are also introduced within universities, businesses, and companies aiming for purpose-driven innovation, as activities that create authentic connections with multiple disciplines, contexts in which learning with others and from others' expertise, sharing knowledge, connecting stakeholders, and empowering the voices of individuals.

Recently, hackathons have been revealed as precious cooperative tools to improve local services, better reconnect the social fabric, unveil the **sense of local pride, restore the connection with the local territory, and make communication more effective**.

In this sense, some of the results achieved with the hackathons realized in **Pollica**, a small rural village in Italy and an **Emblematic Community of the Mediterranean Diet**, are evident. Future Food has designed an educational approach grounded on the ancient Mediterranean philosophy (exploration and discovery by the **five senses**, learning and teaching through **savoir-faire**, and transdisciplinary **sharing per sustaining**), a design model that relies on the implementation of the Prosperity Thinking, and a regeneration toolkit, which encompasses the political, environmental, social, human, cultural, and economic regeneration, some of the outcomes includes:

- the regeneration of the Mediterranean areas and sea through hackathons as part of the Boot Camp in Pollica involving several Climate Shapers;
- a community bakery, a hub for restoring contact with the Cilento nature, and other innovative ideas to connect local actors through training programs, all codesigned by locals through Hack the Village;
- a playroom redesigned by the kids from Pollica to play, recreate, and develop their personalities through the hackathon with children.

These achievements could not have been possible without a multilayer collaboration between public and private companies, local youth and adults, mayors and entrepreneurs: the necessary ingredients to combine tradition and innovation also in the communication field. In this sense, Pollica, a symbol of the truest Mediterranean values and know-how, has become a tangible example of innovating while respecting the Mediterranean Ecological knowledge also through new forms of communication. The **Paideia Digital Academy**, which was launched in 2021 in Pollica, confirms that adventure video making, app design and game design, digital storytelling, virtual experiences of cultural and tourist heritages, and 3D printers can all become crucial layers to protect regenerative food systems and defend the heritage of the past for the future.

To **weave together for a thriving society**, we urge understanding the messages and languages spoken by the different actors on the international stage, of local and global leaders. We urge merging the healthy side of competition, aiming at gradually improving solutions, with cooperation, which represents the most important Sustainable Development Goal. The result, cooperating with a competitor to achieve a common goal (also known as coopetition) is what can lead us toward building learning communities.

This can be done only by trying, succeeding, or failing, and trying again and again. In this sense, food is a great teacher. After all, we have all learned how to fail in the kitchen, and yet we didn't stop cooking. Instead, we have adapted, and reshaped, till we got better results.

Food is the perfect playground for innovation and a vehicle for common language.

Conclusion

Communicating has always been fundamental to us humans. It is what has allowed us to survive as a species and to be able to develop into complex systems. Today, however, we have reached a crucial point in our history as human beings: thanks to the spread of the Internet and social media, we have never communicated as much as we do today. The development of computers and technologies has further expanded this ability to relate, which is one of our basic needs. However, we are now realizing that communication, and therefore access to information, is meaningless without context. It is therefore important to link communication and education. Just as we are taught in school to analyze a newspaper article or an advertisement, we need to do the same with all other forms of communication in order to develop critical thinking skills more broadly. As we are seeing during the current pandemic, the fundamental part of education is not in the passive passing of information but in the creation of critical thinking and reciprocity, of response and appropriation or denial on the part of the learner (i.e., all of us). Today, however, it is no longer enough to have a critical spirit, to truly apply in the field what is often explained only in theory: show don't tell! Secrecy and competition no longer make sense in the era we are living in. Keeping knowledge to oneself is a short-term solution. Now we are called to think in the long term including all the stakeholders and the most participatory knowledge. In a society where access to information has become extremely widespread and personalized, it is critical to include everyone in the development of heartfelt and more critical knowledge. This can lead us to develop a true sense of community by listening to all voices and making knowledge truly participatory. By including everyone and equipping everyone with a sense of responsibility and agency, we will be able to develop not only a sense of action with purpose but also more responsible and therefore more resilient communities.

Open innovation in sustainable corporate communication: a case study from Italian food companies

Marco Bassan

Department of Business Economics, Università degli studi Roma Tre, Rome, Italy

Introduction

Sustainability has emerged as a new marketing model in recent decades, illustrating the strong interconnection between these two disciplines (Kumar et al., 2012; Kemper & Ballantine, 2019), strengthening the long-standing link with the business target.

According to previous studies, the two concepts are incompatible because sustainability leads to reducing consumption, while the main objective of marketing should be to increase it (Jones et al., 2008).

Marketing has often been accused of having worked to increase consumption even at the cost of generating unsustainable impacts (Jones et al., 2008; Peattie & Peattie, 2009; Sandberg & Polsa, 2015). The consequences of marketing appear quite in contrast with the affinities of sustainability (Jones et al., 2008; Lim, 2016; Pereira Heath & Chatzidakis, 2012), both from the environmental point of view, therefore reduction of the excessive use of resources and reduction of waste production and emissions, but also from the point of view social and therefore health and economic sustainability.

Marketing, in particular, is not undertaken to ensure long-term profitability, yet there is so much potential that is sometimes ignored (Gordon et al., 2011). In recent years, scholars have explored the intrinsic concept of marketing in an attempt to adapt the role of marketing to sustainable growth (Desmond & Crane, 2004; Peattie & Peattie, 2009; van Dam & Apeldoorn, 1996).

Several academic papers started to explore a sustainable marketing paradigm (Gordon et al., 2011; Peattie & Peattie, 2009; Cooper & Schindler, 2005); green marketing has become an increasingly crucial factor in a sustainable company strategy, and as a result,

Food Sustainability and the Media
DOI: https://doi.org/10.1016/B978-0-323-91227-3.00008-1

organizations are embracing green marketing strategies to improve business success while reducing the environmental effect of their operations (Vafaei et al., 2019). To put it another way, going "green" is both a necessity and an opportunity for businesses (Souri et al., 2018).

Since marketing will help the company to meet the needs of its customers, it has the potential to promote long-term economic stability (Gordon et al., 2011; Fisk, 2001).

According to recent research (Baldassarre & Campo, 2016), communication is the cornerstone of sustainable marketing. The range of behavior varies from companies that use sustainability solely as a marketing tool, manipulating their data, or adopting a hyperbolic communication style, to companies that see sustainability as a benefit rather than a cost. Indeed, a transparent company organizes its communication to express sustainability most clearly (e.g., reports, website, product packaging) and transform sustainability into a marketing tool (Baldassarre & Campo, 2016).

Sustainability is the most critical hurdle of modern marketing for a company that wants to be a pioneer in its industry (Sheth et al., 2011). Indeed, sustainability elements are key factors of corporate image and reputation(Fombrun, 2005; Hillenbrand & Money, 2007; Pfau et al., 2008).

Companies, on the other hand, are not sure of the right tools to successfully communicate sustainability and often suffer from "marketing myopia" (Levitt, 1960), a circumstance in which firms that do not think widely may lose out to competition, particularly those that do not adapt to changing customer demands and expectations, as well as other marketplace changes (Kamasastry, 2020). Indeed, this attitude can be found in some companies that are not consumer-centered, when it comes to sustainability. Those companies focus on their performance in terms of impact reduction or sustainability-related metrics but do not take enough into account that all of this improvement has to be accepted in some way or verified by the market.

Sustainability is a global concept that can provide a way to communicate with society, broaden the consumer base, lower operating costs, and benefit the environment, as defined in the Sustainable Development Goals by United Nations 2015.

In brief, there is a two-way partnership between marketing and sustainability: on the one hand, sustainability can be used as a marketing topic, and on the other hand, communication practices are crucial to achieving sustainability. Indeed, sustainability represents a cost in the short term for companies if they don't achieve a competitive advantage if the stakeholders recognize it as useful and of preferable value, then the investments made to achieve it turns into an advantage, indeed (Falkenberg & Brunsæl, 2011; Gerstlberger et al., 2014) have all claimed that sustainability is a way to gain a competitive advantage.

The main goal of the relationship between sustainability and communication is to promote a good brand representation in the media by promoting the company's efforts to overcome its natural, social, and economic impacts (Mitchell et al., 2010). Obtaining this positioning requires not only concrete and certified actions but also a communication activity, both to make known the actions carried out but also to engage in advance the stakeholders who will have to help the company to achieve them. So in short, communication can be useful both to achieve results and to make them known and thus obtain a market advantage.

If the connection between sustainability and marketing is established, the need for stakeholder engagement—involving primary and secondary stakeholders such as suppliers, customers, institutions, NGOs, academia, becomes clear: a stakeholder-focused strategy outperforms a market-driven or sustainability-centered strategy (Mena et al., 2019). Indeed, as global sustainability issues are characterized by high complexity and uncertainty, effective communication could be essential to tackle them (Newig et al., 2013).

Through the study of three business cases, our research proposes a new approach to achieving sustainable marketing that includes stakeholder participation. We agree that including stakeholders in the various stages of the sustainable marketing process provides a strategic benefit to companies that wish to incorporate sustainability into their operations and strategy rather than using communication for greenwashing, helping them to frame the sustainable message that needs to be delivered and gain a competitive advantage over competitors. Indeed, stakeholder participation will help organizations overcome corporate social responsibility new challenges and potential emergencies (Miles et al., 2006; Noland & Phillips, 2010).

Sustainable development goals in the food industry

Hunger affects around 820 million individuals globally (or one in every nine persons) (IFAD, UNICEF, WFP, & WHO, 2019). In 2018, an estimated 149 million children under the age of 5 had stunted growth, which is a symptom of malnutrition marked by inadequate height for age. Similarly, the number of overweight children under the age of 5 has increased to 40.1 million, with Asia accounting for nearly half and Africa accounting for a quarter. Because of the food system disruption induced by the COVID-19 pandemic, these circumstances are expected to increase (Headey et al., 2020).

The global food system contributes significantly to greenhouse gas emissions at all stages of the supply chain, from agricultural production to disposal, distribution, retail, domestic food preparation, and waste, all of which play a role (Garnett, 2011). It also involves other important environmental impacts, including loss of biodiversity and water extraction and pollution (Garnett, 2014).

Although the definition of food sustainability encompasses a wide range of factors, methodologies, and interpretations (Garnett, 2011; Aiking & de Boer, 2004), there can be extreme differences between the sustainability awareness of producers and consumers (Aiking & de Boer, 2004), the "Sustainable Consumption in Italy (2021)" research realized by SAP declare that in Italy "In the food industry, 48% of the interviewed would choose with better probability a food product brand recognized as sustainable while 45% considers important for brands to focus on a sustainable supplying because they feel that in this way, the environment will be respected. In conclusion, 56% wishes to monitor the carbon footprint of food companies, while 24% allegedly already does."

Initially consumers, on the other hand, have used "sustainability" as a sort of abbreviation for "green and good" to identify production and consumption systems linked to a broader range of characteristics, such as community-based efforts to build healthy, fair, and local food systems (Kloppenburg et al., 2000).

As the planet faces the challenge of ensuring sustainable food security in the face of population growth, habitat depletion, water scarcity, and climate change, the modern food system is at the center of a global nexus of social, environmental, and economic issues (Freibauer et al., 2011; Garnett, 2014; Gladek et al., 2016; Godfray et al., 2010; IPES-Food, 2015; Sustainable Development, 2009; Searchinger et al., 2014; WWW-UK, 2013). Furthermore, environmental, social, and economic problems in the food system interconnect and magnify each other (Hinrichs, 2014).

There is a global concern that food systems may lead to major ecosystem destruction (IPCC, 2019; Willett et al., 2019). If existing trends hold, the impact of these increases will be felt more intensely in low- and middle-income countries that are still facing malnutrition (HLPE, 2017). There is a need for structural and transformative change, including significant shifts in production and consumption habits. The idea of "sustainable transition" has been coined to support the goal of change into sustainable structures (Lachman, 2013).

The Sustainable Development Goals (SDGs) have ambitious targets, including food security, nutrition, climate change, sustainable use, human dignity, and justice, as the world's main transparency tool for sustainable development in the next 15 years (Fanzo, 2019). They are considered a roadmap for creating a greater and more prosperous future for all (Byerlee & Fanzo, 2019)

International organizations (such as the International Integrated Reporting Council) and global initiatives (such as the Global Compact) have lately urged that businesses provide additional information to stakeholders to help them evaluate the impact of their policies and strategies on sustainability. The introduction of sustainability and integrated reports as a result of this international "trend," which, according to voluntary disclosure theory, increases corporations' accountability for a variety of environmental, social, and economic concerns (Izzo et al., 2020).

Sometimes this is demanded by consumers such as what happened with palm oil, what happened with glyphosate in agriculture, what happened with plastic or packaging for which the market is asking. In other cases, the trade is, therefore, the intermediaries who acquire a positive reputation toward end consumers, need to have suppliers compliant with the SDGs or in any case who demonstrate a commitment to sustainability.

Several companies throughout the world have begun this path by defining and implementing sustainable strategies as fundamental drivers of their mission, ambitions, and business models (Busco et al., 2018).

On the one hand, the SDGs provide a chance for companies to rethink their goals and integrate their business strategies (strengthening stakeholder engagement and better identifying future business opportunities). In this regard, private business groups play a critical role in achieving the SDGs: Organizations, according to UN Secretary-General Ban Ki-moon, are a critical partner in attaining the SDGs because the public sector alone would not be able to meet the challenge (Jensen & Berg, 2012).

The benefits of the SDGs can only be fully realized by businesses if these benefits are externally acknowledged by stakeholders and if business organizations can maintain a trusting relationship.

Sustainability and stakeholder engagement

Sustainability is one of the world's big problems (Ferraro et al., 2015), and it is a problem of such magnitude that it cannot be solved by individual intervention alone. Sustainable growth and climate change can only be achieved through collaboration between actors at all levels (Chakrabarti et al., 2020).

Sustainability is always a team game (Doelle & Sinclair, 2006), for example, within the supply chain, because a company will be able to declare itself sustainable if its raw materials are also sustainable, cooperation, coordination, integration, and collaboration, as well as a recognition of its cross-disciplinary character, represent key factors in the definition of the sustainable supply chain (Frankel et al., 2008).

Due to the seriousness of sustainability issues such as climate change, simply recognizing the (systematic) reality of multiple trade-offs is impossible to address the most pressing challenges of sustainability management. In this context, the view of stakeholder theory varies significantly from strictly instrumental interpretations that measure the presence of an objective relationship between various stakeholder interests (e.g., stakeholder participation and corporate financial performance), to the creation of constructive connections between stakeholders' interests (Hörisch et al., 2014).

The release in 2015 of a comprehensive and detailed strategic plan of objectives and indicators underlying the Sustainable Development Goals (SDGs) marked a watershed moment in the process of aligning not just developing but also developed nations on the path to sustainable development (Pradhan et al., 2017), especially the SDG17 focus on "Strengthen the means of implementation and revitalize the global partnership for sustainable development."

It is crucial to recognize that the sustainable management philosophy must be able to solve both short and long-term challenges, as well as to provide companies with both short and long-term potentials and possibilities (Hörisch et al., 2014). In the social sense of a market, stakeholder theory is about solving (potential) conflicts between money and ethical obligations by building shared interests between the demands of all related stakeholder groups (Parmar et al., 2010).

Stakeholder theory is one of the most applied approaches in global, financial, and sustainability policy science, if not the most widely used (Montiel & Delgado-Ceballos, 2014).

An individual firm must investigate tools and processes that enable different market players and stakeholders to collaborate on value-added projects (Gold et al., 2013).

The relationship of stakeholder engagement and communication process

New stakeholder engagement models for communication are emerging in many corporate sectors, with the most developed initiatives concentrated in the area of environmental sustainability (Aakhus & Bzdak, 2015; Burchell & Cook, 2006; Souri et al., 2018; Spitzeck & Hansen, 2010), for example, show how companies are using dialog techniques to react to criticism of their social and environmental policies. Vogel (2005) looks at how CSR evolved from a more charity-driven strategy to a more bottom-line-driven one and how the instruments and methods for engaging stakeholders changed as a result. Perrini (2006) examines how nonfinancial reporting tools and norms are adjusting to shifting stakeholder expectations and new media circumstances.

To create share value (Patrizia & Carlotta, 2011), the company must communicate with stakeholders to grasp chances for economic development and growth that is socially responsible. A systematic conversation with social agents via meetings and other exchanges ("stakeholder interaction") is required without interfering with or altering the goals of the stakeholders, which often vary (Schmidheiny & Zorraquin, 1998; Patrizia & Carlotta, 2011; Seabright & Kurke, 1997). By collaborating with stakeholders to create a strong, reliable, and credible sustainability path, the businesses cases that we have studied have turned sustainability into a communication competitive advantage (Rossi, 2017), enabling them to respond quickly to a variety of crises while maintaining their distinct and strategic positions on their sustainability issues.

Methodology

Considering that the research hypothesis is focused on contemporary phenomena with real-life context, we implemented a case study methodology to approach the analysis. Indeed, exploring the relationship between marketing, sustainability, and stakeholder engagement is an investigation with the boundaries between phenomenon and context not being clearly evident (Yin, 2013). The research is based on the multiple-case model, which is the most suitable when the context is relevant, and the research area is a new one.

A case study is an empirical investigation that examines current phenomena in its naturalistic setting, particularly when the boundaries between the object of research and the environment are not readily apparent (Ebneyamini & Sadeghi Moghadam, 2018).

The use of a case study methodology allows us to achieve a deeper understanding of the situation as well as to generalize the results to other contexts (Eisenhardt, 1989; Eisenhardt & Graebner, 2007).

We followed the stages of the research design involving: Case selection, Theory Building, Data Collection, Data Analysis, and Conclusion (Patton & Appelbaum, 2003).

Case selection

In this research, the geographical area of analysis was restricted to Italy, where stakeholder involvement is particularly useful considering that the territory is fragmented with over 7900 municipalities. The Italian food industry is highly fragmented, The Hofstede's Model, a six-dimensional model that represents culturally independent preferences for one state of affairs over another, allowing nations (rather than individuals) to be distinguished from one another (Hofstede, 2011) identify that individualism is a characteristic of the Italian culture. Individualism, in contraposition to collectivism, describes the relationships individuals have in each culture. In individualistic societies, individuals look after themselves and their immediate family only, whereas, in collectivistic cultures, individuals belong to groups that look after them in exchange for loyalty (Soares et al., 2007). All these factors lead to the hypothesis that stakeholder engagement can be strongly useful in managing fragmentation in the Italian food sector (Gupta et al., 2015).

The companies included in our analysis were chosen to represent different elements of the value chain: Skretting is a supplier, world leader in aquaculture feed, Mulino Bianco is a biscuit manufacturer, and Coop is an Italian consumers cooperative that manages the largest supermarket chain in Italy. Moreover, the case studies represent different business models: Mulino Bianco and Coop are B2C meanwhile Skretting is a B2B company, "see Table 10.1."

In terms of their commitment to sustainability, all three have been leaders in their sector for many years in facing sustainable challenges. To give some examples: according to

TABLE 10.1   Companies' market description

Company	Market description	Data
Skretting	Skretting is a global leader in providing innovative and sustainable nutritional solutions for the aquaculture industry. Skretting has production facilities in 18 countries on five continents and manufactures and delivers high-quality feeds from hatching to harvest for more than 60 species. The total annual production volume of feed is more than 2.4 million tonnes. The head office is located in Stavanger, Norway	Employees: 3500 worldwide Products: Fish and shrimp feed for more than 60 species Production: 2.4 million tonnes annually (2020)
Barilla	The Barilla family has been at the helm of the company for over 143 years. Barilla Group controls Barilla (multinational pasta maker), Cucina Barilla, Mulino Bianco, Gran Cereale, Pan di Stelle, Pavesi, Voiello, First and Academia Barilla (Italy), Harrys (France), Wasabröd (Sweden), MISKO (Greece), Filiz Makarna (Turkey), Casa Barilla (United States), Yemina and Vesta (Mexico) trademarks	In 2020, the Company had a net sales of 3890 MLN EURO Employees: 8.591
Coop	Coop is a system of Italian consumers' cooperatives that operates the largest supermarket chain in Italy. Its headquarters are located in Casalecchio di Reno, Province of Bologna	14.3 bn of turnover 6.7 million shareholders 8 million weekly customers

Forbes, Coop is one of the five sustainable leaders in Italy; in 2019, Skretting won the *Edie* award in the Sustainability Product Innovation of the Year category; Barilla founded a think tank that since 2014 is an independent foundation: the Barilla Foundation for Food & Nutrition. The company's commitment to sustainability issues has led it to reduce greenhouse gas emissions by 30% per tonne of finished product and reduce water consumption by 21% per tonne of finished product, since 2010.

Data collection

As a primary source, we used interviews and informal speeches with the internal project owners in the companies (Pucci et al., 2020). During the entire process of development of the projects, we have collected information from Table 10.2.

We also used field notes, internal reports, participation in workshops and seminars organized in the companies, as well as multiple visits to local offices. Moreover, each project produced several work-in-progress reports that were used to gather key information about the process. Informal conversations and direct participation in workshops and seminars allowed the research team to gather a deep understanding of the intents, the process, and outcomes through the ability to: gain the trust of informants, clarify uncertainties, and integrate evidence from observations and interviews (Creswell, 2009; Yin, 2013).

TABLE 10.2   Interviewers.

Company	Role
COOP	CSR manager
	Quality Manager
	Communication Manager
BARILLA	Marketing Manager
	Purchasing Director & Sustainable Sourcing Coordination
SKRETTING	General Manager
	Marketing Manager

Data analysis

Starting from the data collected (project goals, stakeholder engagement process, and outcomes), the different projects have been systematized to develop a protocol that can represent a stakeholder engagement process. We analyze the results using a cross-case comparison based on pattern matching (Baxter & Jack, 2010).

The protocol we implemented for each case study followed the proposed process (Ihrig & MacMillan, 2017):

Company sustainable goals

The first step is the initial point from which the project starts. Achieving sustainability is a complex process and since it's mainly related to the longevity of the business and responsibility to external needs (Bateh et al., 2014), some goals that the organization wants to achieve must involve external stakeholders to fulfill the expected goal.

Stakeholder mapping

Once identified, the goals we want to achieve and which stakeholder is involved, to produce an effective strategy is necessary to map these stakeholders identifying their power, interest, and attitude (Murray-Webster & Simon, 2006).

Map stakeholders' tensions

The mapping of stakeholder tensions allows both to identify which are the mutual interests based on which the company aims to create value for all stakeholders involved (Parmar et al., 2010) and to realize that there are always trade-offs and that the most proactive and challenging task for management is to search for means to overcome trade-offs (Hörisch et al., 2014).

Stakeholders' involvement

Finally, to overcome the trade-offs that emerge when interests do not perfectly match, we implemented a design thinking workshop that has been considered a suitable method to promote the integration of different stakeholders (Redante et al., 2019); indeed, overcoming trade-offs and conflicts is where stakeholder engagement is mainly active by creating mutual interests among the demands of all relevant stakeholders (Parmar et al., 2010).

Case study

Mulino Bianco

In recent years, Mulino Bianco has implemented a series of sustainable policies (recyclability of packaging, animal welfare, sustainable agriculture, etc.), ahead of the market, to maintain leadership, as consumer concerns no longer derive from social anxiety, but from those issues revolving around sustainability, especially in the agri-food sector. In line with Barilla's "Good for You, Good for the Planet" project, Mulino Bianco presents in 2019, the "Carta del Mulino," to strengthen its commitment to sustainability starting from its primary ingredient: wheat.

It is an innovative guideline for sustainable agriculture, created in collaboration with the WWF, containing 10 rules designed to increase the quality of products, support the work of farmers, and give backspace to nature by promoting biodiversity, reducing the use of chemicals, and safeguarding pollinating insects (Dipace et al., 2019).

Mulino Bianco defined the "Carta del Mulino" as: "a set of 10 rules for the sustainable cultivation of soft wheat. With the Carta del Mulino we not only bring quality to our products, but we also support the work of farming communities and we promote biodiversity by safeguarding pollinating insects."

Despite the complexity of extending this specification to all Mulino Bianco plants and products, the company has decided to undertake, with the help of WWF Italia, a stakeholder engagement process to define a sustainability formula shared by all.

The stakeholders who have joined the projects over time are: dozens of Italian mills (one of which is owned by Barilla located in Galliate in the province of Novara), hundreds of storage centers scattered throughout the country and internationally, and thousands of farms.

Since 2017—the year in which the idea of formulating the "Carta del Mulino" was born—a series of training meetings and workshops have been organized to understand and reconcile the different needs of all partners (Dipace et al., 2019). Below is a table containing the various meetings held from 2017 to April 2019, the year in which the first biscuit made exclusively with sustainable soft wheat flour was launched on the market see Table 10.3.

The involvement of some stakeholders in the formulation of the "Carta del Mulino" meant that they were more inclined to accept and implement the Guidelines aimed at a more sustainable production. Barilla, given its size and the large number of partners in the company, could have the contractual power to not involve directly the stakeholder and impose the guidelines to the stakeholders. Instead, it understood the potential of a proper stakeholder engagement, thus favoring the creation of an ecosystem of involved and collaborative subjects.

TABLE 10.3   Project roadmap.

	January	First encounter with the mills.
2017	May	Start of collaboration with WWF, University of Bologna, University of Tuscia, and Open Fields
	July	First draft of the Mill Charter
	September	Presentation of the draft of the Charter to the mills, and discussion with the University of Bologna, University of Tuscia, WWF, and Coldiretti
2018	January	Meeting with the breeders
	February	Choice of the ISCC PLUS sustainable cultivation scheme.
	March	Meeting with the mills
	April	Training for mills, stackers, and farmers at the Barilla plants in Novara and Cremona. Software launch for the traceability of the whole process
	June	First harvest and certification
	July	First revision of the Mill Charter. Involvement of 500 farmers for 2018 sowing
	September	Meeting and training with the mills and with the stockmen
	October	Meeting for the training and alignment of RINA auditors
	November	Meeting with the Hungarian, German and Austrian stackers
	December	Meeting with the French stackers
2019	April	Launch of Buongrano from sustainable flour Presentation to the press of Buongrano in the production plant of Castiglione delle Stiviere in the presence of all stakeholders Communication to consumers

The guideline has been realized with the involvement of WWF Italy that represented the stakeholder that can guarantee the use of 100% sustainable and environmental-friendly wheat for all the products.

The "Carta del Mulino" represented for barilla and its stakeholders the tool on which to build a common communication strategy, strengthening the barilla brand as a market leader and allowing small stakeholders to benefit from the communication benefits linked to a shared initiative.

The agency IPRN (International Public Relation Network) realized a communication strategy around the "Carta del Mulino" that promotes the project implementing a "PR First" strategy to raise awareness on the brand purpose 2 months before the marketing launch and then implementing an engaging phase with media and stakeholders taking them among the production lines, where the sustainable biscuits are made, with Paolo Barilla, VP of the Barilla Group.

The Media reported clearly and in detail all the campaign key messages with over 87 media hits among the most important Italian newspaper (La Repubblica, Il Sole 24 Ore, Corriere della Sera), the news rebounded on the social media thanks to the engagement of food and sustainability influencer that enabled 17.000 + social media interactions and a social presence of over 100,000 people.

Skretting

Skretting Italy started a stakeholder engagement initiative in 2017 intending to build community trust by leveraging sustainability. To combat fake news and increase the perceived importance of Italian aquaculture products, the initiative exposed the need to improve and organize the coordination efforts of the various stakeholders in the supply chain (feed manufacturers, fishermen, retailers, and distributors). Through a series of meetings held according to the methods of design thinking and active involvement, it emerged that farmers needed to coordinate their communication activities through the supply chain, to enhance the Italian aquaculture products. Therefore, the need to evolve the consumer's perception of the efforts of the whole supply chain in the continuous improvement of product quality, led, in early 2019, to the drafting of "Acqua in Bocca! Practical guide for not spreading... fish illusions." This communication kit, drafted in collaboration with customers, is designed to allow farmers to tell the value of the products they sell, dispelling fake news and misperceptions.

The handbook was developed in collaboration with stakeholders to bridge the information gap between industry and consumers, answering frequently asked questions about aquaculture in areas such as human nutrition, sustainability, and apiculture.

It was then posted on the website and made available to the public free of charge, to answer questions and provide solid, scientific information to anyone who needs it.

This project has four innovative elements, also considering that Skretting is a company with a B2B approach. These are:

The contact points: operating with a B2B approach, Skretting Italia entrusted customer relations to the relevant agents.

B2B2C approach: it was difficult for the company to involve customers in a path of common dialog, also because they often have different interests or are competitors. Skretting Italia thus decided to identify and exploit a need that was present in all customers and that the company was able to satisfy. It was therefore decided to adopt a B2B2C-type approach, for example, the creation of partnerships between companies and customers, to support and encourage sales to the final consumer and influence their market. In this sense, Skretting Italia's customers find in the common objective of influencing the perceptions of final consumers an opportunity to optimize profit margins.

Attention to the target market: "Acqua in Bocca" is a strategic operation that owes its success to the use of an idea tailored to the target market. This initiative overturned the classic approach to marketing strategies of multinationals, where corporate inputs are translated and adapted to specific local situations. In this case, indeed, the input came from Skretting Italia, becoming a point of reference for the other companies in the group.

Precompetitive level: Skretting Italia decided to collect all the information and difficulties related to aquaculture to start a precompetitive development process common to its customers. To this end, it has made the Acqua in Bocca communication kit publicly available on its website, so that it can be consulted by anyone who wants to transmit and obtain accurate and reliable information about aquaculture.

The case study of Skretting Italia shows how the active involvement of stakeholders (in this case the breeders) allows the identification of specific needs that led to the creation of a tool, such as the communication kit, from which the entire category benefited. Sustainability and its communication contribute to achieving a competitive advantage at sector level and not only for the single player. This advantage can be obtained by increasing the awareness of the final consumer on sustainability issues and how they affect the production chain of the specific sector, in this specific case the aquaculture sector. The impact that these activities have on the image of the sector and of individual companies may be reflected not only on the company's intangible values (e.g., in terms of brand image), but also on tangible factors, such as volumes and profit margins. In the case of Skretting Italia, the data provided by the company show that the strategy adopted is showing the first results for companies in the sector: 46% of the customers involved improved their opinion of Skretting, and 12% declared that they had obtained better commercial results thanks to "Acqua in Bocca."

The communication kit helped the customers of Skretting to frame their key communication messages around the Skretting products, providing information about the sustainable topic of their products as a collection of questions and answers.

The document was divided into five major topics:

Human nutrition.

Food safety.

Sustainability.

Animal wellness.

Innovation—toward 2030.

Coop

Coop has a long-standing commitment to environmental protection (Riboldazzi, 2016), which is reflected in specific measures on Coop products: from point-of-sale control to freight transportation by reducing polluting pollution, from waste reduction to engaging stakeholders in the supply chain. The system's set of principles includes environmental stewardship, which informs the whole supply chain's management requirements.

Furthermore, Coop is committed to raising awareness about the environmental quality of goods, packaging, production practices, and systems, providing customers with the knowledge and environmental education resources needed to better understand the effects of individual and public consumption decisions on the environment.

Coop began the study in 2020 intending to identify and analyze new developments in sustainability issues. It did so using an approach based on conceptual thinking and stakeholder participation. It was able to capture weak signals and identify new patterns using a combination of qualitative and quantitative approaches, such as in-depth interviews, online and offline studies, sensitive analysis, and design thinking workshops.

In this situation, stakeholder participation also helped the firm to recognize industry demands to develop a plan that would enable it to gain a competitive advantage. Coop's previous initiatives have all been linked to targeted and constructive engagement to promote the company's environmental contribution and reinforce its brand value.

The project led to the identification of five trends that reflect consumer attitudes toward sustainability practices. These results provide a starting point for management to implement a business engagement approach based on the trend to better capture the principles that motivate customers to engage in healthy consumption behavior.

The five trends were named as follows:

- Selfish autarchy.
- Tech enthusiasm.
- Pragmatic rationalism.
- Happy degrowth.
- Diversity and sharing.

To process of identifying the five trends is based on the collection and interpretation of weak signals. Indeed, weak signals consist of advanced and imprecise symptoms of impending future problems. Coop involved different stakeholders during the two main phases of the trend identification process: in the research phase with the engagement of the influencers, and the design thinking workshop with the researcher of the Department of Business Economics of University Roma Tre. The collection phase included over 700 signals intercepted, 30 keywords analyzed, 43 interviews, and 5 design thinking workshops. The results of the project were the identification of five major customer values that helped Coop to identify a communication strategy to cluster consumers in the retails.

Results

In the Mulino Bianco case, the engagement of suppliers in the definition of common sustainable goals leads to the creation of the "Carta del Mulino" agreement, resulting in a competitive advantage for the company and all the stakeholders involved throughout the supply chain.

In the Skretting Case, the company wanted to include different customers and given a large number of consumers, it was critical to develop a strategy that, thanks to stakeholder participation, could be seen as beneficial for all sectors to prevent rivalries and therefore obstruction.

In the last example, Coop enlisted the help of a university, as well as an influencer and an opinion maker, to collected the necessary data to develop an effective communication plan in the retails. The involvement of different stakeholders helped the company to clarify costumer's needs and values and to reshape the commercial and marketing strategy in line with the results that emerged from the project.

Discussion and conclusion

The three business cases presented can be brought together in a communication ecosystem framework that can be replicated by food companies to involve different stakeholders at different stages of their sustainable communication strategy.

Stakeholder engagement fosters a definition of sustainable marketing that goes beyond a commercial and product orientation to a broader societal view of its impact and is based on long-term relationships (Peattie & Belz, 2010; Gordon et al., 2011). The three dimensions of sustainable development: economic, social, and environmental, which represent the core of the most recent conceptualization of sustainable marketing (Peattie & Belz, 2010; Martin & Schouten, 2014), are all included when stakeholders are engaged in a communication project for sustainability.

The focal points of the ecosystem are the upstream and downstream partners of the company. The **Mulino Bianco** shows how to interact effectively with a provider and how, through a design thinking process, and the involvement of third-party stakeholders, it is possible to arrive at a common strategy to implement integrated sustainability through the upper part of the chain production. This strategy allowed **Mulino Bianco** to build shared communication messages with stakeholders, reinforcing its positioning toward other stakeholders and competitors.

With a similar approach, **Skretting** included the lower part of the supply chain by involving customers in the development, multiplying the contact points the company has with its corporate customers, and ensuring that customers could relate to the various actors in the production chain. With this strategy, **Skretting** was able to collect real-time market needs to promptly adapt communication strategies.

Finally, **Coop** created a communication ecosystem with the university and influencers to activate internal stakeholders on the topic of sustainability and to gain a strategic competitive advantage over competitors who were less focused on setting a corporate vision toward sustainability, increasing internal awareness on the issues and penetrating sustainability issues within the company DNA.

References

Aakhus, M., & Bzdak, M. (2015). Stakeholder engagement as communication design practice. *Journal of Public Affairs*, 15(2), 188–200. Available from https://doi.org/10.1002/pa.1569.

Aiking, H., & de Boer, J. (2004). Food sustainability: Diverging interpretations. *British Food Journal*, 106(5), 359–365. Available from https://doi.org/10.1108/00070700410531589.

Baldassarre, F., & Campo, R. (2016). Sustainability as a marketing tool: To be or to appear to be? *Business Horizons*, 59(4), 421–429. Available from https://doi.org/10.1016/j.bushor.2016.03.005.

Bateh, J., Heaton, C., Arbogast, G. W., & Broadbent, A. (2014). Defining sustainability in the business setting. *Journal of Sustainability Management (JSM)*, 1(1), 1–4. Available from https://doi.org/10.19030/jsm.v1i1.8386.

Baxter, P., & Jack, S. (2010). *Qualitative case study methodology: Study design and implementation for novice researchers.* Available from https://doi.org/10.46743/2160-3715/2008.1573.

Burchell, J., & Cook, J. (2006). Confronting the "corporate citizen": Shaping the discourse of corporate social responsibility. *International Journal of Sociology and Social Policy*, 26(3–4), 121–137. Available from https://doi.org/10.1108/01443330610657188.

Busco, C., Granà, F., & Izzo, M. F. (2018). Sustainable development goals and integrated reporting. *Sustainable development goals and integrated reporting* (pp. 1–125). Taylor and Francis. Available from https://doi.org/10.4324/9780429027314.

Byerlee, D., & Fanzo, J. (2019). The SDG of zero hunger 75 years on: Turning full circle on agriculture and nutrition. *Global Food Security*, 21, 52–59. Available from https://doi.org/10.1016/j.gfs.2019.06.002.

Chakrabarti, R., Henneberg, S. C., & Ivens, B. S. (2020). Open sustainability: Conceptualization and considerations. *Industrial Marketing Management*, 89, 528–534. Available from https://doi.org/10.1016/j.indmarman.2020.04.024.

Cooper, D.R., & Schindler, P.S. (2005). Marketing research.

Creswell, J. W. (2009). *Research design: Qualitative, quantitative, and mixed methods approaches.* Sage Publications Inc.

Desmond, J., & Crane, A. (2004). Morality Ad the consequences of marketing action. *Journal of Business Research*, 57(11).

Dipace, A., Mirone, L., Mora, G., & Pratesi, C. A. (2019). *Case Study: Mulino Bianco per un agricoltura più sostenibile.*

Doelle, M., & Sinclair, A. J. (2006). Time for a new approach to public participation in EA: Promoting cooperation and consensus for sustainability. *Environmental Impact Assessment Review*, 26(2), 185–205. Available from https://doi.org/10.1016/j.eiar.2005.07.013.

Ebneyamini, S., & Sadeghi Moghadam, M. R. (2018). Toward developing a framework for conducting case study research. *International Journal of Qualitative Methods*, 17(1). Available from https://doi.org/10.1177/1609406918817954.

Eisenhardt, Kathleen M. (1989). Building theories from case study research. *Academy of Management Review*, 14(4), 532–550. Available from https://doi.org/10.5465/amr.1989.4308385.

Eisenhardt, K. M., & Graebner, M. E. (2007). Theory building from cases: Opportunities and challenges. *Academy of Management Journal*, 50(1), 25–32. Available from https://doi.org/10.5465/AMJ.2007.24160888.

Falkenberg, J., & Brunsæl, P. (2011). Corporate social responsibility: A strategic advantage or a strategic necessity? *Journal of Business Ethics*, 99(1), 9–16. Available from https://doi.org/10.1007/s10551-011-1161-x.

Fanzo, J. (2019). Healthy and sustainable diets and food systems: The key to achieving sustainable development goal 2? *Food Ethics*, 4(2), 159–174. Available from https://doi.org/10.1007/s41055-019-00052-6.

Ferraro, F., Etzion, D., & Gehman, J. (2015). Tackling grand challenges pragmatically: Robust action revisited. *Organization Studies*, 36(3), 363–390. Available from https://doi.org/10.1177/0170840614563742.

Fisk, G. (2001). Reflections of George fisk: Honorary Chair of the 2001 Macromarketing Conference. *Journal of Macromarketing*, 21(2), 121–122. Available from https://doi.org/10.1177/0276146701212002.

Fombrun, C. J. (2005). Building corporate reputation through CSR initiatives: Evolving standards. *Corporate Reputation Review*, 8, 7–11.

Frankel, Bolumole, Yemisi, ham, Eltantawy, Paulraj, & Gundlach, G. (2008). The domain and scope of SCM's foundation disciplines—Insights and issues to advance research. *Journal of Business Logistics*, 29. Available from https://doi.org/10.1002/j.2158-1592.2008.tb00066.x.

Freibauer, A., Mathijs, E., Brunori, G., Damianova, Z., Faroult, E., Gomis, J. G., O'brien, L., & Treyer, S. (2011). Sustainable food consumption and production in a resource-constrained world. *EuroChoices*, 10(2), 38–43. Available from https://doi.org/10.1111/j.1746-692X.2011.00201.x.

Garnett, T. (2014). Three perspectives on sustainable food security: Efficiency, demand restraint, food system transformation. What role for life cycle assessment? *Journal of Cleaner Production*, 73, 10–18. Available from https://doi.org/10.1016/j.jclepro.2013.07.045.

Garnett, T. (2011). Where are the best opportunities for reducing greenhouse gas emissions in the food system (including the food chain)? *Food Policy*, 36(1), S23–S32. Available from https://doi.org/10.1016/j.foodpol.2010.10.010.

Gerstlberger, W., Præst Knudsen, M., & Stampe, I. (2014). Sustainable development strategies for product innovation and energy efficiency. *Business Strategy and the Environment*, 23(2), 131–144. Available from https://doi.org/10.1002/bse.1777.

Gladek, E., Fraser., Roemers, G., Munoz, O. S., Hirsch, P., & Kennedy, E. (2016). *The global food system: An analysis*. Metabolic.

Godfray, H. C. J., Crute, I. R., Haddad, L., Muir, J. F., Nisbett, N., Lawrence, D., Pretty, J., Robinson, S., Toulmin, C., & Whiteley, R. (2010). The future of the global food system. *Philosophical Transactions of the Royal Society B: Biological Sciences*, 365(1554), 2769–2777. Available from https://doi.org/10.1098/rstb.2010.0180.

Gold, S., Hahn, R., & Seuring, S. (2013). Sustainable supply chain management in 'Base of the Pyramid' food projects-A path to triple bottom line approaches for multinationals? *International Business Review*, 22(5), 784–799. Available from https://doi.org/10.1016/j.ibusrev.2012.12.006.

Gordon, R., Carrigan, M., & Hastings, G. (2011). A framework for sustainable marketing. *Marketing Theory*, 11(2), 143–163. Available from https://doi.org/10.1177/1470593111403218.

Gupta, J., Pouw, N. R. M., & Ros-Tonen, M. A. F. (2015). Towards an elaborated theory of inclusive development. *European Journal of Development Research*, 27(4), 541–559. Available from https://doi.org/10.1057/ejdr.2015.30.

Headey, D., Heidkamp, R., Osendarp, S., Ruel, M., Scott, N., Black, R., Shekar, M., Bouis, H., Flory, A., Haddad, L., & Walker, N. (2020). Impacts of COVID-19 on childhood malnutrition and nutrition-related mortality. *The Lancet*, 396(10250), 519–521. Available from https://doi.org/10.1016/S0140-6736(20)31647-0.

Hillenbrand, C., & Money, K. (2007). Corporate responsibility and corporate reputation: Two separate concepts or two sides of the same coin? *Corporate Reputation Review*, 10(4), 261–277. Available from https://doi.org/10.1057/palgrave.crr.1550057.

Hinrichs, C. C. (2014). Transitions to sustainability: A change in thinking about food systems change? *Agriculture and Human Values*, 31(1), 143–155. Available from https://doi.org/10.1007/s10460-014-9479-5.

HLPE. (2017). Nutrition and food systems. In: A report by the high level panel of experts on Food Security and Nutrition of the Committee on World Food Security.

Hofstede, G. (2011). Dimensionalizing cultures: The Hofstede model in context. *Online Readings in Psychology and Culture, 2*(1). Available from https://doi.org/10.9707/2307-0919.1014.

Hörisch, J., Freeman, R. E., & Schaltegger, S. (2014). Applying stakeholder theory in sustainability management: Links, similarities, dissimilarities, and a conceptual framework. *Organization and Environment, 27*(4), 328–346. Available from https://doi.org/10.1177/1086026614535786.

IFAD, UNICEF, WFP, & WHO. (2019). *The state of food security and nutrition in the world 2019: Safeguarding against economic slowdowns and downturns.*

Ihrig, M., & MacMillan, I. C. (2017). How to get ecosystem buy-in. Harvard Business Review, 2017(MarchApril). https://hbr.org/download/S001000056303/R1702G-PDF-ENG/R1702G-PDF-ENG

IPES-Food. (2015). The New Science of Sustainable Food Systems: Overcoming Barriers to Food Systems Reform. In: *International Panel of Experts on Sustainable Food Systems* (IPES-Food).

IPCC. (2019). Climate Change and Land: An IPCC special report on climate change, desertification, land degradation, sustainable land management, food security, and greenhouse gas fluxes in terrestrial ecosystems. IPCC.

Izzo, M. F., Ciaburri, M., & Tiscini, R. (2020). The challenge of sustainable development goal reporting: The First evidence from italian listed companies. *Sustainability, 12*(8), 3494. Available from https://doi.org/10.3390/su12083494.

Jensen, J. C., & Berg, N. (2012). Determinants of traditional sustainability reporting versus integrated reporting. An institutionalist approach. *Business Strategy and the Environment, 21*(5), 299–316. Available from https://doi.org/10.1002/bse.740.

Jones, P., Clarke-Hill, C., Comfort, D., & Hillier, D. (2008). Marketing and sustainability. *Marketing Intelligence & Planning, 26*(2), 123–130. Available from https://doi.org/10.1108/02634500810860584.

Kamasastry. (2020). Marketing Myopia—A literature review based contemporary perspective. *Journal of Marketing Vistas, 10*(2), 44–59.

Kemper, J. A., & Ballantine, P. W. (2019). What do we mean by sustainability marketing? *Journal of Marketing Management, 35*(3–4), 277–309. Available from https://doi.org/10.1080/0267257X.2019.1573845.

Kloppenburg, J., Lezberg, S., De Master, K., Stevenson, G. W., & Hendrickson, J. (2000). Tasting food, tasting sustainability: Defining the attributes of an alternative food system with competent, ordinary people. *Human Organization, 59*(2), 177–186. Available from https://doi.org/10.17730/humo.59.2.8681677127123543.

Kumar, V., Rahman, Z., Kazmi, A. A., & Goyal, P. (2012). Evolution of sustainability as marketing strategy: Beginning of new era. *Procedia - Social and Behavioral Sciences, 37*, 482–489. Available from https://doi.org/10.1016/j.sbspro.2012.03.313.

Lachman, D. A. (2013). A survey and review of approaches to study transitions. *Energy Policy, 58*, 269–276. Available from https://doi.org/10.1016/j.enpol.2013.03.013.

Levitt, T. (1960). Marketing Myopia. *Harvard Business Review, 38*(4), 27–47.

Lim, W. M. (2016). A blueprint for sustainability marketing: Defining its conceptual boundaries for progress. *Marketing Theory, 16*(2), 232–249. Available from https://doi.org/10.1177/1470593115609796.

Martin, D. M., & Schouten, J. W. (2014). The answer is sustainable marketing, when the question is: What can we do? *Recherche et Applications En Marketing (English Edition), 29*(3), 107–109. Available from https://doi.org/10.1177/2051570714540497.

Mena, J. A., Hult, G. T. M., Ferrell, O. C., & Zhang, Y. (2019). Competing assessments of market-driven, sustainability-centered, and stakeholder-focused approaches to the customer-brand relationships and performance. *Journal of Business Research, 95*, 531–543. Available from https://doi.org/10.1016/j.jbusres.2018.07.038.

Miles, M. P., Munilla, L. S., & Darroch, J. (2006). The role of strategic conversations with stakeholders in the formation of corporate social responsibility strategy. *Journal of Business Ethics, 69*(2), 195–205. Available from https://doi.org/10.1007/s10551-006-9085-6.

Mitchell, R. W., Wooliscroft, B., & Higham, J. (2010). Sustainable market orientation: A new approach to managing marketing strategy. *Journal of Macromarketing, 30*(2), 160–170. Available from https://doi.org/10.1177/0276146710361928.

Montiel, I., & Delgado-Ceballos, J. (2014). Defining and measuring corporate sustainability. *Organization & Environment, 27*(2), 113–139. Available from https://doi.org/10.1177/1086026614526413.

Sustainable Development Commission (2009). Food security and sustainability: The perfect fit.

Newig, J., Schulz, D., Fischer, D., Hetze, K., Laws, N., Lüdecke, G., & Rieckmann, M. (2013). Communication regarding sustainability: Conceptual perspectives and exploration of societal subsystems. *Sustainability, 5*(7), 2976−2990. Available from https://doi.org/10.3390/su5072976.

Noland, J., & Phillips, R. (2010). Stakeholder engagement, discourse ethics and strategic management. *International Journal of Management Reviews, 12*(1), 39−49. Available from https://doi.org/10.1111/j.1468-2370.2009.00279.x.

Parmar, B. L., Freeman, R. E., Harrison, J. S., Wicks, A. C., Purnell, L., & de Colle, S. (2010). Stakeholder theory: The state of the art. *Academy of Management Annals, 4*(1), 403−445. Available from https://doi.org/10.1080/19416520.2010.495581.

Patrizia, G., & Carlotta. (2011). The role of the sustainability report in capitalistic firm. *Annals of Faculty of Economics, 1*(2), 243−250.

Patton, E., & Appelbaum, S. H. (2003). The case for case studies in management research. *Management Research News, 26*(5), 60−71. Available from https://doi.org/10.1108/01409170310783484.

Peattie, K., & Peattie, S. (2009). Social marketing: A pathway to consumption reduction? *Journal of Business Research, 62*(2), 260−268. Available from https://doi.org/10.1016/j.jbusres.2008.01.033.

Peattie, Ken, & Belz, F.-M. (2010). Sustainability marketing—An innovative conception of marketing. *Marketing Review St. Gallen, 27*(5), 8−15. Available from https://doi.org/10.1007/s11621-010-0085-7.

Pereira Heath, M. T., & Chatzidakis, A. (2012). Blame it on marketing": Consumers' views on unsustainable consumption. *International Journal of Consumer Studies, 36*(6), 656−667. Available from https://doi.org/10.1111/j.1470-6431.2011.01043.x.

Perrini, F. (2006). SMEs and CSR theory: Evidence and implications from an Italian perspective. *Journal of Business Ethics, 67*(3), 305−316. Available from https://doi.org/10.1007/s10551-006-9186-2.

Pfau, M., Haigh, M. M., Sims, J., & Wigley, S. (2008). The influence of corporate social responsibility campaigns on public opinion. *Corporate Reputation Review, 11*(2), 145−154. Available from https://doi.org/10.1057/crr.2008.14.

Pradhan, P., Costa, L., Rybski, D., Lucht, W., & Kropp, J. P. (2017). A systematic study of Sustainable Development Goal (SDG) interactions. *Earth's Future, 5*(11), 1169−1179. Available from https://doi.org/10.1002/2017EF000632.

Pucci, T., Casprini, E., Galati, A., & Zanni, L. (2020). The virtuous cycle of stakeholder engagement in developing a sustainability culture: Salcheto winery. *Journal of Business Research, 119*, 364−376. Available from https://doi.org/10.1016/j.jbusres.2018.11.009.

Redante, R. C., de Medeiros, J. F., Vidor, G., Cruz, C. M. L., & Ribeiro, J. L. D. (2019). Creative approaches and green product development: Using design thinking to promote stakeholders' engagement. *Sustainable Production and Consumption, 19*, 247−256. Available from https://doi.org/10.1016/j.spc.2019.04.006.

Riboldazzi, S. (2016). Corporate governance and sustainability in Italian large-scale retail companies. *European Scientific Journal, ESJ, 12*(16), 1. Available from https://doi.org/10.19044/esj.2016.v12n16p1.

Rossi, A. (2017). Beyond food provisioning: The transformative potential of grassroots innovation around food. *Agriculture, 7*(1), 6. Available from https://doi.org/10.3390/agriculture7010006.

Murray-Webster, R., & Simon, P. (2006). Making sense of stakeholder mapping. *PM World today, 8*(11), 1-5.

Sandberg, M., & Polsa, P. (2015). Efficiency or sufficiency? The (re)construction of discourses about sustainable consumption in marketing research. In: *Proceedings of the 40th annual macromarketing conference*. The Macromarketing Society, Inc.

Schmidheiny, S., & Zorraquin, F. J. (1998). *Financing change: The financial community, eco-efficiency, and sustainable development*. MIT Press.

Seabright, M. A., & Kurke, L. B. (1997). Organizational ontology and the moral status of the corporation. *Business Ethics Quarterly, 7*(4), 91−108. Available from https://doi.org/10.5840/10.2307/3857210.

Searchinger, T., Hanson, C., Ranganathan, J., Lipinski, B., Waite, R., Winterbottom, R., ... Ari, T. B. (2014). Creating a sustainable food future. A menu of solutions to sustainably feed more than 9 billion people by 2050. World resources report 2013-14: interim findings *(Doctoral dissertation)*. World Resources Institute (WRI); World Bank Groupe-Banque Mondiale; United Nations Environment Programme (UNEP); United Nations Development Programme (UNDP); Centre de Coopération Internationale en Recherche Agronomique pour le Développement (CIRAD); Institut National de la Recherche Agronomique (INRA).

Sheth, J. N., Sethia, N. K., & Srinivas, S. (2011). Mindful consumption: A customer-centric approach to sustainability. *Journal of the Academy of Marketing Science, 39*(1), 21−39. Available from https://doi.org/10.1007/s11747-010-0216-3.

Soares, A. M., Farhangmehr, M., & Shoham, A. (2007). Hofstede's dimensions of culture in international marketing studies. *Journal of Business Research*, *60*(3), 277–284. Available from https://doi.org/10.1016/j.jbusres.2006.10.018.

Souri, M. E., Sajjadian, F., Sheikh, R., & Sana, S. S. (2018). Grey SERVQUAL method to measure consumers' attitudes towards green products - A case study of Iranian consumers of LED bulbs. *Journal of Cleaner Production*, *177*, 187–196. Available from https://doi.org/10.1016/j.jclepro.2017.12.105.

Spitzeck, H., & Hansen, E. G. (2010). Stakeholder governance: How stakeholders influence corporate decision making. *Corporate Governance*, *10*(4), 378–391. Available from https://doi.org/10.1108/14720701011069623.

Vafaei, S. A., Azmoon, I., & Fekete-Farkas, M. (2019). The impact of perceived sustainable marketing policies on green customer satisfaction. *Polish Journal of Management Studies*, *19*(1), 475–491. Available from https://doi.org/10.17512/pjms.2019.19.1.36.

van Dam, Y. K., & Apeldoorn, P. A. C. (1996). Sustainable marketing. *Journal of Macromarketing*, *16*(2), 45–56. Available from https://doi.org/10.1177/027614679601600204.

Vogel, D. J. (2005). Is there a market for virtue? The business case for corporate social responsibility. *California Management Review*, *47*(4), 3–45. Available from https://doi.org/10.1177/000812560504700401.

WWW-UK. (2013). A 2020 vision for the global food system.

Willett, W., Rockström, J., Loken, B., Springmann, M., Lang, T., Vermeulen, S., ... Murray, C. J. (2019). Food in the Anthropocene: the EAT–Lancet Commission on healthy diets from sustainable food systems. *The lancet*, *393*(10170), 447–492.

Yin, R. K. (2013). *Case study research: Design and methods*. SAGE Publications.

Index

Note: Page numbers followed by "*f*" and "*t*" refer to figures and tables, respectively.

A

ABC World News Report, 109
Abstract conceptualization, 200
Access, 85, 85*t*, 97
Access to organic food, 85, 85*t*
"Acqua in Bocca", 245
Action Hub, 171–172
Action Pan, 172
Active experimentation, 200
AdHealth, 40
Advertisement framing of health, 143–144
AFNs. *See* Alternative food networks
Agency, 78–79, 97–98
    Eurobarometer, 87, 88*t*
Agri-business CSR initiatives, 108
Agricultural research, 44
Agriculture, 68
Agri-food, 103, 221
    businesses, 17
    contributions, 13
    corporate social responsibility, 106–108
    COVID-19 on, 125–127
    industry, 19, 28–29, 44
    organizations, soft regulation of, 108
    policymaking, 44
AGROVOC, 17–18
AI adoption, 19
Alternative food networks (AFNs), 66–67, 128–133
    media coverage of, 132
    viability of, 131
Alternative meat, 114–117
Alt-meat, 115
Animal agriculture, 103
Animal-based food production practices, 112
Animal-based food systems, 101
Animal food products, 135
    widespread consumption of, 145
Animal products
    alternatives, 113–114
    communicating about
        alternative meat, 114–117
        animal product alternatives, 113–114
        circular agriculture, 119–121

        dairy and plant mylks, 117–118
        farm animal welfare, 112–113
        food waste reduction, 118–119
        problem with, 108–112
Animal source food (ASF), 112, 114–115
Antimicrobial resistance, 101
Application Programming Interface (API), 58
AquaBounty, 116–117
AR-based wearable food-monitoring system, 17
Artificial intelligence, 137
Aspiration, Boot Camps, 228–229
Asset Vulnerability Framework (AVF), 198–199
Attending events, 172
Attitude–behavior gap, 36, 97
Augmented Reality (AR) technology, 16
Automated food recognition methods, 17
Availability, 79–84, 80*t*, 81*t*, 82*t*, 83*t*, 84*t*, 96
Avian influenza, 137

B

"Bad news", 140
Barilla Foundation, 196, 241*t*, 242*t*
B2B2C approach, 245
Belief systems, 221
Benefit message appeals, 119
Benzopyrene, 140
Beverage brands, 34
"Big data" analysis, 142
Big Food
    brands, 33–34
    companies, 115
    corporations, 37
    marketing ubiquity, 33–34
Biochar, 102
Biological products, 62–63
Biomedical vocabularies, 18
Biosafety legislation, 142–143
Biosafety regulations, 144
Biotech foods, 141–142
Biotechnology, 141–142
Bioterrorism, 137
Bitcoin Mining Council, 68–69
Blended learning, 67

Body mass index (BMI), 61
Bogs, 105
Boot Camps, 228–231
    action, 229
    aspiration, 228–229
    inspiration, 228
Bottled water, 183
Bovine Spongiform Encephalopathy (BSE), 109
Branding, food, 33–37
Brand reputation, 21
British GM foods, 143
Bubbles, social media, 111
Buddhist Compassion Relief Foundation in Singapore, 114
Buddy system, 61
"Business as usual" approach, 125

C
Canadian GM foods, 143
Canadian Yellowknife Farmers Market, 127
CAP, 87
    awareness of, 88t
    benefits, 89t
    budget, 93t
    contribution, 92t
    objectives, 90t
    performance, 91t
Carbon capture technologies, 104
Carbon dioxide equivalence, 13
Carbon footprint, 4–5
    information, 21
Carbon mitigation, 104
Caring, 7–8
"Carta del Mulino", 243, 247
Case selection, 240–241
CEASE framework, 201–202
Celebrate variety, 182
Cellular agriculture, 115
    far-reaching effects of, 116
Cellular food agriculture, 115
Censorship, 71
Central Oregon Locavore, 66–67
Challenge-based learning, 232
Champion plant-based food, 181
Charity-driven strategy, 239
Cheese, 182
Chef engagement, 171–172
Chefs' Manifesto, 163, 169
    building network and guiding principles, 170–171
    chef engagement, 171–172
    chefs bridging gap, 167–168
    chefs, spreading message, 172–175
    food messaging

Internet, 164–165
    short history of, 165
    global food crisis, 168–169
    new platform, impact of, 166
    scientists or credible health practitioners, 166–167
Chef Manifesto Action Plan, 5–6, 163
Chefs, 165, 168
    bridging, 167–168
    generating pathways for, 171–172
    spreading message, 172–175
Chemophobia, 98
Chinese government agencies, 135
Chrome browser extension, 40
Chronic risk communications, 136
Chronic Wasting Disease (CWD), 110
Circular agriculture, 119–121
    strategies, 119
Circular food systems, 120
Citizen-driven food waste recovery initiatives, 120
Citizen-driven initiatives, 120
Civic journalism, 130
Clean eating, 31–32
Climate adaptation strategies, 104
Climate change impacts, media framing of, 13, 101–103, 221, 224
    agricultural impacts of, 104–105
    agri-food corporate social responsibility, 106–108
    animal products, communicating about, 108–121
        alternative meat, 114–117
        animal product alternatives, 113–114
        circular agriculture, 119–121
        dairy and plant mylks, 117–118
        farm animal welfare, 112–113
        food waste reduction, 118–119
        problem with, 108–112
    communicating food risks, 133–144
        food risk perceptions, 137–141
        food safety framing, 133–137
        genetically modified foods, 142–144
        nanofoods and other biotech foods, 141–142
    food security, 121–133
        COVID-19 impacts on agri-food, 125–127
        food insecurity framing, 121–125
        food riots, 127–128
        local food and alternative food networks, 128–133
    food–water–energy nexus, 105–106
    media framing of, 102
Climate crisis, 225
Climate League, Food for, 223–228
"Climate Menus", 228
Climate mitigation strategies, 11
Climate-related food insecurity, 127

Climate Shapers, 229, 231
Climate-smart agricultural methods, 104
Climate-smart agriculture news coverage, 105
Climate-Smart Cities, 230–231
Climate-smart eating, 225
Climate-Smart Farms, 230–231
Climate-Smart Kitchens, 230–231
Climate-Smart Oceans, 230–231
Climate suppers, 228–229
CO_2 emissions, 178–179
Cognitive dissonance, 138
Cognitive risk perception, 138
Cognitive structures, 203–204
Collective action frames, 124
Commercialization, 12–13
Communication, 221
    ecosystem, 248
    food insecurity issues, 102
    process, 239–240
    strategy, 6
Communicative ecology, 59
Community, 7–8, 222
Community-based organizations, 130
Company sustainable goals, 242
Comprehensibility, 185
Comprehensive and Progressive Agreement for Trans-
    Pacific Partnership (CPTPP), 68–69
Concrete experience, 200
Confirmatory bias, 138
Conscious capitalism, 106–107
Consumer apps, 43
Consumer attitudes, 116
Consumer behavior, 35, 236
Consumer decisions, 107
Consumer risk perceptions, 18
Consumers, 111
Consumption of local food, 132–133
Controversial food, normalizing, 113
"Cookery the Australian Way", 164
Coop, 240, 241t, 242t, 246–248
Coopetition, 233
Corporate communication, 64–66
Corporate social responsibility (CSR), 66, 107
    communication, 106–107
    obligations, 107
Coupling variety, 182
Covid pandemic, 68
COVID-19 pandemic, 98, 102, 146, 168–169, 227–228
    on agri-food, 125–127
COVID washing, 41
Craft broad-reaching, 135
Credibility, 166
Credible expert organizations, 144

Croatia's first organic food store bio&bio framed
    environmental claims, 15
Cross-disciplinary teams, 201
CSR. See Corporate social responsibility
Cultural entrepreneurship, 132
Cultural factors, 109
Culture, 133
Cultured meat, 114–115
    consumer acceptance of, 115–116
    marketing, 116
Current neural networks, 16–17
Customer level
    engagement activities, 184–185, 185f
    intermediate evaluation stage I, 186–188, 189t

D
Daily Telegraph, 138–139
Dairy farm practices, public perceptions of, 117
Dairy farm systems, 117
Dairy mylks, 117–118
Dairy products, 182
Dairy-related food safety, 135
Dairy system efficiency, 117
Data analysis
    PUSH-UFWH design thinking workshop, 209, 210t,
        211t, 212t
    sustainable corporate communication, 242–243
        company sustainable goals, 242
        map stakeholders' tensions, 242
        stakeholder mapping, 242
        stakeholders' involvement, 243
Decentralization, 170, 175
Deep learning technologies, 17
Deliberative democratic processes, 124
Delicacy campaign, 63–64
De-meatification mechanism, 113
Descriptive message appeals, 119
Design thinking, 197–203
    design thinking, 215–216
    empathy, 217
    experience of receiving nutritionally unbalanced
        meal, 214
    as experiential learning activity, 201–202
    "learning how to think," process of, 216
    participants, 214
    PUSH and UFWH, 197–203
        experiential learning, 200–201
        experiential learning activity, design thinking as,
            201–202
        motivation and perceived ability, experiential
            learning on, 202–203
        theoretical background and definition of key
            concepts, 198–200

Design thinking (*Continued*)
  PUSH-UFWH design thinking workshop, 203–213
    data analysis, 209, 210*t*, 211*t*, 212*t*
    data collection, 207–209
    meal distribution, 209–210
    methodology, 204–209, 205*f*, 206*f*, 207*f*
    participants, 206, 208*t*
    survey, 213
    theoretical framework, 203–204
  theoretical background and definition of key
      concepts, 198–200
  transformational learning, 215
  workshop leaders, 215
Design thinking workshop, 198
Diet-related health problems, 43
Diets, 169, 178
  influencers, 20
Digital Boot Camps, 230
Digital connectivity, 66
Digital marketing platforms, 40
Digital photography, 14
Direct collaborations, 172
Direct participation, 241
Disability-Adjusted Life-Years (DALYs), 3
Disbudding, 117
Disproportionate junk food advertising, 40
Disruptive food technologies, 12
Diversification strategies, 126–127
Domestic biotechnology, 144
Domestic production, 79–82
Dominant storytelling strategies, 225
Doomsday predictions, 122
Dorst's claim, 201
Drastic action, 174
Dread risks, 140
Drink recognition methods, 17
Drones, 125
Drought, 104
Dutch livestock sector, 29

E

East European food systems, 124
Eating seasonally, 83
E-cigarettes, 36–37
Edible food, 145
Educated millennials, 144
Education, 221–222
Emblematic Community of the Mediterranean Diet,
      232–233
Emergency, 203
Emotions, 214
Empathy, 202, 214
Empower individuals, 230

Empower organizations, 231
Endorsement, 68–69
Endorsement network, 68–69
Energy intake, 182
Engagement activities, 183–185, 184*f*
  customer level, 184–185, 185*f*
  food service level, 183–184
Environmental sustainability, 37
EOE model, 217
Equitable Food Initiative (EFI), 108
Erosion of organic food price premiums, 129
Ethical outrage, 168–169
Eurobarometer surveys, 77
  access, 85, 85*t*
  agency, 87, 88*t*
  availability, 79–84, 80*t*, 81*t*, 82*t*, 83*t*, 84*t*
  food safety, 89–96
  getting data on food security using, 78–79
  stability, 86, 86*t*, 87*t*
  sustainability, 87–89, 88*t*, 89*t*, 90*t*, 91*t*, 92*t*, 93*t*, 94*t*
  utilization, 85–86, 86*t*
European Commission, 57, 77, 177–178
European diets, 177–178
European food products, 94*t*
European Food Safety Authority (EFSA), 89
European Green Deal, 177–178
European Member States, 88*t*
European Union (EU), 4, 77, 140
  regulations, 141
eWOM, 66
Excess energy intake, 182
Expectation assimilation, 61
Experiential learning, 200–201
  design thinking as, 201–202
  motivation and perceived ability, 202–203
Experiential learning theory, 203
Exploitative overpurchasing, 125
Extended Theory of Planned Behavior, 118

F

Facebook, 33–34, 40, 64–65
Facebook Advertising platform, 60–61
Fair trade, 129
Fake news, 19, 21–22
FallingFruit.org, 67
False science news, 20
FAO. *See* Food and Agriculture Organization
FAO E-learning Academy, 228
Farm animal welfare, 112–113
Farm food producer organizations, 105
Farming, 68, 103
Farm to Fork, 4, 57
Favorability, 142

FCL. *See* Food for Climate League
FDA Egg Rule, 136
Federal Trade Commission (FTC), 66
FFI. *See* Future Food Institute
Field crops, 104
Finnish customers, 133
"First-foods systems" campaign, 34
Fitbit wearable device, 61
Flash Eurobarometer, 77
Flooding, 104
Florida's 2018 Red Tide, 133–134
Food
  characteristics, 79, 80*t*, 97
  for climate league, 223–228
  on social media, 29–42
    food branding, 33–37
    junk food marketing, 37–42
    uses and gratifications, 29–33
Food-101, 59–60
Food access, 78–79
Food ads, 35
Food allergen, 138–139
Food and Agriculture Organization (FAO), 77,
    224–226, 228–229
Food and Climate Shaper Boot Camp, 231
Food and Climate Shaper Digital Boot Camp, 230
Food and drink brands, 42–43
Food availability, 78–79, 97
  impact of humans on, 79, 80*t*
Food Bank, 96–97
Food bias, 22
Food biotechnology, 141, 145
Food branding, 33–37
Food companies, 166
  and consumers, 103
Food consumption
  on social media and health, 59–61
  uncontrolled or unmonitored, 18–19
Food contamination, 136
Food converges, 221
Food culture, social media, 32
  analysis methods, 58–59
  case study, 68–70
  and individual, 59–64
    food consumption on, 59–61
    food waste, 63–64
    sustainable consumer choice, 62–63
  and organizations, 64–68
    corporate communication, 64–66
    double-edged sword of, 66
    grassroots organization, 66–68
Food deserts, 60
Food discussions, 12

Food ecologies and economies, 11
Food experience, 14
Food for Climate League (FCL), 224–225
Food governance, 124
Food hygiene, 94–96
Food identity, 223
Food imagery, 14–16
Food imports, 82
  perception on, 81*t*
Food industry, 103
  sustainable development goals in, 237–238
Food information, 14
  technologies, 16–19
Food innovation, 138–139
Food insecurity, 198–200, 202
  framing, 121–125
  issues, communications, 102
  media coverage of, 130
  newspaper coverage of, 122
Food journalism, role of, 12–14
Food justice, 131–132
  research, 129
Food labeling, 36
Food messaging
  Internet, 164–165
  short history of, 165
Food misinformation, 19–22
Food neophobia, 115–116
Food object recognition systems, 17–18
Food packaging, 183
Food photo posting, 16
Food policy framing, 107–108
Food politics, 131–132
Food porn, 14
Food processing, 182–183
Food producers, 39
Food production, 80*t*
Food products, diversification of, 126–127
Food purchasing, 136
Food recalls, 136
Food-related communication, 62
Food-related CSR, 107
Food-related GHG emissions, 4
Food-related health hazards, 137
Food-related keywords, 126
Food-related stories, 23
Food-related tweets, 30–31
Food retailers, 110
Food riots, 127–128
  media accounts of, 128
Food risk, 136
  climate change, 133–144
    food risk perceptions, 137–141

Food risk (*Continued*)
    food safety framing, 133–137
    genetically modified foods, 142–144
    nanofoods and other biotech foods, 141–142
    perceptions, 137–141
Food rumors, on social media, 21
Food safety, 6–7, 89–96, 98–99
    concern about, 94t, 95t
    EU level, 94t
    framing, 133–137
    labels, 36
    organizations, 135
    risk, 136
Food security, 6–7, 77, 98–99, 221
    climate change, 121–133
        COVID-19 impacts on agri-food, 125–127
        food insecurity framing, 121–125
        food riots, 127–128
        local food and alternative food networks, 128–133
    Eurobarometer surveys, 77
        access, 85, 85t
        agency, 87, 88t
        availability, 79–84, 80t, 81t, 82t, 83t, 84t
        food safety, 89–96
        getting data on food security using, 78–79
        stability, 86, 86t, 87t
        sustainability, 87–89, 88t, 89t, 90t, 91t, 92t, 93t, 94t
        utilization, 85–86, 86t
    farmers' responsibility, 88t
    as international threat, 80t
    perceived features, 87t
    perception of, 86, 97–98
    progress in, 123
Food selfies, 15
Food self-provisioning (FSP) practices, 128
Food service level
    engagement activities, 183–184
    intermediate evaluation stage I, 186, 187t
Food sovereignty, 129
Foodstagramming, 15
Food supply chains, 12
    corporatization and globalization of, 110–111
Food system, 3–4, 97–98
    European Union, 4
    stakeholders, 110–111
"Food tech justice" approach, 115
Food traceability systems, 18
Food waste, 63–64
    reduction, 118–119
Foodwatch, 130
Food–water–energy (FWE) nexus, 105–106
Four-item green equity measure, 133
Fourth Industrial Revolution digitization, 19

Framed mindful food consumption, 114
Framing, media, of food sustainability, 22–26
From Waste to Delicacy campaign, 63–64
Future Food, 222, 228
Future Food Ecosystem, 231
Future Food Institute (FFI), 221–222, 228–229

G
Gain-framed messages, 104
Gamified approach, 180
Gardening messages, 128
Gathering chefs, 171
Genetically modified (GM) foods, 20, 142–144
    asynchronous approvals of, 144
    British and Canadian GM foods, 143
    domestic biotechnology and biosafety regulations, 144
    interview sources influence, 143
    longitudinal study of, 142
    media attention shapes, 142
    science, 143
    social capital, 144
German food retail and industry sector, 112
German nonprofit organization, 130
Gesis database, 79
GHG emissions. *See* Greenhouse gas (GHG) emissions
GHG-intensive foods, 11
Global climate, 11
Global culture, 225
Global food crisis, 168–169
Global food ecosystems, 221–223
    Boot Camps, 228–231
        action, 229
        aspiration, 228–229
        inspiration, 228
    Food for Climate League, 223–228
    hackathon, 231–233
    value-based communication, 223
Global food insecurity, 143
Global food production, 11–12
Global food sustainability, 3–8, 11, 103, 197, 199–200, 203, 216
    food imagery, 14–16
    food information technologies, 16–19
    food misinformation, 19–22
    media framing of, 22–26
    role of food journalism, 12–14
    social media, food on, 29–42
        food branding, 33–37
        junk food marketing, 37–42
        uses and gratifications, 29–33
Global food system, 11, 237
Global Reporting Initiative guidelines, 108

Global warming, 105
Glyphosate, 238
Good After COVID-19 movement, 223
Good food grammars, 31–32
Good hydration, 183
Goody's five-phase framework, 22–23
Google, 224
Google Glass front-facing camera, 17
Google Trends, 126
Government health institution, 139
GPS-enabled geolocation, 59–60
Grassroots organization, 66–68
GreenApes app, 178–179, 184–185
Greenhouse gas (GHG) emissions, 3–4, 101, 169, 177
    source of, 103
Green marketing, 66, 121
Greenpeace Thailand, 68–69
Greenwashing, 41
Greta Thunberg, 68–69
Gulf Cooperation Council region, 119

H

Hackathon, 231–233
Hashtags, 35
Health communicators, 20
Health policies, 110
Healthy foods, 35–36
HealthyLifestyle4All campaign, 57
Heart disease, 110
Heavy social media, 41
Hemolytic Uremic Syndrome (HUS), 137
Higher education institutions, 196
High-resolution photography, 59–60
Hofstede's Model, 240
Hospitality, 167–168
Household food waste reduction campaigns, 118
Human-centered approach, 232
Humanistic nongovernmental organization, 114
Hungarian media, 143
Hunger, 77
Hypercompetition, 223
Hyperspecialization, 223

I

iBuffet, 17
ICT. *See* Information and communication
    technology
Identification With All Humanity (IWAH), 121
iLog, 18–19
In-depth one-on-one interview, 226
Individual attitudes, 104
Industrialized animal food production, 101–102
Industrial production, 112

Informal conversations, 241
Information and communication technology (ICT), 19,
    43, 125
Information seeking, 32, 134
    on social media, 30
Information sharing, on social media, 29–31
Initiative Tierwohl campaigns, 112
Injunctive message appeals, 119
Innovation, 7–8, 222
    ingredients of, 223
Inspiration, Boot Camps, 228
Instagram, 15, 32–34, 60, 165
    characteristics of, 31
Integral ecological regeneration, 230–231
Interconnectedness, 203
Intergovernmental Panel on Climate Change (IPCC),
    102, 223–224
International expert groups, 110–111
International food security, 86*t*
International organizations, 238
International Public Relation Network (IPRN), 244
Internet, 166
    food messaging, 164–165
Internet-mediated democratization of expertise, 20
Internet of People (IoP), 16–17
Internet of Things (IoT) technologies, 17
Interviewers, 242*t*
Inverted policy approach, 137
IPCC. *See* Intergovernmental Panel on Climate
    Change
Italy's Minister of Health, 137

J

Junk food marketing, 37–42
Justice, three-dimensional theory of, 130

K

Kolb's theory of experiential learning, 7, 200
Kristen Alley Swain, 5
Kuhn's scientific knowledge paradigms, 15
Kyoto protocol, 13

L

Labeling perception, 83
Language, 223
Leafsnap, 67
Learning experiences, 180
Learning process, 200
Liaison Entre Actions pour le Development de
    l'Economie Rurale (LEADER), 129
LIFE Climate Smart Chefs project, 5–6
Livestock consumption, 108–109

Livestock food producer organizations, 105
Livestreaming environment, 35
Local discoveries, 228—229
Local food, consumption of, 132—133
Local Food Loop, 66—67
Local food networks, 128—133
Love Food Hate Waste campaign, 63—64
Low-involvement barriers, 36

M
Machine learning, 137
Mad cow disease, 109
Malaysian dessert, 14
Malnutrition, 169
Manageability, 185
Manal Club, 173
Manufactured alternative meats, 114—115
Map stakeholders' tensions, 242
Market-driven journalism, 127—128
Market-driven strategy, 236—237
Marketing, 235
Marketing myopia, 236
McDonalds, 65—66
Meal distribution, 209—210
Meat consumption, 111
"Meat Free Mondays" campaign, 113
Meat production, 181—182
Meat-selling practices, 111
Meat substitute, 114—115
Media amplification, 43
Media attention, 142
Media corporatization, 12—13
Media coverage, 13—14, 113, 122
    of AFNs, 132
    of food insecurity, 130
    of food-related CSR, 107
    and social media framing shape, 22—23
Media drive food system transformation, 131—132
Media framing, 42
    of food sustainability, 22—26
Media organizations, 126
Mediterranean region, 3—4
Mercury deposition, 135
Methane, 177—178
Methylation, 135
Mezirow's transformational learning, 7
Microblog Weibo, 134
Migrant communities, 164
Millennials and Generation Z, 224, 227
Millennium Development Goals, 121
Ministry of Health, 29
"60 Minutes" investigation, 112
Mixed-method approach, 96

Mobile food apps, 17—18, 43
Modularity, 203
Monsanto claims, 143
Monsanto Protection Act, 65
MRIO model, 4—5
Mulino Bianco, 240, 243—244, 244t, 247—248
Mylks, 117

N
Nanofoods, 141—142
    consumer acceptance of, 141
Narrative storytelling, 12
National food security, 86t
National Health Service, 29
Natural ecosystems, 224
Natural Language Processing (NLP), 58—59, 68
Negative food product perceptions, 132—133
Negative footprint illusion, 21
Negative spillover, 104
New platform, impact of, 166
News articles, 126
News coverage, 146
News media, 102, 127, 139—140
Newspaper coverage, 138
NLP. *See* Natural Language Processing
Noncommunicable diseases, 3
Nongovernmental organizations (NGOs), 111
North Carolina University, 199—200
Norwegian Institute of Public Health, 135
"Novel protein" production, 115
Nutritional knowledge, trainers' lack of, 166

O
Obesity prevalence, 38
Olio, 66—67
One Health Platform, 137
One Planet Business for Biodiversity (OP2B),
    225—226
One-way forms of communication, 231
Online apps, 38—39
Online disinformation tactics, 20
Online fake news, 20—21
Online food media, 29
Online forums, 30
Online images, 42
Online news stories, 23
Online social interaction, 33
Open Food Facts, 18
Opinion leaders and policymakers, 13
Organic food, 36
Organic food price premiums, 129
Organic food production, 82, 134
    perception of, 82t, 83t

Organization for Economic Cooperation and Development, 122
Organizations, social media, 64–68
  corporate communication, 64–66
  double-edged sword of, 66
  grassroots organization, 66–68
Otago Farmers' Market, 133
Overt vegetarian framing, absence of, 114
Ownership, 170–171

P
Pacific islands, 121–122
Paideia Digital Academy, 233
Pandemic-related stories, 126
Parasocial relationships, 34
Paris Agreement, 107
  climate change targets, 103
Participants, PUSH-UFWH design thinking workshop, 206, 208t
Patriotism, 138
Peatlands, 105
Peer-to-peer networks, 16–17, 231
Perceived ability, experiential learning on, 202–203
Perceived food risks, 125
Perceived food safety, of European food products, 94t
Perceived knowledge, 138
Perception
  on food imports, 81t
  food risk, 137–141
  of food security, 86, 98
  of national and international food security, 86t
  of organic food production, 82t, 83t
Perennial crops, 104
Personal dietary change, 113
Personal income level, 126
Persuasive messages, 104
Philippine mainstream media, 140
Photovoice, 14
Pink slime frame, 109
Planet boundaries, 178
Planet-centered approach, 232
Plant-based diets, 101–102, 178
Plant-based milk, 117
  alternatives, 117
  diffusion, 118
Plant mylks, 117–118
Politicization, 140
Politicized framing, 114
Pollinators, 123
Population growth, and climate change, 124
Postworkshop survey, 213
Predictive modeling, 18
Preferences, purchasing food, 83t, 84t

Presidents United to Solve Hunger (PUSH) Leaders Forum, 197–203
  experiential learning, 200–201
  experiential learning activity, design thinking as, 201–202
  motivation and perceived ability, experiential learning on, 202–203
  theoretical background and definition of key concepts, 198–200
Price, 85
Private climate mitigation interventions, 107–108
Private partnerships, 226
Problem-based learning, 232
Processed foods, omnipresent marketing of, 40
Proecological attitudes, 106
Proenvironmental consumers, 34, 113
Proenvironmental frame, 114
Professional organizations, 138–139
Project roadmap, 244t
"Protein-rich" products, 38
Prototyping, 229
Psychological process, 138
Public attitudes, 112
Public buy-in, for food waste reduction, 118
Public climate mitigation interventions, 107–108
Public communication, 18, 146
Public partnerships, 226
Purpose-driven research, 222
PUSH Leaders Forum. See Presidents United to Solve Hunger (PUSH) Leaders Forum
PUSH-UFWH design thinking workshop, 203–213
  data analysis, 209, 210t, 211t, 212t
  data collection, 207–209
  meal distribution, 209–210
  methodology, 204–209, 205f, 206f, 207f
  participants, 206, 208t
  survey, 213
  theoretical framework, 203–204

R
Red and processed meat (RPM), 110–111
Reddit platform, 61
Red meat, 110
"Red Queen" effect, 38
"Reduce waste" messages, 118–119
Reflective observation, 200
Regenerative tracks, 230–231
Registered dietitians, 106
Reimagine Education, 230
Restaurant, 168
Rhetorical frames, 111
Ripple effect, 18
Robotic vehicles, 125

S

Safe and Healthy High-Quality Food, 87–88
Safety, 78–79
Salmonella, 140
Salt producers, French association of, 140
Salutogenic approach, 180
Salutogenic Model of Health (SMH), 180
Science-based design rationale, 179–180
Science-based SU-EATABLE principles, 180–183
SDGs. *See* Sustainable development goals
Selective media exposure, 138
Selfish overpurchasing, 125
Sensationalism, 43
Share value, 239–240
Shiga Toxin-producing *Escherichia coli*, 134
Short-termism, 221–222
Silo approach, 223
Skretting, 240, 241t, 242t, 245–246, 248
Slow Food's Terra Madre, 133
Slow journalism, 12–13
Small-scale dairy systems, 117
Small-scale food producers, 34
Smart e-health systems, 17
Smart phones, 165
SNAP. *See* Supplemental Nutrition Assistance Program
Social Amplification of Risk Framework (SARF), 139
Social capital, 144
Social countermarketing (SCM) strategies, 111
Social diffusion of information, 140
Social media, 5, 15, 19–20, 172, 196
    channels, 20
    conversations, 32
    diet influencers, 20
    eating and cooking experiences on, 16
    expression of opinions on, 31–32
    food on, 29–42
        food branding, 33–37
        junk food marketing, 37–42
        uses and gratifications, 29–33
    food risks and benefits, 20–21
    food rumors on, 21
    information seeking on, 30
    information sharing on, 29–31
    key gratification of, 33
    peer behavior on, 30
Social media, food culture and food sustainability on
    analysis methods, 58–59
    case study, 68–70
    and individual, 59–64
        food consumption on, 59–61
        food waste, 63–64
        sustainable consumer choice, 62–63

    and organizations, 64–68
        corporate communication, 64–66
        double-edged sword of, 66
        grassroots organization, 66–68
Social media framing
    media coverage and, 22–23
    of sustainable food, 114
Soft skills, 232
Soil, 106
Soil and Water Conservation Society, 123
Soil carbon, 105
Soil carbon management, 102
Solutions-oriented news coverage, 21–22
Special Waves, 94
Speciesism, 22
Sporting events, 37–38
Stability, 78–79
    Eurobarometer, 86, 86t, 87t
Stakeholder engagement, 239–240, 247–248
    sustainability and, 238–239
Stakeholder mapping, 242
Stakeholders' involvement, 243
Stakeholder theory, 239
Story frames, 134
SU-EATABLE criteria, 7
SU-EATABLE LIFE project, 177–178
    design and evaluation-stage I, 183–188
        engagement activities, 183–185, 184f
        intermediate evaluation stage I, 185–188, 186t
    design process, 179–183
        science-based design rationale, 179–180
        sustainable diets, science-based SU-EATABLE
            principles for, 180–183
    intermediate evaluation stage I, 185–188, 186t
        customer level, 186–188, 189t
        food service level, 186, 187t
    stage II, 188–191
        adaptation, focus areas for, 190t
        cafeteria experience, 190
        communication style, 190
        digital experience, 190
        implementing toolkits and fostering ownership,
            189–190
        increasing flexibility and reducing effort, 190
        interruption of experiments, 188
        know our audiences, 190
        learning experience, 190
        learnings from stage I, 190t
Sufficient supply, 164
Superfood, 44
Supplemental Nutrition Assistance Program (SNAP),
    202

Sustainability, 78–79, 87–89, 88*t*, 89*t*, 90*t*, 91*t*, 92*t*, 93*t*, 94*t*, 172, 197–203, 229–230, 235
  design thinking, 216
    process, 215
  experience of receiving nutritionally unbalanced meal, 214
  "learning how to think," process of, 216
  messages, 22
  participants, 214
  PUSH and UFWH, 197–203
    experiential learning, 200–201
    experiential learning activity, design thinking as, 201–202
    motivation and perceived ability, experiential learning on, 202–203
    theoretical background and definition of key concepts, 198–200
  PUSH-UFWH design thinking workshop, 203–213
    data analysis, 209, 210*t*, 211*t*, 212*t*
    data collection, 207–209
    meal distribution, 209–210
    methodology, 204–209, 205*f*, 206*f*, 207*f*
    participants, 206, 208*t*
    survey, 213
    theoretical framework, 203–204
  and stakeholder engagement, 238–239
  theoretical background and definition of key concepts, 198–200
  transformational learning, 215
  workshop leaders, 215
Sustainability-centered strategy, 236–237
Sustainability communication scholarship, 44
Sustainable agriculture projects, 195
Sustainable consumer choice, 62–63
Sustainable Consumption and Production (SCP), 57
"Sustainable Consumption in Italy (2021)", 237
Sustainable corporate communication, 235–237
  academic papers, 235–236
  case selection, 240–241
  case study, 243–244
    Mulino Bianco, 243–244, 244*t*
  companies, 236
  Coop, 246–247
  data analysis, 242–243
    company sustainable goals, 242
    map stakeholders' tensions, 242
    stakeholder mapping, 242
    stakeholders' involvement, 243
  data collection, 241, 241*t*
  food industry, sustainable development goals in, 237–238
  global initiatives, 238
  international organizations, 238
  marketing, 235
  methodology, 240
  recent research, 236
  Skretting, 245–246
  stakeholder engagement, 236–237
  stakeholder engagement and communication process, relationship of, 239–240
  sustainability and stakeholder engagement, 238–239
Sustainable development goals (SDGs), 7–8, 77, 107, 163, 168–169, 178, 196–197, 203, 222, 233, 236, 238
  in food industry, 237–238
Sustainable diet, 86*t*, 196
  science-based SU-EATABLE principles for, 180–183
Sustainable dietary models, 178
Sustainable food
  campaigns, 109
  production, 63
  social media framing of, 114
Sustainable food-related behaviors, 13
Sustainable Restaurant Association, 179, 191

T
Tagliatelle, 61
Technical jargon, 170–171
Television cooking, 13–14
Thematic online sources, 136
Theory of Planned Behavior, 15–16
Third-party certification, 129
Three-dimensional theory of justice, 130
Thriving society, 233
"Through marketing", 42
Timor-Leste, 164
Trade-offs, 239
  FWE, 106
Traffic-light labels, 36
Transformational learning theory, 200, 203–204
Transgenic plants, 116–117
Tweets, 114
Twitter, 64–65, 134
  group, 114
  network, 132
Twitter Streaming API, 68

U
UFWH. *See* Universities Fighting World Hunger
Ultraprocessed food product marketing, 41
UN Agenda 2030, 201–202, 222
UN Food and Agriculture Organization, 122
Unhealthy food consumption, 39
Unified Medical Language System, 18
United Nations, 4, 236
United Nations Climate Change Conference, 68–69

United Nations Sustainable Development Goals, 103, 121
Universities Fighting World Hunger (UFWH), 197–203
  experiential learning, 200–201
  experiential learning activity, design thinking as, 201–202
  motivation and perceived ability, experiential learning on, 202–203
  theoretical background and definition of key concepts, 198–200
UN SDGs, 171
UN World Food Program, 173–174
Upcycled foods, 119
Urban farming, 66–67
Urban food activists, 132
Urban food strategies (UFSs), 130
URLs, 68
U.S. Centers for Disease Control, 110
U.S. Department of Agriculture, 198
U.S. Food and Drug Administration, 116–117
U.S. Food Safety Modernization Act, 136
U.S. National Academies of Sciences, Engineering, and Medicine, 144
U.S. newspapers, 130
Utilization, 78–79, 97–98
  Eurobarometer, 85–86, 86t

V

Value-based communication, 223, 229
Veganism, 114
Vegans, 117
Vegetable crops, 104
Vegetarianism, 114
Vegetarians, 117

Virtual reality (VR), 16
Visual design, 15–16
Vitamin-rich foods, 123–124
Voiceless Animal Cruelty Index (VACI), 112
Vygotsky's zone of proximal development, 200–201

W

Waste, food reduction, 118–119
Waste-to-green energy technology, 120
Waste Week campaign, 63–64
Water safety, 183
Web-based platforms, 38–39
WHO-Europe nutrient profile model, 40
Wicked problem, 11
Wikipedia, 134
Win–win solution, 120
Word-of-mouth (WOM), 65
Workshop theoretical framework, 203–204
World Association of Soil and Water Conservation, 123
World Bank, 122
World Environment Day, 67–68
World Health Organization, 19, 42, 108
World Resources Institute, 225–226
World Summit on Sustainable Development in Johannesburg, 57
World Wide Web, 164

Y

Yield gains, 4
Young adults, 40
Young Italians, 196
YouTube, 35

Printed in the United States
by Baker & Taylor Publisher Services